Advances in Geological Science

Series Editors

Junzo Kasahara, Tokyo University of Marine Science and Technology, Tokyo, Japan
Shizuoka University, Shizuoka, Japan

Michael Zhdanov, University of Utah, Utah, USA

Tuncay Taymaz, Istanbul Technical University, İstanbul, Türkiye

Studies in the twentieth century uncovered groundbreaking facts in geophysics and produced a radically new picture of the Earth's history. However, in some respects it also created more puzzles for the research community of the twenty-first century to tackle. This book series aims to present the state of the art of contemporary geological studies and offers the opportunity to discuss major open problems in geosciences and their phenomena. The main focus is on physical geological features such as geomorphology, petrology, sedimentology, geotectonics, volcanology, seismology, glaciology, and their environmental impacts. The monographs in the series, including multi-authored volumes, will examine prominent features of past events up to their current status, and possibly forecast some aspects of the foreseeable future. The guiding principle is that understanding the fundamentals and applied methodology of overlapping fields will be key to paving the way for the next generation.

Michael S. Zhdanov

Advanced Methods of Joint Inversion and Fusion of Multiphysics Data

Springer

Michael S. Zhdanov
Department of Geology and Geophysics
University of Utah
Salt Lake City, UT, USA

ISSN 2524-3829 ISSN 2524-3837 (electronic)
Advances in Geological Science
ISBN 978-981-99-6721-6 ISBN 978-981-99-6722-3 (eBook)
https://doi.org/10.1007/978-981-99-6722-3

This Springer imprint is published by the registered company Springer Nature Singapore Pte Ltd.
The registered company address is: 152 Beach Road, #21-01/04 Gateway East, Singapore 189721, Singapore

Paper in this product is recyclable.

Introduction

The mathematical inverse theory is dedicated to developing methods of reconstructing the properties of the objects generating the observed data. This problem arises in many fields of science and for different applications. For example, in earth science, the geological and geophysical data collected on the ground are used to determine the earth's internal structure and to find the locations of the minerals, oil and gas, water, and other natural resources. In medical science, the recordings of various medical sensors are used to determine the conditions of the internal organs of the human body. In astronomy, the data collected by optical and radio telescopes provide the basis for reconstructing the physical properties of distant stars and galaxies. This list of applications can be expanded to many other fields of science and engineering.

In many applications, researchers collect different types of data representing the same object of investigation. For example, in medical imaging, various imaging techniques, e.g., X-ray, ultrasound, magnetic resonance imaging (MRI), etc., are used to study the internal organs of the human body. In earth science applications, multiple physical field data, e.g., gravity, magnetic, electromagnetic, seismic, etc., are collected to study the earth's internal structure. In astronomy, optical and radiotelescopes are used to study electromagnetic radiation from the stars and galaxies, as well as observations of neutrinos, cosmic rays, or gravitational waves.

One common feature of all these problems is that the observed data are known, but the sources generating these data are unknown and have to be recovered from the data. At the same time, many of these applications use multiphysics data to study the same object of interest while providing information about different physical properties of the target. Therefore, combining all available information about the target makes sense to produce the most reliable solution to the inverse problem. This concept is similar to how living organisms, including humans, study the environment. They do not rely on one specific sensor but evaluate the information provided by multiple organs of sense, e.g., eyes, ears, nose, or tongue. In other words, they see, hear, smell, taste, or feel the surrounding world. The advantage discovered by nature is that if the information provided by one sensor is incomplete, it could be complemented by the signals received by other sensors. The joint inversion and fusion of multiphysics data serve the same purpose of integrating information from multiple sensors to produce the most reliable reconstruction of the objects generating the observed data.

The field of "data fusion" or "information fusion" has been developed extensively over the last decades, specifically with application to multisensor data analysis in navigation, surveillance, and remote sensing. The classical data fusion techniques are based on data association, state of the target estimation, and probability estimation (decision-based) fusion. In this book, I use the term "fusion" in a more general sense as an integration of all available information about the target with the goal of the most reliable reconstruction of the target from the observed multiphysics data. The joint inversion methods provide a mathematical framework for integrated analysis and fusion of multiphysics data which can be used in solving this problem.

In this book, I describe the state of the art in the field of joint inversion and present those advanced methods which emerged over the last decade.

The book is organized into five parts. In Part I, I review some of the mathematical concepts we use in this book, including the elements of probability theory and functional analysis. These concepts are critical to understanding the probabilistic and deterministic approaches to the inverse theory.

Part II discusses the foundations of the inverse theory, starting with the principles of regularized inversion and considering different methods of inverse problem solutions. The regularization theory is critical for developing effective methods of inversion. It is based on the simple idea introduced in the pioneering work of Tikhonov that the nonuniqueness and instability in the inverse problem solution can be overcome by bringing a priori information about the inverse model. This can also be achieved by using complementary data as the source of a priori information.

Part III, central part of the book, focuses on the mathematical algorithms of joint inversion. I begin this part by formulating the multimodal inverse problem and the joint inversion based on analytical and statistical relationships between different model parameters. Then, the methods of joint inversion based on structural similarities, joint focusing, and joint minimum entropy are introduced. This group of methods incorporates the a priori information in the form of structural constraints on the inverse problem solution.

Chapters 12 and 13 introduce the theory of Gramian spaces and corresponding stabilizing functionals, which play an important role in the formulation of the general principles of joint inversion. Gramian constraints make it possible to enforce the relationships between model parameters without a priori knowledge of the specific form of these relationships. Gramian spaces provide the convenient quadratic metric in the model spaces, which simplifies the minimizations of the Gramian constraints for arbitrary multimodal parameterization.

Chapter 14 focuses on the processing and fusion of multiphysics data as applied to digital image restoration. This problem is of great importance in biomedical, geophysical, astronomical, high-definition television, remote sensing, and other applications. I demonstrate that the methods of joint inversion of multiphysics data developed in this book can be effectively used for solving image restoration and deblurring problems.

In Part IV of the book, I discuss the idea of using artificial intelligence (AI) algorithms in the solution of the inverse problem. The AI algorithms work through a process called "machine learning." The concept of building machines possessing "artificial intelligence" was introduced long ago; however, this concept has become practical only recently with the rapid scaling up of computing power during the last decade. The application of AI algorithms in solving the inverse problem is usually based on using the training data that helps the algorithm to learn. I demonstrate that this process can be significantly improved by using knowledge-based neural networks, which rely both on the training data and a priori information about the laws governing the data prediction (forward modeling operators).

The final Part V presents the case histories of joint inversion of the potential field data used in geophysical applications. The gravity and magnetic geophysical data are widely used in the exploration of mineral resources and regional geological studies. Chapter 17 provides an overview of the principles of modeling and inversion of gravity and magnetic data, including their gradients (gravity and magnetic tensors). In Chap. 18, I present several case histories illustrating the effectiveness of joint inversion of potential field data in geophysical exploration.

In conclusion, I would like to thank my associates and graduate students in the Consortium for Electromagnetic Modeling and Inversion (CEMI) at the University of Utah, who contributed to developing different aspects of the joint inversion methods.

I am deeply grateful to my wife, Olga, for her encouragement and continuing support during the work on this book.

Salt Lake City, UT, USA
March 2023

Michael S. Zhdanov

Contents

Part II Foundations of Inverse Theory

Part I
Mathematical Background

Chapter 1
Introduction to Inversion Theory

Abstract This chapter introduces the definitions of forward and inverse problems of mathematical physics and concepts of the well-posed and ill-posed problems. It also reviews the main principles of ill-posed inverse problem solutions based on the regularization theory developed in the pioneering research by Tikhonov. The notions of sensitivity and resolution, which are important in understanding the regularization principles, are also introduced. Finally, two different main approaches to formulating the inverse problem, one based on the deterministic concept and the other using the framework of the Bayesian probability theory, are considered as well.

Keywords Inverse problem · Well-posed problem · Ill-posed problem · Regularization theory

1.1 Formulation of Forward and Inverse Problems

Scientific experiments are usually based on measuring the observable data in the lab or natural environment and determining the parameters of the object of investigation based on these measurements. A similar problem arises in many biological, physical, and geophysical studies where the observer collects various physical field data emitted by the target and tries to draw a conclusion about the properties of the target by analyzing these data. In mathematical physics, predicting the observed data (measured fields) caused by the target is called *a forward problem*. The opposite problem of extracting information about the target from the observed data is called *an inverse problem*.

One can express forward and inverse problems symbolically using the following mathematical notations:

$$\mathbf{d} = A(\mathbf{m}). \tag{1.1}$$

In Eq. (1.1), symbol $\mathbf{m}$ represents a model characterizing the structure and properties of the target, e.g., geological structure of the earth in geological applications, or

M. S. Zhdanov, *Advanced Methods of Joint Inversion and Fusion of Multiphysics Data*, Advances in Geological Science,
https://doi.org/10.1007/978-981-99-6722-3_1

internal structure of the brain or other body tissue in medical applications. Symbol **d** denotes the observed data, e.g., geophysical data in geological applications or medical imaging data in medical applications. The symbol A stands for the set of rules, represented by physical laws and/or mathematical equations, which relate the given model to the observed data. In mathematics, symbol A is often called *an operator*, which transforms model, **m**, into the data, **d**.

For example, if data, **d**, represent the observed gravity field, and model, **m**, describes a given distribution of the density of the target, then the operator, A, is the mathematical representation of Newton's law. Similarly, if data, **d**, represent the observed electromagnetic field, and model, **m**, describes a given distribution of the electrical properties of the target, then operator, A, is the mathematical representation of the laws of electromagnetism (e.g., Maxwell's equations).

In a case when model **m** is known, and the goal is to find the data, **d**, generated by this model, Eq. (1.1) is called a *forward modeling problem*. Contrary, we call Eq. (1.1) an *inverse modeling problem*, if the data, **d**, are known, but the model, **m**, generating this data is unknown.

One of the major difficulties with the solution of the inverse problems is related to the fact that the size of the set forming the unknown model parameters usually exceeds the size of the set of the observed data significantly. In other words, the inverse problems are hugely underdetermined. For example, in geophysical applications, the data are collected on a small area on the earth's surface or in a borehole. Still, the inversion aims to recover the physical properties of the large volume of the rock formations underground. In medical imaging, the sensors are placed on or around the patient's body, while the goal is to produce images of the internal organs. In addition, the observed data always have some instrumental noise, which also strongly affects the inversion. Solving these undetermined and uncertain problems requires developing specialized mathematical methods. These methods constitute the body of inversion theory, which development was stimulated by the needs of applied science, first and foremost by geophysical exploration problems.

The history of the development of inversion theory goes back to the beginning of the twentieth century.

Almost a century ago, French mathematician Hadamard (1902) formulated three critical questions related to the solution of any inverse problem:

(1) Does the solution exist?

(2) Is it unique?

(3) Is it stable?

The question of the solution's existence is associated with the mathematical formulation of the inverse problem. From the physical point of view, there should be some particular solution, if we study a real physical target. However, from the mathematical point of view, there could be a situation when we cannot fit the observed data by a specific class of models. For example, let us assume that the real target represents a body composed of 20 rectangular bricks of different densities; however, we are looking for a solution described by five rectangular bricks only. In this situation, an exact solution may not exist.

The following formulae illustrate the question of the uniqueness of the solution. Assume that we have two different models, $\mathbf{m}_1$ and $\mathbf{m}_2$, which generate the same data $\mathbf{d}_0$:

$$A(\mathbf{m}_1) = \mathbf{d}_0, \ A(\mathbf{m}_2) = \mathbf{d}_0.$$

In this case, it is impossible to distinguish these two models from the given data. That is why the question of uniqueness is so important in inversion.

The last question of solution stability is a critical one in inversion theory as well. In fact, the observed data are always contaminated by some noise $\delta\mathbf{d}$. The question is whether the difference in the responses for different models is larger than the noise level. For example, let two different models, $\mathbf{m}_1$ and $\mathbf{m}_2$, generate two different data sets, $\mathbf{d}_1$ and $\mathbf{d}_2$, which can be expressed schematically as follows:

$$A(\mathbf{m}_1) = \mathbf{d}_1, \text{ and } A(\mathbf{m}_2) = \mathbf{d}_2.$$

Assume also that these two models are very different, while the data difference is within the noise level ε:

$$\|\delta\mathbf{m}\| = \|\mathbf{m}_1 - \mathbf{m}_2\| > C,$$

$$\|\delta\mathbf{d}\| = \|\mathbf{d}_1 - \mathbf{d}_2\| < \varepsilon, \ C >> \varepsilon,$$

where symbol $\|\ldots\|$ denotes some norm or measure of the difference between two models and two data sets, and C is some large number. We will discuss a rigorous definition of norm later.

It is also impossible to distinguish these two models from the observed data in this situation.

Hadamard realized the importance of these three questions for inverse problem solutions. He suggested that a particular mathematical problem was formulated correctly if and only if all three questions posed above had positive answers. In other words, the mathematical problem was *well-posed,* if its solution did exist, was unique, and was stable.

A problem was *ill-posed,* according to Hadamard (1902), if at least one of the conditions listed above would fail. In other words, the problem was ill-posed if the solution did not exist, or was not unique, or if it was not stable. Furthermore, Hadamard considered that an ill-posed mathematical problem was not physically and mathematically meaningful (that was why one could call it an "ill" problem).

Many mathematicians of the early twentieth century followed Hadamard's view. They did not consider ill-posed problems as being the legitimate subject of the research. The rapid development of applied computational mathematics in the middle of the twentieth century made it clear, however, that ill-posed problems appeared in the majority of the practical computational problems. The researchers have to solve these problems in many natural science applications. It was also subsequently found that Hadamard's opinion was incorrect: ill-posed problems were physically and mathematically meaningful and could be solved.

The foundations of the theory of ill-posed problems were formulated by Russian mathematician A. N. Tikhonov in the middle of the twentieth century (Tikhonov 1943; Tikhonov and Arsenin 1977). Since then, this theory has been developed in hundreds of research papers and monographs (e.g., Lavrent'ev et al. 1986; Isakov 1993; Morozov 1993; Hansen 2010; Muller and Siltanen 2012; Neto and Neto 2013).

This chapter reviews the main principles of ill-posed inverse problem solutions. A detailed exposition of this subject can be found in Zhdanov (2002, 2015).

The solution of the inverse problem consists in determining such a model $\mathbf{m}^{pr}$ (predicted model) that generates predicted data $\mathbf{d}^{pr}$, that represent well the observed data $\mathbf{d}$. In other words, the goal is to find a model generating the predicted data being close enough to the observed data (usually within the accuracy of observations).

In order to find a mathematical solution to this problem, one should define the notion of a "distance" between two data sets that will help us to evaluate the accuracy of the inverse problem solution. In other words, we need to introduce geometry to measure the distance between the actual and predicted data. The concept of a distance between two functions or two data sets is provided by the mathematical theory of function spaces, reviewed in Chap. 3.

1.2 Concept of Well-Posed and Ill-Posed Inverse Problems

In this section, we provide rigorous mathematical definitions of well-posed and ill-posed problems. Consider a mathematical model of some phenomenon characterized by a multidimensional vector of the model parameters, $\mathbf{m}$, which belongs to some set of models M. Let $\mathbf{d}$ be a multidimensional vector of observed data attributed to this phenomenon, which belongs to a set of data D. We assume that data, $\mathbf{d}$, and model, $\mathbf{m}$, are related by a corresponding physical or mathematical law described by Eq. (1.1).

In a case when the model parameters, $\mathbf{m}$, are known, formula (1.1) represents a forward modeling problem of finding data, $\mathbf{d}$, and A is a *forward modeling operator*. Formula (1.1) becomes an equation of the inverse problem for $\mathbf{m}$, when the data and forward modeling operator are known. Our focus is on studying this inverse problem.

We generally expect that the true data $\mathbf{d}$ are not directly observable. Instead, we have only some approximate values, $\mathbf{d}_\delta$. The following formula provides an estimate of the accuracy of the observed data: $\|\mathbf{d} - \mathbf{d}_\delta\|_D \leq \delta$. In the last formula, δ is the noise level of the data. Symbol $\|\ldots\|_D$ represents some norm ("distance" between vectors $\mathbf{d}$ and $\mathbf{d}_\delta$) in the set of data, D. Solution of the inverse problem consists of the following principal steps:

(1) Finding a model, $\mathbf{m}_\delta \in M$, with the attributes (data) such that $\|\mathbf{d}-A(\mathbf{m}_\delta)\|_D \leq \delta$ under the condition that $\mathbf{m}_\delta \to \mathbf{m}$ as $\delta \to 0$.
(2) Determining, if possible, the errors of the approximate solutions within the adopted model; that is, the estimation of the distance between $\mathbf{m}_\delta$ and $\mathbf{m}$ using an appropriate metric.

The mathematical difficulty in solving inverse problems is that the inverse operator, A^{-1}, may not exist or be continuous over the domain $AM \subset D$, where AM is formed by all possible data, attributed to all available models from the model set, M.

In the beginning of this chapter, we have already introduced the concept of ill-posed and well-posed problems. We can now provide rigorous mathematical definitions of these problems. A well-posed problem must meet the following requirements:

(1) The solution $\overline{\mathbf{m}}$ of equation $\mathbf{d} = A(\mathbf{m})$ exists over the entire set D.
(2) The solution $\overline{\mathbf{m}}$ is unique.
(3) The solution is stable; that is, small perturbations of $\mathbf{d}$ cause only small perturbations of the solution, $\overline{\mathbf{m}}$.

The approximate solution of an inverse problem can be written as follows, $\mathbf{m}_\delta = A^{-1}\mathbf{d}_\delta$, inasmuch as $\mathbf{m}_\delta \rightarrow \overline{\mathbf{m}}$ as $\delta \rightarrow 0$. The problem (1.1) is ill-posed if at least one of the conditions listed above fails. As a consequence, the element $\mathbf{m}_\delta = A^{-1}\mathbf{d}_\delta$ (which may not even exist!) is not an approximate solution inasmuch as $\mathbf{m}_\delta$ can deviate from the exact solution $\overline{\mathbf{m}}$ arbitrarily even for small δ.

Attempts to solve ill-posed problems were undertaken well before the development of a coherent general approach. As a result, simple ill-posed problems were solved intuitively. However, intuitive techniques work when we consider relatively simple models only. Present capabilities in measuring and processing experimental data make it necessary to consider very complex models. At present, through the use of regularization theory, we can find highly effective numerical algorithms to solve a wide range of inverse problems for complex targets. This theory is based on the concept of a regularizing algorithm or operator. The regularizing operator establishes a relationship between each pair $(\mathbf{d}_\delta, \delta)$ and an element $\mathbf{m}_\delta \in M$ such that $\mathbf{m}_\delta \rightarrow \overline{\mathbf{m}}$ as $\delta \rightarrow 0$. Regularizing algorithms have been developed for many ill-posed problems and implemented in computer codes. These permitted the automated processing of the experimental data.

1.3 Regularized Solution of the Ill-Posed Problem

The following fundamental idea serves as a basis of Tikhonov's regularization theory. One can solve an ill-posed problem by substituting a family of well-posed problems for the original ill-posed problem. The requirement is that the well-posed solutions should provide an asymptotic approximation to the solution of the corresponding ill-posed problem.

We can explain this major concept of regularization theory more rigorously using mathematical notations. Let us assume that the inverse problem described by Eq. (1.1) is ill-posed. The main idea of the regularization method is to consider, instead of one ill-posed inverse problem (1.1), a family of well-posed problems,

$$\mathbf{d} = A_\alpha(\mathbf{m}), \tag{1.2}$$

which approximates the original inverse problem in some sense (Strakhov 1968, 1969a, b). The scalar parameter $\alpha > 0$ is called *a regularization parameter*. Family of well-posed problems (1.2) delivers a solution to the ill-posed problem (1.1) if the following condition holds:

$$\mathbf{m}_\alpha \rightarrow \mathbf{m}, \text{ if } \alpha \rightarrow 0. \tag{1.3}$$

The critical question of the regularization theory is how to find this family of the well-posed problem. Over the years, since Tikhonov conceived the concept of the regularized solution, many different methods were developed to construct this family. Most of these methods are based on introducing some a priori information about the properties of the required solution. For example, one can search for a target model with a smooth distribution of physical properties. Another example would be a solution with the bounded variations of the parameters within the given boundaries. In both examples, introducing additional constraints on the possible solution of the inverse problem results in limiting the class of possible solutions and, eventually, in making the ill-posed problem well-posed.

In this book, we will discuss the multiple ways of regularization using various types of constraints imposed on the solution of the inverse problem.

1.4 Sensitivity and Resolution of Data Inversion

We formulate the notions of sensitivity and resolution, which are important in understanding the regularization principles. These notions can be readily introduced in a case when operator, A, of forward modeling problem is a linear one. A rigorous definition of a linear operator will be given in Chap. 3; however, for the purpose of this section, it is enough to understand that the linear operator acts similarly to a linear function. In other words, the application of the linear operator to a sum of two vectors should be equal to the sum of the results obtained by applying the same operator to each vector separately.

1.4.1 Sensitivity

One can describe any forward problem by operator Eq. (1.1). Let us consider some given model $\mathbf{m}_0$ and corresponding data $\mathbf{d}_0$. For the sake of simplicity, we assume that in some vicinity of point $\mathbf{m}_0$, operator $A = A_{m_o}$ is a linear operator. Then we have

$$A_{m_o}(\mathbf{m} - \mathbf{m}_0) = A_{m_o}\mathbf{m} - A_{m_o}\mathbf{m}_0 = \mathbf{d} - \mathbf{d}_0,$$

or

$$A_{m_o}(\Delta\mathbf{m}) = \Delta\mathbf{d}, \tag{1.4}$$

where

$$\Delta \mathbf{m} = \mathbf{m} - \mathbf{m}_0 \text{ and } \quad \Delta \mathbf{d} = \mathbf{d} - \mathbf{d}_0$$

are the perturbations of the model parameters and of the data, respectively.

Following work by Dmitriev (1990), we can now give a corresponding definition of sensitivity.

Definition 1.1 The sensitivity S_{m_o} of the data is determined by the ratio of the norm of the perturbation of the data to the norm of the perturbation of the model parameters.

The maximum sensitivity is given by the following formula:

$$S_{m_o}^{max} = \sup \left\{ \frac{\|\Delta d\|}{\|\Delta m\|} \right\} = \sup \left\{ \frac{\|A_{m_o}(\Delta m)\|}{\|\Delta m\|} \right\} = \|A_{m_o}\| \,, \tag{1.5}$$

where symbol $\sup \varphi$ denotes the least upper bound or *supremum* of the variable φ. We will find in Chap. 3 that $S_{m_o}^{max}$ is equal to the norm of operator A_{m_o}.

If we know $S_{m_o}^{max}$, according to (1.4) and (1.5), we can determine the variations of the model that can produce the variations of the data greater then the errors of observations, δ:

$$\|\mathbf{m} - \mathbf{m}_0\| \geq \delta / S_{m_o}^{max}. \tag{1.6}$$

Therefore, the data are sensitive to those perturbations of the model parameters only that exceed the level $\delta / S_{m_o}^{max}$. Thus, one cannot distinguish any other variations of the model from the data.

1.4.2 Resolution

Let us assume now that in some vicinity of the model $\mathbf{m}_0$ the following inequality is satisfied:

$$\|A_{m_o}(\Delta \mathbf{m})\| \geq k \|\Delta \mathbf{m}\|,$$

for any $\Delta \mathbf{m}$, where $k > 0$ is some constant. It will be shown in Chap. 3 (Theorem 3.41) that, in this case, there exists a linear and bounded inverse operator $A_{m_o}^{-1}$. It means that the solution of the inverse problem in the vicinity of the model $\mathbf{m}_0$ can be written as follows:

$$\mathbf{m} = \mathbf{m}_0 + A_{m_o}^{-1}(\mathbf{d} - \mathbf{d}_0). \tag{1.7}$$

The same expression can be written for data $\mathbf{d}_\delta$, observed with some noise $\mathbf{d}_\delta = \mathbf{d} + \delta \mathbf{d}$:

$$\mathbf{m}_\delta = \mathbf{m}_0 + A_{m_o}^{-1}(\mathbf{d}_\delta - \mathbf{d}_0). \tag{1.8}$$

From (1.7) and (1.8) we have

$$\mathbf{m}_\delta - \mathbf{m} = A_{m_o}^{-1}(\mathbf{d}_\delta - \mathbf{d}). \tag{1.9}$$

Now we can determine the maximum possible errors in the solution of the inverse problem for the given level of the errors in the observed data, equal to $\delta = \|\delta\mathbf{d}\|$:

$$\Delta_{max} = \sup_{\|\mathbf{d}_\delta - \mathbf{d}\| = \delta} \|\mathbf{m}_\delta - \mathbf{m}\| = \sup_{\|\mathbf{d}_\delta - \mathbf{d}\| = \delta} \|A_{m_o}^{-1}(\mathbf{d}_\delta - \mathbf{d})\| = \|A_{m_o}^{-1}\|\delta, \tag{1.10}$$

where by Theorem 3.41 of Chap. 3,

$$\|A_{m_o}^{-1}\| \leq \frac{1}{k}. \tag{1.11}$$

Based on the last formulae, we can determine the resolution of the data inversion. Two models, $\mathbf{m}_1$ and $\mathbf{m}_2$, in the vicinity of the model $\mathbf{m}_0$, can be resolved if the following condition is satisfied:

$$\|\mathbf{m}_1 - \mathbf{m}_2\| \geq \Delta_{max} = \|A_{m_o}^{-1}\|\delta = \frac{\delta}{R_{m_0}}. \tag{1.12}$$

The value

$$R_{m_0} = \frac{1}{\|A_{m_o}^{-1}\|} \tag{1.13}$$

is the *measure of resolution* of the given inverse problem. It follows from (1.11) and (1.13) that

$$R_{m_0} \geq k\ . \tag{1.14}$$

The smaller the norm of the inverse operator, the bigger the resolution, R_{m_0}, and the closer to each other are models that can be resolved. For example, in the case of unbounded inverse operator $A_{m_o}^{-1}$ with the norm going to infinity, the resolution goes to zero, $R_{m_0} = 0$. Therefore, the maximum possible errors in the determination of $\mathbf{m}$ are infinitely large. We have this case for the ill-posed problem precisely.

1.5 Deterministic and Probabilistic Approaches to the Formulation of the Inverse Problem

There are two different main approaches to the formulation of the inverse problem. The first approach is based on the deterministic concept, which considers the data and model parameters characterized by specific functions or vectors with certain (maybe unknown) values. This approach was developed in the works of Lanczos (1961), Backus and Gilbert (1967), Backus (1970a, b, c), Marquardt (1963, 1970), Tikhonov and Arsenin (1977), and others.

In the second probabilistic approach, the observed data and model parameters are treated as realizations of some random variables. This approach was introduced in the pioneering papers of Foster (1961), Franklin (1970), Jackson (1972), Tarantola and Valette (1982), Tarantola (1987, 2005), among others.

The probabilistic approach often formulates the inverse problem solution in the framework of the Bayesian probability theory. This allows bringing the a priori information about the models to reduce the ambiguity and instability of inversion. In addition, the probabilistic approach enables the use of statistical methods to evaluate the uncertainty and a posteriori probability of the inverse problem solutions.

It can be demonstrated, however, that both these approaches result in similar numerical solutions of the inverse problem (Zhdanov 2002, 2015; Menke 2018). For example, in the framework of deterministic Tikhonov regularization, one can bring the a priori information into the solution by imposing specific constraints on the parameters of the models. At the same time, deterministic or probabilistic interpretation of the observed data and model parameters emphasizes different aspects of the inversion algorithms. This also helps understand better the properties of the inversion parameters.

In this book, I will consider both approaches to inversion. The probabilistic and deterministic methods of inverse problem solutions are inextricably intertwined. Chapter 2 will present the elements of the probability theory, while in Chap. 3, I will discuss the concepts of functional model and data spaces which serve as the foundation of the deterministic approach.

References and Recommended Reading to This Chapter

Backus GE (1970a) Inference from inadequate and inaccurate data, I. Proc Natl Acad Sci 65:1–7

Backus GE (1970b) Inference from inadequate and inaccurate data, II. Proc Natl Acad Sci 65:281–287

Backus GE (1970c) Inference from inadequate and inaccurate data, III. Proc Natl Acad Sci 67:282–289

Backus GE, Gilbert TI (1967) Numerical applications of a formalism for geophysical inverse problems. Geophys J R Astr Soc 13:247–276

Dmitriev VI, Editor in chief (1990) Computational mathematics and techniques in exploration geophysics (in Russian). Nedra, Moscow, 498 pp

Foster M (1961) An application of the Wiener-Kolmogorov smoothing theory to matrix inversion. J Soc Ind Appl Math 9:387–392

Franklin JN (1970) Well-posed stochastic extensions of ill-posed linear problems. J Math Anal Appl 31:682–716

Hadamard J (1902) Sur les problèmes aux derivées partielles et leur signification physique, vol 13. Princeton University Bulletin, pp 49–52. Reprinted in his Oeuvres (1968) vol III. Centre National de la Recherche Scientifique, Paris, pp 1099-1105

Hansen PC (2010) Discrete inverse problems: insight and algorithms. SIAM Press

Hjelt S-E (1992) Pragmatic inversion of geophysical data. Springer, Berlin, Heidelberg, New York, 262 pp

Jackson DD (1972) Interpretation of inaccurate, insufficient and inconsistent data. Geophys J Roy Astron Soc 28:97–110

Isakov V (1993) Uniqueness and stability in multi-dimensional inverse problem. Inverse Probl 6:389–414

Lanczos C (1961) Linear differential operators. D. van Nostrand Co.

Lavrent'ev MM, Romanov VG, Shishatskii SP (1986) Ill-posed problems of mathematical physics and analysis. Translations of mathematical monographs, vol 64. American Mathematical Society, Providence, Rhode Island, 290 pp

Marquardt DW (1963) An algorithm for least-squares estimation of nonlinear parameters. J Soc Ind Appl Math 11:431–441

Marquardt DW (1970) Generalized inverses, ridge regression, biased linear estimation, and nonlinear estimation. Technometrics 12:591–612

Menke W (2018) Geophysical data analysis: discrete inverse theory, 4th edn. Elsevier

Morozov VA (1993) Regularization methods for ill-posed problems. CRC Press

Muller JL, Siltanen S (2012) Linear and nonlinear inverse problems with practical applications. SIAM Press

Neto FDM, Neto AJ (2013) An introduction to inverse problems with applications. Springer

Strakhov VN (1968) Numerical solution of incorrect problems representable by integral equations of convolution type (in Russian). DAN SSSR 178(2):299

Strakhov VN (1969a) Theory of approximate solution of the linear ill-posed problems in a Hilbert space and its application in applied geophysics, part I (in Russian). Izvestia AN SSSR, Fizika Zemli, No 8, pp 30–53

Strakhov VN (1969b) Theory of approximate solution of the linear ill-posed problems in a Hilbert space and its application in applied geophysics, part II (in Russian). Izvestia AN SSSR, Fizika Zemli, No 9, pp 64–96

Tarantola A (1987) Inverse problem theory. Elsevier, Amsterdam, Oxford, New York, Tokyo, 613 pp

Tarantola A (2005) Inverse problem theory and methods for model parameter estimation. SIAM, 344 pp

Tarantola A, Valette B (1982) Generalized nonlinear inverse problem solved using the least squares criterion. Rev Geophys Space Phys 20:219–232

Tikhonov AN (1943) On the stability of inverse problems (in Russian). Doklady AN SSSR 39(5):195–198

Tikhonov AN, Arsenin VY (1977) Solution of ill-posed problems. V. H. Winston and Sons

Zhdanov MS (2002) Geophysical inverse theory and regularization problems. Elsevier, 628 pp

Zhdanov MS (2015) Inverse theory and applications in geophysics. Elsevier, 704 pp

Chapter 2
Elements of Probability Theory

Abstract In this chapter, we review the foundations of the probability theory, which are needed to develop a probabilistic approach to inverse problem solutions. These include basic formulas and notations from probability theory, properties of discrete and continuous random variables, and the concept of Shannon's entropy. We also study linear correlations between random variables and describe the properties of the correlation coefficient and the covariance matrix of multiple random variables. These properties play a fundamental role in the joint inversion of multiphysics data sets.

Keywords Probability theory · Random variables · Correlation coefficient · Covariance matrix · Entropy

We have discussed in Chap. 1 that there exist two approaches to inverse problems based on probabilistic and deterministic concepts. This chapter reviews the foundation of the probability theory, which is needed to develop a probabilistic approach to inverse problem solutions. The interested reader can find more details in the textbooks on the probability theory (e.g., Ross 2010).

2.1 Basic Formulas and Notations from Probability Theory

2.1.1 Discrete Random Variables

In the framework of the probabilistic approach to the solution of the inverse problem, we can treat the data, d, as a random variable because we do not know the data before conducting a physical experiment and the data are always contaminated by random noise. The actually observed data, d_{obs} can be considered as the *realization* of the random variable d in a given experiment. In a general case, the value of the observed data is determined by the results of the physical experiment; therefore, we can assign probabilities to the possible values of the random data.

In practice, the data are measured at discrete observation points and at discrete-time moments. Therefore, we can represent the data as a discrete random variable

M. S. Zhdanov, *Advanced Methods of Joint Inversion and Fusion of Multiphysics Data*, Advances in Geological Science,
https://doi.org/10.1007/978-981-99-6722-3_2

that can take on a finite number of possible values, $d = \xi = (\xi_1, \xi_2, \ldots \xi_n)$, with probabilities $(p(\xi_1), p(\xi_2), \ldots, p(\xi_n))$, respectively, where the function, $p(\xi)$, is called *a probability mass function*. This function is nonnegative, and it satisfies the following condition:

$$\sum_{i=1}^{n} p(\xi_i) = 1, \tag{2.1}$$

since ξ must take on one of the values ξ_i.

We can use the probability mass function to determine the *expectation*, or *mean value* of a random variable, as follows:

$$\langle \xi \rangle = \sum_{i=1}^{n} \xi_i \, p(\xi_i). \tag{2.2}$$

Given the mean value and the probability mass function, one can introduce a variance, σ^2, of the random variable, ξ, which characterizes the variations of ξ from its mean value:

$$\sigma^2 = \langle (\xi - \langle \xi \rangle)^2 \rangle = \sum_{i=1}^{n} (\xi_i - \langle \xi \rangle)^2 p(\xi). \tag{2.3}$$

In practice, we usually work with some samples of distribution of the random variable. In this case, we can calculate the statistical estimate of the mean value of the unknown distribution. For example, the mean value estimation of the sample of the normally distributed random variable χ, can be calculated as follows (for the definition of the normal or Gaussian distribution, see Sect. 2.1.3 below):

$$\langle \chi \rangle = \frac{1}{n} \sum_{i=1}^{n} \chi_i. \tag{2.4}$$

The estimate of the variance, σ_χ^2, of the normally distributed random variable, χ, takes the following form:

$$\sigma_\chi^2 = \langle (\chi - \langle \chi \rangle)^2 \rangle = \frac{1}{n} \sum_{i=1}^{n} (\chi_i - \langle \chi \rangle)^2 = \frac{1}{n} \sum_{i=1}^{n} (\chi_i^2 - 2\chi_i \langle \chi \rangle + \langle \chi \rangle^2)$$

$$= \frac{1}{n} \sum_{i=1}^{n} \chi_i^2 - 2\frac{1}{n} \sum_{i=1}^{n} \chi_i \langle \chi \rangle + \frac{1}{n} \sum_{i=1}^{n} \langle \chi \rangle^2 =$$

$$= \frac{1}{n} \sum_{i=1}^{n} \chi_i^2 - 2\langle \chi \rangle \langle \chi \rangle + \langle \chi \rangle^2 == \frac{1}{n} \sum_{i=1}^{n} \chi_i^2 - \langle \chi \rangle^2. \tag{2.5}$$

2.1.2 Continuous Random Variables

In many applications, it is convenient to consider continuous random variables. Let us assume that the random data d can accept any real number in the experiment. Despite the random character of the data, d, their properties can be fully described by the *probability density function* (PDF), $P(d)$. The probability that the measurement is between d and $d + \Delta d$ is determined by the value of $P(d)\Delta d$, which is an analog of the probability mass function introduced above (see Fig. 2.1):

$$p(d) = P(d)\Delta d.$$

In particular, we have for a random variable, d, the following equality:

$$\int_{-\infty}^{+\infty} P(d)\Delta d = 1, \tag{2.6}$$

where Δd denotes a differential of d, and integration is done over all possible values of the random variable, d, from $-\infty$ to $+\infty$. Formula (2.6) represents a simple fact that the random variable will take at least one value in the experiment.

The following expression determines the *mean value* of the continuous random variable d:

$$\langle d \rangle = \int_{-\infty}^{+\infty} d\ P(d)\Delta d. \tag{2.7}$$

A *variance* σ^2, of the random variable, d, describes the deviation of d from its mean value. It is calculated by the following formula:

$$\sigma^2 = \langle (d - \langle d \rangle)^2 \rangle = \int_{-\infty}^{+\infty} (d - \langle d \rangle)^2 P(d)\Delta d. \tag{2.8}$$

The square root of the variance, σ, is called the *standard deviation* of data d.

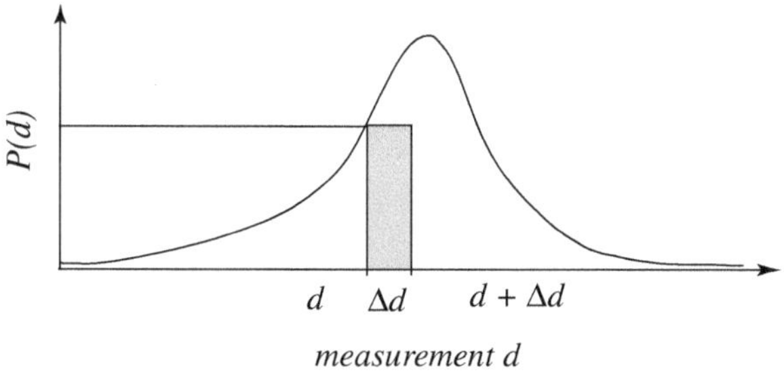

Fig. 2.1 The shaded area of the probability distribution gives the probability that the datum will fall between d and $d + \Delta d$

The most likely measurement d_{ML} is the one with the highest probability:

$$P(d_{ML}) = \max\{P(d)\}. \tag{2.9}$$

The value d_{ML} is called the *maximum likelihood point*.

2.1.3 Standard Probability Density Distributions

In the solution of the inverse problems, we can consider several standard probability density distributions. The simplest one is *uniform distribution*. Random variable d is said to have a uniform distribution over the interval $[d_{\min}, d_{\max}]$, if it has the following PDF:

$$P_U(d) = \begin{cases} 1/(d_{\max} - d_{\min}), & d_{\min} \le d \le d_{\max} \\ 0, \quad d < d_{\min} \text{ or } & d > d_{\max} \end{cases}. \tag{2.10}$$

Formula (2.10) indicates that the random variable d has an equal probability of taking any value between its maximum and minimum values.

One of the most widely used types of probability density distributions is the *normal* or *Gaussian distribution*, described by the following formula:

$$P_G(d) = \frac{1}{(2\pi)^{\frac{1}{2}}\sigma} \exp\left[-\frac{(d - \langle d \rangle)^2}{2\sigma^2}\right]. \tag{2.11}$$

This distribution has a mean $\langle d \rangle$ and variance σ^2. Figure 2.2 shows two typical Gaussian distributions with zero mean, $\sigma = 1$ for curve A, and $\sigma = 2$ for curve B. One can see that the smaller variance corresponds to the narrower and sharper probability distribution, while the bigger variance describes the wider and smoother distribution.

The importance of the Gaussian distribution comes from the Central Limit Theorem (CLT) of the probability theory. This theorem states that the normal distribution can approximate well the probability density distribution of a large sum of the ran-

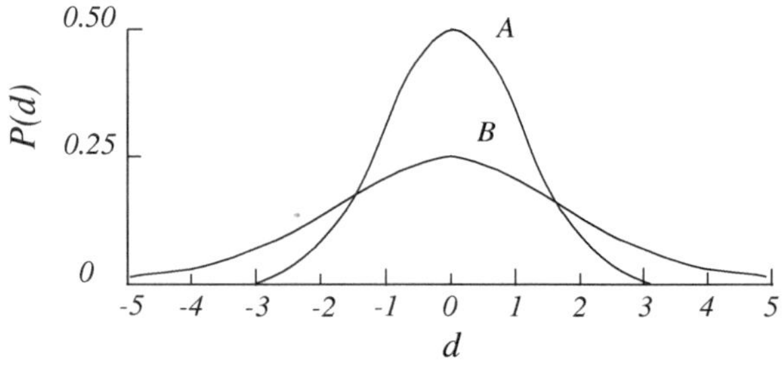

Fig. 2.2 Typical Gaussian distributions with zero mean and $\sigma_1 = 1$ for curve A, and $\sigma_2 = 2$ for curve B. Variance σ_1^2 corresponds to the sharp probability distribution, while variance σ_2^2 describes the smooth distribution

dom variables. One can always consider the observed data as the superposition of a large number of independent random variables. Therefore, we can apply the CLT to the observed data.

Another widely used probability density function is *lognormal distribution*. The random variable, d, is lognormally distributed if its natural logarithm, $\ln d$, is normally (Gaussian) distributed. The following formula describes the probability density function, $P_{LN}(d)$, of lognormal distribution:

$$P_{LN}(d) = \frac{1}{(2\pi)^{\frac{1}{2}}\sigma_{\ln d}} \exp\left[-\frac{(\ln d - \langle \ln d \rangle)^2}{2\sigma_{\ln d}^2}\right], \tag{2.12}$$

where $\sigma_{\ln d}$ is the standard deviation of $\ln d$.

Note that, the random variable that is lognormally distributed may have positive real values only. It is clear also that if random variable, $\tilde{d}$, has a Gaussian distribution, then exponential function, $d = \exp \tilde{d}$, has a lognormal distribution.

One can also consider an *exponential distribution*. With the probability density function, $P_{\exp}(d)$, defined as follows:

$$P_{\exp}(d) = \frac{1}{(2)^{\frac{1}{2}}\sigma} \exp\left[-\frac{|d - \langle d \rangle|}{\sigma}\right]. \tag{2.13}$$

It is helpful to note that the plot of the exponential distribution has much longer tails compared to the plot of the Gaussian distribution with the same variance, σ, and mean value, $\langle d \rangle$.

2.2 Multiple Random Variables

Let us assume that we have the column vector $\mathbf{d} = [d_1, d_2, d_3, \ldots, d_N]^T$ of the observed data (where the upper subscript "T" denotes transposition). In this case, the mean value of the datum, d_i, is determined by the following formula:

$$\langle d_i \rangle = \int_{d_N=-\infty}^{+\infty} \int_{d_2=-\infty}^{+\infty} \cdots \int_{d_1=-\infty}^{+\infty} d_i \; P(\mathbf{d}) \Delta d_1 \Delta d_2 \ldots \Delta d_N. \tag{2.14}$$

It can be demonstrated that the mean value of the product of independent random variables is equal to the product of mean values of each random variable:

$$\langle d_1 d_2 \ldots d_N \rangle = \langle d_1 \rangle \langle d_2 \rangle \ldots \langle d_N \rangle. \tag{2.15}$$

For example, for two independent random variables, d_1 and d_2, we have

$$\langle d_1 d_2 \rangle = \langle d_1 \rangle \langle d_2 \rangle. \tag{2.16}$$

We have already established that random variables d_1 and d_2 are dependent if equality (2.16) does not hold. In this case, we can use the difference between the left and right parts of Eq. (2.16) as the measure of their dependence. This measure is called the *covariance*:

$$\mathrm{cov}(d_1, d_2) = \langle d_1 d_2 \rangle - \langle d_1 \rangle \langle d_2 \rangle. \tag{2.17}$$

It is easy to show that the mean value is a linear function:

$$\langle \alpha_1 d_1 + \alpha_2 d_2 \rangle = \alpha_1 \langle d_1 \rangle + \alpha_2 \langle d_2 \rangle, \tag{2.18}$$

where α_1 and α_2 are some scalar coefficients.

Considering property (2.18), we can transform expression (2.17) for the covariance as follows:

$$\mathrm{cov}(d_1, d_2) = \langle (d_1 - \langle d_1 \rangle)(d_2 - \langle d_2 \rangle) \rangle. \tag{2.19}$$

It is clear from formula (2.17) that the covariance of two independent random variables is always equal to zero.

Finally, we should note that in the case of the normally distributed discrete random variables, ξ and χ, one can use the following statistical estimate of the covariance:

$$cov(\xi, \chi) = \frac{1}{n-1} \sum_{i=1}^{n} (\xi_i - \langle \xi \rangle)(\chi_i - \langle \chi \rangle). \tag{2.20}$$

2.2.1 Linear Correlation Between Two Random Variables

The covariance serves as a measure of probabilistic dependence of two random variables without specifying the form of this dependence. At the same time, we are often interested in the linear dependence between different observed data in practical applications. The measure of linear dependence between two random variables is provided by the *correlation coefficient*, η, determined by the following formula:

$$\eta(d_1, d_2) = \frac{\mathrm{cov}(d_1, d_2)}{\sigma_1 \sigma_2}, \tag{2.21}$$

where σ_1 and σ_2 are the standard deviations of d_1 and d_2, respectively.

The correlation coefficient satisfies the following conditions:

$$-1 \le \eta \le 1. \tag{2.22}$$

If the correlation coefficient between two random variables, d_1 and d_2, equals 1 or -1, these variables are linearly dependent. This means that there exist two coefficients, c_1 and c_2, satisfying the condition $c_1^2 + c_2^2 > 0$, and the linear combination of the random variables d_1 and d_2 with these coefficients is equal to constant,

$$c_1 d_1 + c_2 d_2 = const, \tag{2.23}$$

with probability equal to 1.

The closer the absolute value of the correlation coefficient to 1 is, the more accurately the relationship between d_1 and d_2 can be described by the linear formula (2.23).

In a general case, the relationship between the observed data, d_1 and d_2, is typically nonlinear. However, finding a linear function that best approximates this relationship is always possible. The line that best fits this linear relationship is known as a *least-squares regression*. The following equation describes this line:

$$d_2^{(p)} = \langle d_2 \rangle + \eta \frac{\sigma_2}{\sigma_1} \left(d_1 - \langle d_1 \rangle \right), \tag{2.24}$$

where $d_2^{(p)}$ denotes the values of the random variable d_2 predicted by the linear regression.

The deviation of the actual dependence between d_2 and d_1 from the linear regression (2.24) can be estimated by the *error of linear approximation*, ε, calculated by the following formula:

$$\varepsilon^2 = \langle d_2 - d_2^{(p)} \rangle^2 = \sigma_2^2 \left(1 - \eta^2 \right). \tag{2.25}$$

The linear approximation error, ε, goes to zero if and only if the correlation coefficient $\eta = \pm 1$.

2.2.2 Linear Correlation Between Multiple Random Variables

We consider again a set of N random variables, $d_1, d_2, d_3, \ldots, d_N$. For dependent data, we can introduce some measure of their dependence by calculating the *covariance*:

$$\mathrm{cov}(d_i, d_j) = \langle (d_i - \langle d_i \rangle)(d_j - \langle d_j \rangle) \rangle =$$

$$\int_{d_N=-\infty}^{+\infty} \int_{d_2=-\infty}^{+\infty} \cdots \int_{d_1=-\infty}^{+\infty} (d_i - \langle d_i \rangle)(d_j - \langle d_j \rangle)\, P(\mathbf{d}) \Delta d_1 \Delta d_2 \ldots \Delta d_N. \tag{2.26}$$

Note that the covariance of a datum with itself is just the variance:

$$\mathrm{cov}(d_i, d_i) = \sigma_i^2. \tag{2.27}$$

Thus, for a column vector $\mathbf{d} = [d_1, d_2, d_3, \ldots, d_N]^T$ of the observed data, we can introduce a *covariance matrix* as follows:

$$\boldsymbol{\sigma}_N = \begin{bmatrix} \sigma_{11} & \sigma_{12} & \ldots & \sigma_{1N} \\ \sigma_{21} & \sigma_{22} & \ldots & \sigma_{2N} \\ \ldots & \ldots & \ldots & \ldots \\ \sigma_{N1} & \sigma_{N2} & \ldots & \sigma_{NN} \end{bmatrix} = [\sigma_{ij}] = [\mathrm{cov}(d_i, d_j)]. \tag{2.28}$$

In the next section, we show that the determinant of the covariance matrix, $|\sigma_{ij}|$ is always nonnegative, and it is equal to zero if and only if the random variables, $d_1, d_2, d_3, \ldots, d_N$, are linearly dependent. Thus, the determinant of the covariance matrix provides some measure of linear dependence between the random variables.

In a general case, the relationship between random variable, d_k, and variables $d_1, d_2, d_3, \ldots, d_{k-1}$, can be nonlinear. However, in the case of multiple random variables, as we did for two variables, one can find a linear formula that best approximates the true dependence between the random variables.

We consider a linear function of random variables defined by the following formula:

$$L(d_1, d_2, d_3, \ldots, d_{k-1}) = c_0 + c_1 d_1 + c_2 d_2 + \cdots + c_{k-1} d_{k-1},$$

where $c_0, c_1, c_2, \ldots, c_{k-1}$ are some scalar coefficients defined under the minimum condition for the mean value of the square difference between random variable d_k and $L(d_1, d_2, d_3, \ldots, d_{k-1})$:

$$\langle [d_k - L(d_1, d_2, d_3, \ldots, d_{k-1})]^2 \rangle = \min. \tag{2.29}$$

Using these coefficients, we can write the equation of multidimensional regression as follows:

$$d_k^{(p)} = \langle d_k \rangle + \sum_{q=1}^{k-1} c_q \left(d_q - \langle d_q \rangle \right), \tag{2.30}$$

where $d_k^{(p)}$ denotes the values of the random variable d_k predicted by the linear regression.

The deviation of empirical dependence between random variable, d_k, and variables $d_1, d_2, d_3, \ldots, d_{k-1}$, from multidimensional regression (2.30) can be characterized by the approximation error, ε, calculated as follows:

$$\varepsilon^2 = \langle \left[d_k - d_k^{(p)} \right]^2 \rangle = \frac{|\boldsymbol{\sigma}_k|}{|\boldsymbol{\sigma}_{k-1}|}, \tag{2.31}$$

where $|\boldsymbol{\sigma}_k|$ is the determinant of the covariance matrix of a set of random variables $d_1, d_2, d_3, \ldots, d_{k-1}, d_k$; and $|\boldsymbol{\sigma}_{k-1}|$ is the determinant of the covariance matrix of random variables $d_1, d_2, d_3, \ldots, d_{k-1}$.

In practice, it is convenient to use another parameter to characterize the degree of the linear relationship between $(k - 1)$ independent random variables, $d_1, d_2, d_3, \ldots, d_{k-1}$, and a single dependent variable, d_k. This parameter is called *a multiple correlation coefficient*, R, which is calculated as follows:

$$R^2 \left(d_k \left| d_1, d_2, d_3, \ldots, d_{k-1}\right.\right) = 1 - \frac{|\boldsymbol{\sigma}_k|}{\sigma_k^2 \left|\boldsymbol{\sigma}_{k-1}\right|}. \tag{2.32}$$

The multiple correlation coefficient is a generalization of the standard correlation coefficient for the multiple regression study. Substituting expression (2.32) into (2.31), we find the relationship between the approximation error, ε, and the multiple correlation coefficient, R:

$$\varepsilon^2 = \sigma_k^2 \left(1 - R^2\right). \tag{2.33}$$

One can see that formula (2.33) is a complete analog for multiple random variables of expression (2.25) for two random variables. This formula shows that the absolute value of the multiple correlation coefficient, R, similar to the standard correlation coefficient, η, is always less or equal to 1, because $\varepsilon^2 \geq 0$:

$$-1 \leq R \leq 1. \tag{2.34}$$

The dependent variable, d_k, is linearly connected with independent random variables, $d_1, d_2, d_3, \ldots, d_{k-1}$, if and only if $R^2 = 1$ and $\varepsilon^2 = 0$.

In the opposite case, when random variable d_k is completely independent of the random variables, $d_1, d_2, d_3, \ldots, d_{k-1}$, the multiple correlation coefficient is equal to zero, and the approximation error is equal to the standard deviation σ_k of d_k:

$$R = 0, \ \varepsilon = \sigma_k. \tag{2.35}$$

Formula (2.35) shows that, in this case, variables $d_1, d_2, d_3, \ldots, d_{k-1}$ provide no information about the behavior of the random variable d_k.

Thus, the multiple correlation coefficient estimates the quality of the prediction of the dependent variable by a linear combination of the independent variables.

2.2.3 Properties of the Covariance Matrix

In this section, we consider the properties of the covariance matrix, $\boldsymbol{\sigma}_N$, of a random vector-column, $\mathbf{d}$, formed by independent random variables, $d_1, d_2, d_3, \ldots, d_{N-1}, d_N$. According to formula (2.19), this matrix can be represented in the following form:

$$\boldsymbol{\sigma}_N = \langle(\mathbf{d} - \langle\mathbf{d}\rangle)(\mathbf{d} - \langle\mathbf{d}\rangle)^T\rangle, \tag{2.36}$$

where $\langle\mathbf{d}\rangle$ is the mean value of the random vector $\mathbf{d}$.

The covariance matrix is symmetric,

$$\boldsymbol{\sigma}_N = \boldsymbol{\sigma}_N^T, \tag{2.37}$$

because

$$\sigma_{ij} = \text{cov}(d_i, d_j) = \text{cov}(d_j, d_i) = \sigma_{ji}.$$

Its diagonal elements are equal to the variances of the scalar components of vector **d** and therefore all positive:

$$\sigma_{ii} = \text{cov}(d_i, d_i) = \sigma_i^2 > 0. \tag{2.38}$$

The symmetric $[N \times N]$ matrix, $\mathbf{B}_N$, is called positive definite if the following inequality holds for any nonzero vector **x** from Euclidean space E_N (for definition of E_N see Chap. 3):

$$\mathbf{x}^T \mathbf{B}_N \mathbf{x} > 0. \tag{2.39}$$

We can prove that the covariance matrix, $\boldsymbol{\sigma}_N$, has this property. Let us consider arbitrary nonzero vector-column $\mathbf{x} \in E_N$ and calculate the square of the product of vector-raw $((\mathbf{d} - \langle \mathbf{d} \rangle))^T$ and vector-column **x:**

$$\left[(\mathbf{d} - \langle \mathbf{d} \rangle)^T \mathbf{x}\right]^2 = \left[(\mathbf{d} - \langle \mathbf{d} \rangle)^T \mathbf{x}\right]^T \left[(\mathbf{d} - \langle \mathbf{d} \rangle)^T \mathbf{x}\right]$$

$$= \mathbf{x}^T (\mathbf{d} - \langle \mathbf{d} \rangle)(\mathbf{d} - \langle \mathbf{d} \rangle)^T \mathbf{x} = \left(\sum_{i=1,2,\ldots,N} (d_i - \langle d_i \rangle) x_i \right)^2 > 0, \tag{2.40}$$

where x_i $(i = 1, 2, \ldots, N)$ are the scalar components of nonzero vector **x**.

The mean value of expression $\left\{\mathbf{x}^T (\mathbf{d} - \langle \mathbf{d} \rangle)(\mathbf{d} - \langle \mathbf{d} \rangle)^T \mathbf{x}\right\}$ can be written using the definition of the covariance matrix (2.36) as follows:

$$\left\langle \mathbf{x}^T (\mathbf{d} - \langle \mathbf{d} \rangle)(\mathbf{d} - \langle \mathbf{d} \rangle)^T \mathbf{x} \right\rangle = \mathbf{x}^T \left\langle (\mathbf{d} - \langle \mathbf{d} \rangle)(\mathbf{d} - \langle \mathbf{d} \rangle)^T \right\rangle \mathbf{x} = \mathbf{x}^T \boldsymbol{\sigma}_N \mathbf{x}. \tag{2.41}$$

From the last formula and inequality (2.40), it follows at once that

$$\mathbf{x}_N^T \boldsymbol{\sigma}_N \mathbf{x} > 0, \tag{2.42}$$

and covariance matrix, $\boldsymbol{\sigma}_N$, is positive definite.

We can apply the spectral decomposition to the symmetric matrix (Golub and Van Loan 2013) and write covariance matrix in the following form:

$$\boldsymbol{\sigma}_N = \mathbf{V} \Lambda \mathbf{V}^T, \tag{2.43}$$

where $\mathbf{V} = (\mathbf{v}_1, \mathbf{v}_2, \ldots, \mathbf{v}_N)$ is an $N \times N$ matrix formed by the eigenvectors, and Λ is an $N \times N$ diagonal matrix, formed by the eigenvalues of the covariance matrix, $\lambda_1, \lambda_2, \ldots, \lambda_N$, respectively. Since the covariance matrix is symmetric and positive definite, all its eigenvalues are real and positive numbers (Golub and Van Loan 2013).

The determinant of the covariance matrix can be calculated as follows:

$$\det(\boldsymbol{\sigma}_N) = \det(\mathbf{V}) \det(\boldsymbol{\Lambda}) \det\left(\mathbf{V}^T\right) = \det(\boldsymbol{\Lambda}) = \lambda_1 \lambda_2 \ldots\ldots \lambda_N, \tag{2.44}$$

because $\mathbf{V}$ is an orthogonal matrix, $\det(\mathbf{V}) = 1$, and $\boldsymbol{\Lambda}$ is a diagonal matrix.

Considering that all eigenvalues are positive numbers, $\lambda_i > 0,\ i = 1, 2, \ldots, N$; we find the determinant of the covariance matrix is always positive for independent random variables:

$$\det(\boldsymbol{\sigma}_N) = |\boldsymbol{\sigma}_N| > 0. \tag{2.45}$$

This determinant is equal to zero if and only if the random variables, $d_1, d_2, d_3, \ldots, d_{N-1}, d_N$, are linearly dependent. In this case, according to formulas (2.32) and (2.33), the multiple correlation coefficient R is equal to 1, and the linear approximation error, ε, reduces to zero. Thus, the determinant of the covariance matrix can be used as a measure of linear dependence between the random variables.

2.2.4 *Joint Probability Distribution of Multiple Random Variables*

The joint probability density $P(\mathbf{d})$ of multiple random variables, $d_1, d_2, d_3, \ldots, d_N$, determines the probability that the first datum is between d_1 and $d_1 + \Delta d_1$, the second datum between d_2 and $d_2 + \Delta d_2$, the third datum between d_3 and $d_3 + \Delta d_3$, etc. For example, if the data are independent, then the product of the individual distributions represents the joint distribution as follows:

$$P(\mathbf{d}) = P(d_1) P(d_2) P(d_3) \ldots\ldots P(d_N). \tag{2.46}$$

The random variables $d_1, d_2, d_3, \ldots, d_N$ are called *dependent* if Eq. (2.46) does not hold. Estimating $P(d)$ is quite challenging if the data are dependent (correlated). However, similar to the case of one random variable, according to the Central Limit Theorem, a large sum of multiple random variables can be characterized by the Gaussian joint distribution function defined by the following formula (Menke 2018):

$$P(\mathbf{d}) = \frac{|\boldsymbol{\sigma}_N|^{-\frac{1}{2}}}{(2\pi)^{\frac{N}{2}}} \exp\left[-\frac{1}{2}(\mathbf{d} - \langle\mathbf{d}\rangle)^T \boldsymbol{\sigma}_N^{-1} (\mathbf{d} - \langle\mathbf{d}\rangle)\right], \tag{2.47}$$

where $\langle \mathbf{d} \rangle$ is the mean value of vector $\mathbf{d}$, and $\boldsymbol{\sigma}_N = [\sigma_{ij}]$ is the covariance matrix introduced in expression (2.28).

The set of linear functions of Gaussian random variables can also be characterized by the Gaussian distribution of the same form provided by expression (2.47).

2.3 Shannon's Entropy

2.3.1 Concept of Entropy of a Discrete Random Variable

Shannon introduced the concept of entropy in probability theory in his famous paper "A mathematical theory of communication" (Shannon 1948). It was proposed to provide a measure of uncertainty in the results of some physical experiments. We will find below that Shannon's entropy is defined as a functional of the probability distributions.

For simplicity, we begin with the concept of entropy for a discrete random variable. Let us again represent the data obtained in some physical experiment as a discrete random variable that can take on a finite number of possible values, $d = \xi = (\xi_1, \xi_2, \ldots, \xi_n)$, with probabilities $(p(\xi_1), p(\xi_2), \ldots, p(\xi_n))$.

It is obvious that there is an uncertainty in what specific value will be measured as a result of an experiment. This uncertainty depends on the probability, $p(\xi_i)$, associated with the particular results of the measurement, $d = \xi_i$. The goal is to introduce the function $u(p)$, that will estimate the amount of this uncertainty. We call $u(p)$ the *uncertainty function*. Considering that probability, $p = p(\xi)$, is usually a positive number that cannot exceed $1, 0 < p \leq 1$, the uncertainty function is defined on the semi-segment (0, 1]. We can now calculate the expectation or average value of the uncertainty function over all possible values of the random variable ξ using formula (2.2) as follows:

$$\langle u(\mathbf{p}) \rangle = \sum_{i=1}^{n} p(\xi_i) u[p(\xi_i)], \tag{2.48}$$

where $\mathbf{p}$ is a vector formed by probabilities, $\mathbf{p} = (p(\xi_1), p(\xi_2), \ldots, p(\xi_n))$.

It is important to emphasize that the uncertainty in the results of measurements exists before conducting the experiment only. After the measurements are done, all uncertainty is completely removed, and the proper information about the data is obtained. It is also obvious that if there was little uncertainty in the results of the experiment, e.g., if we knew a priori that the data should be equal to a specific number (with some small variations), then the experiment itself would not provide any useful information about the data. Oppositely, if the uncertainty was large, then conducting the experiment would result in gaining significant new information about the data. Thus, we can conclude that the uncertainty can be equally treated as the amount of information supplied by the physical experiment. Hence, we arrive at the interpretation of the average uncertainty in the information theory as the measure of information contained in the data, $H(\mathbf{p})$,

$$H(\mathbf{p}) = \langle u(\mathbf{p}) \rangle = \sum_{i=1}^{n} p(\xi_i) u[p(\xi_i)]. \quad (2.49)$$

Function $H(\mathbf{p})$ must satisfy several requirements in order to serve as a reasonable measure of information (Aczel and Daroczy 1975). For example, it can be demonstrated that the uncertainty function $u(p)$, is proportional to the logarithm of probability (e.g., Khinchin 1957):

$$u(p) = -\lambda \log p, \quad (2.50)$$

where λ is a positive coefficient, which in information theory is selected as $1/\log 2$.

Substituting expression (2.50) into (2.49), we arrive at the following formula:

$$H(\mathbf{p}) = \langle u(\mathbf{p}) \rangle = -\sum_{i=1}^{n} p(\xi_i) \log_2 [p(\xi_i)]. \quad (2.51)$$

Expression (2.51) provides a definition of the famous Shannon's entropy as the measure of uncertainty. Entropy defined by formula (2.51) possesses many useful properties (for details, see, for example, Aczel and Daroczy 1975; Cover and Thomas 1991). The most important property of Shannon's entropy is nonnegativity:

$$H(\mathbf{p}) \geq 0. \quad (2.52)$$

2.3.2 The Entropy of a Continuous Random Variable

We now extend Shannon's definition of entropy to the continuous random variables introduced above. Let us d is a continuous random variable with the probability density function, $P(d)$.

By analogy with formula (2.51), we can define entropy, $H(d)$, as follows:

$$H(d) = -\int_{-\infty}^{+\infty} P(d) \ln P(d) \Delta d. \quad (2.53)$$

Note that the last expression is often called a *differential entropy* because its properties are not exactly the same as of Shannon's entropy for discrete random variables. For example, differential entropy may not satisfy inequality (2.52).

One can see from the definition, formula (2.53), that the differential entropy is uniquely determined by the corresponding probability distribution of the random variable. Moreover, by imposing different conditions on the differential entropy, one can determine the probability distribution which satisfies these conditions. This result is known as the *maximum entropy principle* (Kapur 1989; Kapur and Kesovan 1992).

The following standard probability distributions can be derived based on the maximum entropy principle.

(1) The uniform probability distribution (2.10) provides the maximum value of the differential entropy.

(2) The Gaussian probability distribution defined by the following formula:

$$P_G(d) = \frac{1}{(2\pi)^{\frac{1}{2}}\sigma} \exp\left[-\frac{(d - \langle d \rangle)^2}{2\sigma^2} \right], \tag{2.54}$$

maximizes the differential entropy (2.53) subject to the constraints

$$\int_{-\infty}^{+\infty} d\, P(d) \Delta d = \langle d \rangle, \tag{2.55}$$

and

$$\int_{-\infty}^{+\infty} (d - \langle d \rangle)^2 P(d) \Delta d = \sigma^2. \tag{2.56}$$

Imposing other constraints, one can arrive at different standard probability distributions.

References and Recommended Reading to This Chapter

Aczel J, Daroczy Z (1975) On measures of information and their characterizations. Academic, New York

Cover TM, Thomas JA (1991) Elements of information theory. Wiley, New York

Golub GH, Van Loan CF (2013) Matrix computations, 4th edn. The Johns Hopkins University Press, Baltimore and London, 753 pp

Kapur JN (1989) Maximum-entropy models in science and engineering. Wiley Eastern Limited, New Delhi

Kapur JN, Kesavan HK (1992) Entropy optimization principles with applications. Academic, New York

Khinchin AYa (1957) Mathematical foundations of information theory. Dover, New York

Menke W (2018) Geophysical data analysis: discrete inverse theory, 4th edn. Elsevier

Ross S (2010) A first course in probability, 8th edn. Printice Hall, Upper Saddle River, New Jersey, p 07458

Shannon CE A mathematical theory of communication. Bell Sys Tech J 27:379–423, 623–656

Chapter 3
Vector Spaces of Models and Data

Abstract The mathematical theory of functional spaces plays a critical role in inversion theory. This chapter introduces the concept of a mathematical space and describes the different types of spaces, including multi-dimensional Euclidean, metric, linear vector, Hilbert, and Gramian spaces. The fundamental properties of all these spaces are discussed in detail. The definitions and properties of operators and functionals acting in mathematical spaces are also considered. The chapter concludes with a review of the major principles of variational calculus.

Keywords Euclidean space · Vector space · Hilbert space · Gramian space · Operators · Functionals

In this chapter, I review the key ideas and principles of the mathematical theory of functional spaces. These ideas help us understand how to introduce distance in describing multiphysics data and models and develop the mathematical framework of the inversion theory.

I begin this chapter by describing the most fundamental space—Euclidean space, which is a straightforward generalization of the conventional 3D physical space into multi-dimensional space. I will move then to introducing the metric and linear vector spaces. This discussion will be concluded by constructing the Hilbert space, which is the generalization of the 3D physical space to the case of infinite dimensions.[1]

3.1 Concept of a Space

The main description of conventional three-dimensional physical space is provided by its geometrical properties. These properties were first described in detail in the classical book of the great Greek mathematician Euclid (Fig. 3.1), who introduced the five fundamental geometrical axioms. One of the most critical geometrical concepts of conventional physical space is the distance between two points and the triangle

[1] In presenting the theory of functional spaces, I will closely follow Appendix A of Zhdanov (2015).

M. S. Zhdanov, *Advanced Methods of Joint Inversion and Fusion of Multiphysics Data*, Advances in Geological Science,
https://doi.org/10.1007/978-981-99-6722-3_3

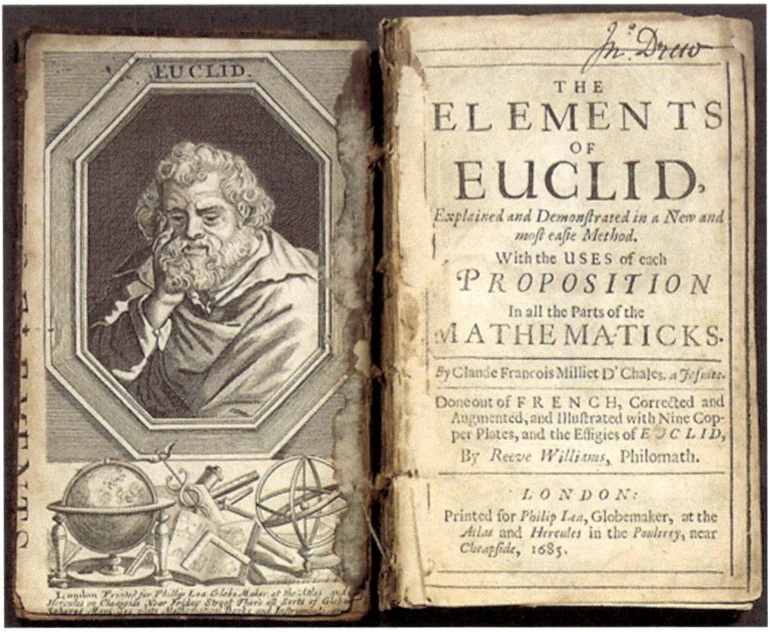

Fig. 3.1 Euclidian geometry

inequality theorem. According to this theorem, one side of a triangle can never be greater than the sum of the lengths of the other two sides. Thus, for example, if A, B, and C be the three vertices of a triangle (Fig. 3.2), then the length of one side, $|AB|$, is always shorter or equal to the sum of the lengths of other sides, $|AC|$ and $|CB|$. In other words, the shortest distance between the two distinct points is always a straight line:

$$|AB| \leq |AC| + |CB| . \tag{3.1}$$

One can write inequality (3.1) using vector notations as follows:

$$|\mathbf{a} + \mathbf{b}| \leq |\mathbf{a}| + |\mathbf{b}| , \tag{3.2}$$

where vectors **a**, **b**, and $(\mathbf{a} + \mathbf{b})$ represent three sides of the triangle ACB, as shown in Fig. 3.2.

These geometrical ideas, developed for the conventional physical space, can be expanded by introducing a more general concept of a mathematical space. This means that we can use as an element of a space not only geometrical points but also any mathematical objects, such as vectors or functions. In the last case, the corresponding space is called a function space. One of the most fundamental properties of mathematical space is the geometrical property, which allows us to consider a distance between

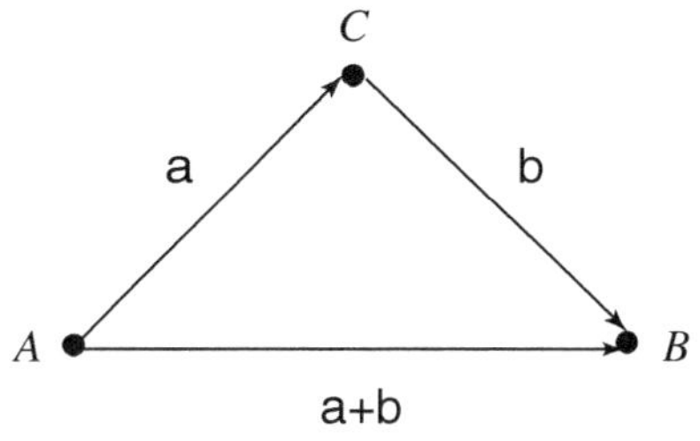

Fig. 3.2 Triangle inequality

any two elements (vectors) of the mathematical space. This property is fundamental in many physical and mathematical applications. The geometry assigned to much more complicated objects than just geometrical points makes it possible to measure the distance between complex physical and mathematical objects, which otherwise have no apparent geometrical representation (e.g., between different mathematical functions or digital data sets).

To illustrate the significance of the geometrical concept in inversion theory, let us recall the formulation of the inverse problem provided above as a solution of the following operator equation:

$$\mathbf{d} = A(\mathbf{m}), \tag{3.3}$$

where $\mathbf{m}$ is some function (or a vector) describing the model parameters, and $\mathbf{d}$ is a data set, which can also be characterized as a function of the observation point (in the case of continuous observations) or as a vector (in the case of discrete observations). The solution of the inverse problem consists in determining such a model, $\mathbf{m}^{pr}$ (predicted model), which generates the predicted data, $\mathbf{d}^{pr}$, fitting the observed data, $\mathbf{d}$, well. In practical applications, we do not want to fit the observed data exactly because they always contain some noise, which we should not fit. Therefore, we are looking for some predicted data to be close enough to the observed data (usually within the accuracy of observations). But what does "close enough" mean? How can we measure the closeness of the two data sets?

We can answer this question by introducing some "distance" between the two data sets, which would allow us to evaluate the accuracy of the inverse problem solution. The mathematical theory of vector spaces provides us with the means of solving this problem.

In the next section, we consider the most fundamental mathematical vector space—Euclidean space.

3.2 Euclidean Space

3.2.1 Vector Operations in Euclidean Space

Conventional physical space has three dimensions. Any point in this space can be represented by three Cartesian coordinates (x_1, x_2, x_3). The natural generalization of three-dimensional (3D) physical space is the n dimensional Euclidean space E_n (or R_n), which can be described as the set of all possible vectors of order n:

$$\mathbf{a} = (a_1, a_2, a_3, \ldots \ldots a_n),$$

where the scalars $a_1, a_2, a_3, \ldots \ldots a_n$ are usually real numbers.

By analogy with the length of the vector in 3D physical space, we can introduce a *norm of the vector* $\|\mathbf{a}\|$ as follows:

$$\|\mathbf{a}\| = \sqrt{a_1^2 + a_2^2 + a_3^2 + \cdots \cdots + a_n^2}. \tag{3.4}$$

It is easy to check that the norm introduced above satisfies the conditions

$$\|\mathbf{a}\| > 0 \text{ if } \mathbf{a} \neq 0, \ \ \|\mathbf{a}\| = 0 \text{ if } \mathbf{a} = 0, \tag{3.5}$$

$$\|\lambda\mathbf{a}\| = |\lambda| \, \|\mathbf{a}\|, \tag{3.6}$$

$$\|\mathbf{a} + \mathbf{b}\| \leq \|\mathbf{a}\| + \|\mathbf{b}\|. \tag{3.7}$$

The last inequality is called a *triangle inequality*. In 3D physical space it has a very simple geometrical sense explained above (Fig. 3.2). In a general case of n dimensional Euclidean space, the triangle inequality comes from the Cauchy inequality, which we will discuss below.

We can introduce also an operation on two vectors, called the *inner (dot) product,* as follows:

$$\mathbf{a} \cdot \mathbf{b} = \sum_{i=1}^{n} a_i b_i. \tag{3.8}$$

Obviously, a norm can be determined as a square root of the dot product of the vector with itself:

$$\|\mathbf{a}\| = \sqrt{\mathbf{a} \cdot \mathbf{a}}. \tag{3.9}$$

By analogy with conventional 3D physical space, we can say that two vectors **a** and **b** in Euclidean space are orthogonal if:

$$\mathbf{a} \cdot \mathbf{b} = 0.$$

The following vectors play a similar role in the space E_n as the vectors of the Cartesian basis in 3D space:

$$\mathbf{e}_1 = (1, 0, 0, \ldots\ldots 0),\ \mathbf{e}_2 = (0, 1, 0, \ldots\ldots 0),\ \mathbf{e}_3 = (0, 0, 1, \ldots\ldots 0), \ldots\ldots.$$

$$\mathbf{e}_n = (0, 0, 0, \ldots\ldots 1).$$

We will call these vectors, $\mathbf{e}_1, \mathbf{e}_2, \ldots\ldots \mathbf{e}_n$, *a basis* of Euclidean space. Any vector $\mathbf{a} \in E_n$ can be represented as a linear combination of the basis vectors

$$\mathbf{a} = a_1\mathbf{e}_1 + a_2\mathbf{e}_2 + \cdots\cdots\cdots + a_n\mathbf{e}_n = \sum_{i=1}^{n} a_i\mathbf{e}_i, \tag{3.10}$$

where numbers $a_1, a_2, a_3, \ldots\ldots a_n$ are the scalar components of vector $\mathbf{a}$. Evidently

$$\mathbf{e}_i \cdot \mathbf{e}_k = 0;\ \text{if}\ i \neq k,\ \mathbf{e}_k \cdot \mathbf{e}_k = 1. \tag{3.11}$$

We can write Eq. (3.11) in a short form:

$$\mathbf{e}_i \cdot \mathbf{e}_k = \delta_{ik} = \begin{cases} 1, & i = k \\ 0, & i \neq k \end{cases},$$

where symbol δ_{ik} is called a symmetric Kronecker symbol. Using this symbol one can find that a_k can be treated as the projection of the vector $\mathbf{a}$ on the basis vector $\mathbf{e}_k$:

$$\mathbf{a} \cdot \mathbf{e}_k = \sum_{i=1}^{n} a_i (\mathbf{e}_i \cdot \mathbf{e}_k) = \sum_{i=1}^{n} a_i\delta_{ik} = a_k.$$

Using the dot product operation, we can prove the *Cauchy inequality*:

$$\mathbf{a} \cdot \mathbf{b} \leq \|\mathbf{a}\| \|\mathbf{b}\|, \tag{3.12}$$

which in scalar form can be written as follows:

$$\left(\sum_{i=1}^{n} a_i b_i\right)^2 \leq \left(\sum_{i=1}^{n} a_i^2\right)\left(\sum_{i=1}^{n} b_i^2\right). \tag{3.13}$$

Proof Let us introduce a non-negative function $\varphi(x)$:

$$\varphi(x) = \sum_{i=1}^{n} (a_i x + b_i)^2 \geq 0.$$

Opening the brackets in the last equation, we obtain

$$\left(\sum_{i=1}^{n} a_i^2\right) x^2 + 2\left(\sum_{i=1}^{n} a_i b_i\right) x + \sum_{i=1}^{n} b_i^2 \geq 0. \tag{3.14}$$

Inequality (3.14) means that the equation

$$\left(\sum_{i=1}^{n} a_i^2\right) x^2 + 2\left(\sum_{i=1}^{n} a_i b_i\right) x + \sum_{i=1}^{n} b_i^2 = 0 \tag{3.15}$$

has only one real root, or no real roots at all, which is possible only if its discriminant is non-positive:

$$\left(\sum_{i=1}^{n} a_i b_i\right)^2 - \left(\sum_{i=1}^{n} a_i^2\right)\left(\sum_{i=1}^{n} b_i^2\right) \leq 0, \tag{3.16}$$

which is exactly the Cauchy inequality (3.13).

3.2.2 *Linear Transformations (Operators) in Euclidean Space*

Suppose that for any $\mathbf{a} \in E_n$ we can assign, according to a certain rule, some element $\mathbf{a}' \in E_n$. We call this rule *an operator* A : $\mathbf{a}' = A(\mathbf{a})$. Operator A is called linear if for any vectors $\mathbf{a}_1, \mathbf{a}_2 \in E_n$ and any scalars $\alpha_1, \alpha_2 \in E_1$ we have

$$A(\alpha_1 \mathbf{a}_1 + \alpha_2 \mathbf{a}_2) = \alpha_1 A(\mathbf{a}_1) + \alpha_2 A(\mathbf{a}_2). \tag{3.17}$$

Let us find the relationships between the components of the vectors $\mathbf{a}$ and $\mathbf{a}'$:

$$\mathbf{a}' = A(\mathbf{a}) = A\left(\sum_{i=1}^{n} a_i \mathbf{e}_i\right) = \sum_{i=1}^{n} a_i A(\mathbf{e}_i). \tag{3.18}$$

At the same time, by applying operator A to the basis vector $\mathbf{e}_i$ we obtain a new vector $\mathbf{e}_i'$, which in turn can be decomposed in terms of the basis vectors:

$$A(\mathbf{e}_i) = \mathbf{e}_i' = \sum_{k=1}^{n} A_{ki} \mathbf{e}_k. \tag{3.19}$$

Substituting Eq. (3.19) into (3.18), we obtain

$$\mathbf{a}' = A(\mathbf{a}) = \sum_{i=1}^{n} a_i \sum_{k=1}^{n} A_{ki} \mathbf{e}_k. \tag{3.20}$$

On the other hand, we can also express vector $\mathbf{a}'$ in the same basis by

$$\mathbf{a}' = A(\mathbf{a}) = \sum_{k=1}^{n} a'_k \mathbf{e}_k. \tag{3.21}$$

Comparing (3.20) with (3.21), we find a very important relationship between the scalar components of the vectors $\mathbf{a}$ and $\mathbf{a}'$:

$$a'_k = \sum_{i=1}^{n} A_{ki} a_i.$$

The matrix $[A_{ki}]$ is called *the matrix of operator* A. It describes the transformation of the scalar components of the vector by the linear operator A.

3.2.3 *Norm of the Operator*

We can calculate a norm of the vector $\mathbf{a}'$ as follows

$$\|\mathbf{a}'\|^2 = \sum_{i=1}^{n} a_i'^2 = \sum_{i=1}^{n} \left(\sum_{k=1}^{n} A_{ik} a_k \right)^2. \tag{3.22}$$

From the Cauchy inequality we have

$$\left(\sum_{k=1}^{n} A_{ik} a_k \right)^2 \leq \left(\sum_{k=1}^{n} A_{ik}^2 \right) \left(\sum_{k=1}^{n} a_k^2 \right) = M_i \|\mathbf{a}\|^2, \tag{3.23}$$

where $M_i = \sum_{k=1}^{n} A_{ik}^2$ are some constants. Substituting Eq. (3.23) into (3.22), we obtain

$$\|\mathbf{a}'\|^2 \leq M^2 \|\mathbf{a}\|^2, \text{ where } M^2 = \sum_{i=1}^{n} M_i = \sum_{i=1}^{n} \sum_{k=1}^{n} A_{ik}^2 \tag{3.24}$$

or

$$\|\mathbf{a}'\| \leq M \|\mathbf{a}\|. \tag{3.25}$$

Let us introduce the definition

Definition 3.1 The norm of an operator A is the minimum value of all possible M that satisfy inequality (3.25):

$$\|A\| = \min \{M > 0, \ \|A(\mathbf{a})\| \leq M \|\mathbf{a}\|\}, \ \|A\| \leq M. \tag{3.26}$$

Thus we have

$$\|A(\mathbf{a})\| \leq \|A\| \, \|\mathbf{a}\| \leq M\|\mathbf{a}\|. \tag{3.27}$$

Based on the last formula, we can write an equivalent expression for the norm of operator A as follows

$$\|A\| = \sup_{\|\mathbf{a}\| \neq 0} \|A(\mathbf{a})\| / \|\mathbf{a}\|. \tag{3.28}$$

Taking into account this definition and inequality (3.24), we can write

$$\|A\| \leq \sqrt{\sum_{i=1}^{n} \sum_{k=1}^{n} A_{ik}^2}. \tag{3.29}$$

Note that the expression on the right-hand side of inequality (3.29) is called the *Frobenius norm of the matrix,* $\|A\|_F$:

$$\|A\|_F = \sqrt{\sum_{i=1}^{n} \sum_{k=1}^{n} A_{ik}^2}.$$

Definition 3.2 A linear operator A is called a bounded operator if it has a bounded norm:

$$\|A\| < \infty$$

It is easy to show that a linear bounded operator is a continuous operator, i.e., that the small variations of the argument of the operator will result in a small variation of its values. Clearly, from inequality (3.27) we have

$$\|A(\mathbf{a}) - A(\mathbf{b})\| = \|A(\mathbf{a} - \mathbf{b})\| \leq \|A\| \|\mathbf{a} - \mathbf{b}\|.$$

Therefore, if $\|\mathbf{a} - \mathbf{b}\| < \delta = \varepsilon / \|A\|$, then $\|A(\mathbf{a}) - A(\mathbf{b})\| < \varepsilon$.

In conclusion, note that in a finite-dimensional space, any linear operator is bounded and continuous.

3.2.4 Linear Functionals

A functional in Euclidean space is a rule that unambiguously assigns a single real number to an element in the space E_n. The functional is linear if for any vectors $\mathbf{a}_1, \mathbf{a}_2 \in E_n$ and any scalars $\alpha_1, \alpha_2 \in E_1$ we have

$$f(\alpha_1 \mathbf{a}_1 + \alpha_2 \mathbf{a}_2) = \alpha_1 f(\mathbf{a}_1) + \alpha_2 f(\mathbf{a}_2). \tag{3.30}$$

Consider as an example the following linear functional

$$f(\mathbf{a}) = \mathbf{a} \cdot \mathbf{l}\,, \tag{3.31}$$

where $\mathbf{a}$ is an arbitrary vector, and $\mathbf{l}$ is some fixed vector.

The remarkable fact is that any linear functional can be represented in the form (3.31). To prove this, we introduce scalars $l_i = f(\mathbf{e}_i),\ i = 1, 2, 3, \ldots, n$. Then

$$f(\mathbf{a}) = f\left(\sum_{i=1}^{n} a_i \mathbf{e}_i\right) = \sum_{i=1}^{n} a_i f(\mathbf{e}_i) = \sum_{i=1}^{n} a_i l_i = \mathbf{a} \cdot \mathbf{l},$$

which is exactly Eq. (3.31).

3.2.5 Norm of the Functional

Consider the Cauchy inequality

$$\mid f(\mathbf{a}) \mid = \mid \mathbf{a} \cdot \mathbf{l} \mid \leq \|\mathbf{a}\|\ \|\mathbf{l}\|. \tag{3.32}$$

Let us introduce some large enough constant $L > 0$, which satisfies the following inequality for all $\mathbf{a} \in E_n$:

$$\mid f(\mathbf{a}) \mid \leq L\|\mathbf{a}\|. \tag{3.33}$$

Definition 3.3 The norm of the functional f is the minimum value of all possible L that satisfy inequality (3.33): $\|f\| = \min\{L > 0,\ |f(\mathbf{a})| \leq L\|\mathbf{a}\|\},\ \|f\| \leq L.$

On the other hand we have

$$|f(\mathbf{l})| = \mathbf{l} \cdot \mathbf{l} = \|\mathbf{l}\|\ \|\mathbf{l}\| \leq L\ \|\mathbf{l}\|.$$

So the minimum value of L that satisfies (3.33) is the norm of the constant vector $\mathbf{l}$:

$$L \geq \|\mathbf{l}\|.$$

Therefore we have established that the norm of the functional is equal to the norm of the vector $\mathbf{l}$ given by its representation (3.31):

$$\|f\| = \|\mathbf{l}\|.$$

3.3 Metric Space

The *metric space* is the simplest and, at the same time, the most important mathematical space that contains geometry (in the sense that there is a distance between any two elements of this space).

3.3.1 Definition of the Metric Space

A metric space is a set, M, of elements, $\{\mathbf{h}\}$, for each two of which the non-negative number $\mu(\mathbf{h}, \mathbf{g})$ can be defined, called the distance between the two elements, $\mathbf{h}$ and $\mathbf{g}$, or metric. Moreover, the metric has to satisfy the following conditions:

$$\mu(\mathbf{h}, \mathbf{g}) = 0 \quad \text{if and only if } \mathbf{h} = \mathbf{g}, \qquad (i)$$

$$\mu(\mathbf{h}, \mathbf{g}) = \mu(\mathbf{g}, \mathbf{h}), \qquad (ii)$$

$$\mu(\mathbf{h}, \mathbf{g}) \leq \mu(\mathbf{h}, \mathbf{q}) + \mu(\mathbf{q}, \mathbf{g}), \text{ for any } \mathbf{h}, \mathbf{g}, \mathbf{q} \in M. \quad (iii)$$

The last inequality is called *the triangle inequality* by analogy with the conventional triangle inequality (3.1) of Euclidian geometry. Indeed, it also states that the distance between any two elements (points) in the metric space, $\mathbf{h}$ and $\mathbf{g}$, is shorter or equal to the sum of distances between the given elements and the third element, $\mathbf{q}$ (Fig. 3.3).

One important property of the metric space is that we can introduce a concept of convergence of a sequence of elements in this space.

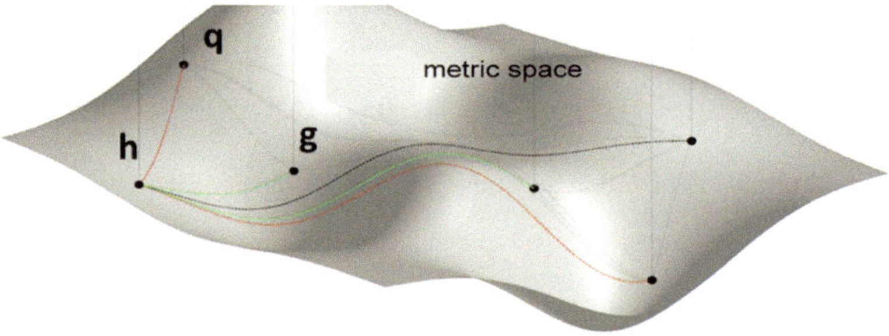

Fig. 3.3 Triangle inequality in the metric space: $\mu(\mathbf{h}, \mathbf{g}) \leq \mu(\mathbf{h}, \mathbf{q}) + \mu(\mathbf{q}, \mathbf{g})$

3.3.2 Convergence, Cauchy Sequences, and Completeness

We begin with several definitions.

Definition 3.4 In a metric space an infinite sequence of elements $\mathbf{f}_1, \mathbf{f}_2, \mathbf{f}_3, \ldots$ is said to converge to element $\mathbf{g}$ if as $k \to 0$, the distance between $\mathbf{f}_k$ and $\mathbf{g}$ tends to zero: $\mu(\mathbf{f}_k, \mathbf{g}) \to 0$.

Definition 3.5 Any sequence in which the distance between any two elements tends to zero, $\mu(\mathbf{f}_k, \mathbf{f}_j) \to 0$, as $k, j \to \infty$, is called a Cauchy sequence.

One can prove that any convergent sequence is a Cauchy sequence. In fact, from the triangle inequality, we can write

$$\mu(\mathbf{f}_k, \mathbf{f}_j) \leq \mu(\mathbf{f}_k, \mathbf{g}) + \mu(\mathbf{g}, \mathbf{f}_j) \to 0, \quad \text{as } k, j \to \infty.$$

On the other hand, there exist Cauchy sequences of elements that do not converge to an element in the metric space. For example, let us consider as a metric space the internal part of the geometric 3D ball B without a boundary. We can introduce series of points $\mathbf{s}_1, \mathbf{s}_2, \mathbf{s}_3, \ldots$, which converge to the element $\mathbf{s}_0$ located at the boundary. Obviously, the set $\mathbf{s}_1, \mathbf{s}_2, \mathbf{s}_3, \ldots$ forms a Cauchy sequence, but it converges to the element $\mathbf{s}_0$ outside our metric space B. From this point of view, we can call B an incomplete metric space.

Now we give a rigorous mathematical definition.

Definition 3.6 A metric space is said to be incomplete if there are Cauchy sequences in it that do not converge to an element of this metric space. Conversely, a space M is complete if every Cauchy sequence converges to an element of the space.

We give below several additional definitions which play an important role in inversion theory.

Definition 3.7 A subset C of the elements of the metric space M is called compact if any sequence of elements $\mathbf{f}_1, \mathbf{f}_2, \mathbf{f}_3, \ldots$ from C contains a convergent sequence, which converges to an element $\mathbf{g} \in C$.

Definition 3.8 A subset N of the elements of the metric space M is called bounded if $\mu(\mathbf{f}, \mathbf{g}) \leq R = const$ for any $\mathbf{f}, \mathbf{g} \in N$.

For example, the metric 3D ball B introduced above is a bounded subset of the physical 3D space.

There are several important theorems about compact sets of elements.

Theorem 3.9 *Any compact set is bounded.*

Theorem 3.10 *For any subset S of Euclidean space E to be compact, it is necessary and sufficient that S be bounded.*

Thus we can see that a metric space contains one fundamental property of the conventional space: there is a distance between any two points. However, the metric space is very amorphous; it has no rigid geometrical structure. We would like to have more specific geometrical properties in many applications than just a distance between two points. This goal can be reached by introducing a new operation on the elements of an abstract mathematical space, which is a generalization of the vector operations in the conventional geometric space.

3.4 Linear Vector Spaces

3.4.1 Vector Operations

A *linear vector space* is a set, L, containing elements (vectors) that can be related by two operations, addition and scalar multiplication, satisfying the conditions

$$\mathbf{f} + \mathbf{g} = \mathbf{g} + \mathbf{f},$$

$$\mathbf{f} + (\mathbf{g} + \mathbf{h}) = (\mathbf{f} + \mathbf{g}) + \mathbf{h},$$

$$\mathbf{0} \in L; \ \mathbf{f} + \mathbf{0} = f,$$

$$(\alpha + \beta)\mathbf{f} = \alpha\mathbf{f} + \beta\mathbf{f},$$

$$\alpha(\beta\mathbf{f}) = (\alpha\beta)\mathbf{f},$$

$$\alpha(\mathbf{f} + \mathbf{g}) = \alpha\mathbf{f} + \alpha\mathbf{g},$$

where

$$\alpha, \beta \in E_1, \ \mathbf{f}, \mathbf{g} \in L,$$

and element $\mathbf{0}$ is called a zero element of the linear vector space.

We now give several definitions, which largely determine the properties of linear spaces.

Definition 3.11 A *linear subspace* of L is a subset of L that forms a linear vector space under the rules of addition and scalar multiplication defined for L.

Definition 3.12 A *linear combination* of elements $\mathbf{f}_1, \mathbf{f}_2, \mathbf{f}_3, \ldots \mathbf{f}_n$ is any vector of the form $\alpha_1\mathbf{f}_1 + \alpha_2\mathbf{f}_2 + \alpha_3\mathbf{f}_3 + \cdots + \alpha_n\mathbf{f}_n$.

Definition 3.13 Elements $\mathbf{f}_1, \mathbf{f}_2, \mathbf{f}_3, \ldots \mathbf{f}_n$ are linearly dependent if it is possible to find a linear combination of them whose value is zero element and not all the scalars of the combination are zero.

Definition 3.14 Elements $\mathbf{f}_1, \mathbf{f}_2, \mathbf{f}_3, \ldots \mathbf{f}_n$ are linearly independent if a linear combination of them is equal to zero if and only if all the scalars of the combination are zero.

Definition 3.15 The linear space L is called *finite dimensional* if there is a finite number of linearly independent elements $\mathbf{e}_1, \mathbf{e}_2, \mathbf{e}_3, \ldots \mathbf{e}_n$, a linear combination of which can determine any element of L.

The elements $\mathbf{e}_1, \mathbf{e}_2, \mathbf{e}_3, \ldots \mathbf{e}_n$ form a *basis* of L.

Thus, we can see that the linear vector space contains another very important Euclidean space property: the basis. However, there is no distance in a linear vector space. Therefore, it would be advantageous to combine these two properties of the Euclidean space, a distance, and a basis, within one space. This space is called *a normed linear space.*

3.4.2 Normed Linear Spaces

A *normed linear space* is a linear space, N, in which for any vector $\mathbf{f}$ there corresponds a real number, denoted by $\|\mathbf{f}\|$ and called the *norm* of $\mathbf{f}$, in such a manner that

$$\|\mathbf{f}\| \geq 0, \text{ and } \|\mathbf{f}\| = 0 \text{ if and only if } \mathbf{f} = \mathbf{0}, \ (i)$$

$$\|\alpha \mathbf{f}\| = \mid \alpha \mid \|\mathbf{f}\|, \qquad (iii)$$

$$\|\mathbf{f} + \mathbf{g}\| \leq \|\mathbf{f}\| + \|\mathbf{g}\|, \qquad (ii)$$

where $\alpha \in E_1$. A normed linear space can be made a metric space if we introduce a metric by the following formula

$$\mu(\mathbf{f}, \mathbf{g}) = \|\mathbf{f} - \mathbf{g}\|.$$

We introduce a special type of linear space by the following definition:

Definition 3.16 *A Banach space B* is a complete linear normed space.

This means that every Cauchy sequence in a Banach space converges to an element of this space.

Thus we can see that a normed linear vector space contains both a basis and a distance. It has two important properties of Euclidean space, but not all of its properties. One property, which is still missing, is the analog of the dot product of two vectors in the conventional geometric space. This property is very important because it actually provides the possibility not only to determine the distance between two points but also to characterize the direction from one point to another in abstract mathematical space. Therefore, the geometrical properties of the space become more rigid. We introduce the space with these properties below.

3.5 Hilbert Spaces

The Hilbert space extends the geometrical properties of 3D physical or Euclidean spaces to mathematical spaces with any finite or infinite dimension. This extension is based on the inner product operation, which allows the measurement of both the distance and angles in this space, thus making it a complete geometrical analog of the conventional 3D physical space.

3.5.1 *Inner Product*

Let us introduce a linear vector space L^I in which for every pair of elements $\mathbf{f}, \mathbf{g}$, we define a functional, called *the inner product* $(\mathbf{f}, \mathbf{g})$, with the properties

$$(\mathbf{f}, \mathbf{g}) = (\mathbf{g}, \mathbf{f}) \qquad \text{(symmetry)}, \tag{3.34}$$

$$(\mathbf{f} + \mathbf{g}, \mathbf{h}) = (\mathbf{f}, \mathbf{h}) + (\mathbf{g}, \mathbf{h}) \quad \text{(linearity)}, \tag{3.35}$$

$$(\alpha\mathbf{f}, \mathbf{g}) = \alpha(\mathbf{f}, \mathbf{g}) \qquad \text{(linearity)}. \tag{3.36}$$

where $\alpha \in E_1$, $\mathbf{f}, \mathbf{g} \in L^I$.

This functional must also be positive definite, i.e.,

$$(\mathbf{f}, \mathbf{f}) > 0 \tag{3.37}$$

and

$$(\mathbf{f}, \mathbf{f}) = 0 \text{ if and only if } \mathbf{f} = \mathbf{0}. \tag{3.38}$$

The operation of the inner product can be treated as an analog of the dot product in Euclidean space.

Evidently the space L^I comes equipped with the norm:

$$\|\mathbf{f}\| = \sqrt{(\mathbf{f}, \mathbf{f})}.$$

The linear normed space L^I equipped with the inner product is called *a pre-Hilbert space*. In order to obtain a Hilbert space we require that the space L^I be complete, in other words, every Cauchy sequence of elements from L^I must converge to an element of this space. So we arrive at the following definition.

Definition 3.17 *A Hilbert space H* is a complete linear normed space whose norm arises from the inner product defined above.

We can now prove a very important inequality, which is the generalization of the Cauchy inequality for Euclidean space.

Theorem 3.18 (Schwarz inequality) *If* $\mathbf{x}$ *and* $\mathbf{y}$ *are any two vectors in a Hilbert space, then*

$$(\mathbf{x}, \mathbf{y}) \leq \|\mathbf{x}\| \; \|\mathbf{y}\| \, . \tag{3.39}$$

Proof When $\mathbf{y} = \mathbf{0}$ the result is clear, for both sides vanish. When $\mathbf{y} \neq \mathbf{0}$ the inequality (3.39) is equivalent to

$$\left(\mathbf{x}, \frac{\mathbf{y}}{\|\mathbf{y}\|}\right) \leq \|\mathbf{x}\|.$$

We have therefore prove only that if $\|\mathbf{y}\| = 1$, then

$$(\mathbf{x}, \mathbf{y}) \leq \|\mathbf{x}\| \quad \text{for all} \quad \mathbf{x}. \tag{3.40}$$

To prove the last inequality we note that:

$$0 \leq \|\mathbf{x} - (\mathbf{x}, \mathbf{y})\mathbf{y}\|^2 = (\mathbf{x} - (\mathbf{x}, \mathbf{y})\mathbf{y}, \mathbf{x} - (\mathbf{x}, \mathbf{y})\mathbf{y})$$

$$= (\mathbf{x}, \mathbf{x}) - 2(\mathbf{x}, \mathbf{y})^2 + (\mathbf{x}, \mathbf{y})^2 \|\mathbf{y}\|^2 = \|\mathbf{x}\|^2 - (\mathbf{x}, \mathbf{y})^2,$$

since $\|\mathbf{y}\| = 1$, from which statement (3.40) follows at once.

Thus, Hilbert space crowns the construction of different functional spaces. Figure 3.4 illustrates the hierarchy of functional spaces.

The simplest is a metric space, which possesses only geometrical properties—a distance (a metric) between any two points. A linear vector space has algebraic properties—addition and multiplication of the vectors. A normed vector space combines these geometrical and algebraic properties. However, the geometry is still very amorphous because there is no way to introduce direction or "angle" between two vectors in this space.

The Hilbert space is the richest with geometrical properties. One can consider not only a distance between any two vectors from a Hilbert space, but also an angle, φ, between any two vectors, $\mathbf{f}$ and $\mathbf{g}$, determined by the following formula:

$$\cos \varphi = \frac{(\mathbf{f}, \mathbf{g})}{\|\mathbf{f}\| \; \|\mathbf{g}\|}. \tag{3.41}$$

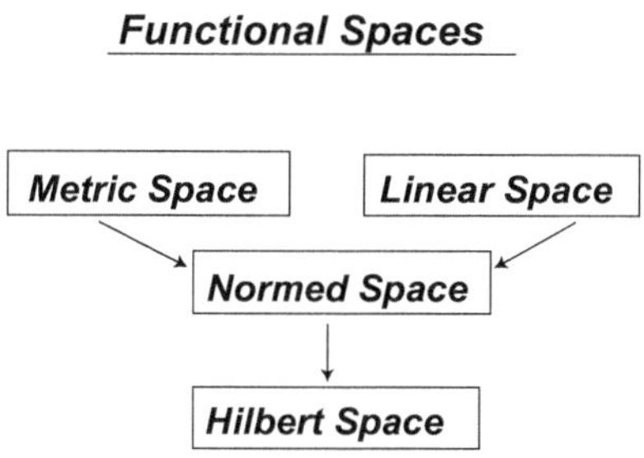

Fig. 3.4 The hierarchy of functional spaces

It follows from the Schwarz inequality that

$$\frac{|(\mathbf{f}, \mathbf{g})|}{\|\mathbf{f}\| \; \|\mathbf{g}\|} \leq 1.$$

Therefore, for any **f** and **g**, expression (3.41) determines some angle φ, and $0 \leq \varphi \leq \pi$.

If $(\mathbf{f}, \mathbf{g}) = 0$, than we have from (3.41) that $\varphi = \pi/2$, so that we can make the following definition.

Definition 3.19 Two elements **f** and **g** of H are orthogonal if $(\mathbf{f}, \mathbf{g}) = 0$.

The geometrical structure of the Hilbert space makes it possible to build a basis in the Hilbert space, similar to the orthogonal basis in the Euclidean space. We will introduce a basis by the following sequence of definitions and theorems.

Definition 3.20 A finite or countable set of elements $\{e_i\}$ of a Hilbert space H is called an *orthonormal set* if

$$\mathbf{e}_i \perp \mathbf{e}_j \; for \; i \neq j \tag{3.42}$$

$\|\mathbf{e}_i\| = 1 \; for \; every \; i.$

Definition 3.21 An orthonormal set of elements $\{e_i\}$ is said to be *complete* if

$$(\mathbf{x}, \mathbf{e}_i) = 0 \text{ (for any } i\text{) if and only if } \mathbf{x} = \mathbf{0}. \tag{3.43}$$

Theorem 3.22 *If* $\{\mathbf{e}_i\}$ *is an orthonormal set in a Hilbert space* H*, and if* x *is an arbitrary vector in* H*, then*

$$\mathbf{x} - \sum_i (\mathbf{x}, \mathbf{e}_i)\mathbf{e}_i \perp \mathbf{e}_j$$

for each j.

Proof

$$(\mathbf{x} - \sum_i (\mathbf{x}, \mathbf{e}_i)\mathbf{e}_i, \mathbf{e}_j) = (\mathbf{x}, \mathbf{e}_j) - \sum_i (\mathbf{x}, \mathbf{e}_i)(\mathbf{e}_i, \mathbf{e}_j) = (\mathbf{x}, \mathbf{e}_j) - (\mathbf{x}, \mathbf{e}_j) = 0,$$

from which equation the theorem statement follows at once.

Theorem 3.23 *If* $\{\mathbf{e}_i\}$ *is an orthonormal and complete set in a Hilbert space* H*, and if* **x** *is an arbitrary vector in* H*, then*

$$\mathbf{x} = \sum_i (\mathbf{x}, \mathbf{e}_i)\mathbf{e}_i \tag{3.44}$$

and

$$\|\mathbf{x}\|^2 = \sum_i (\mathbf{x}, \mathbf{e}_i)^2. \tag{3.45}$$

Proof From Theorem 3.22, $\left(\mathbf{x}-\sum_i (\mathbf{x}, \mathbf{e}_i)\mathbf{e}_i\right)$ is orthogonal to $\{\mathbf{e}_i\}$, so Eq. (3.43) for orthonormal and complete set implies that

$$\mathbf{x}-\sum_i (\mathbf{x}, \mathbf{e}_i)\mathbf{e}_i = \mathbf{0},$$

or equivalently, that

$$\mathbf{x} = \sum_i (\mathbf{x}, \mathbf{e}_i)\mathbf{e}_i. \tag{3.46}$$

By the joint continuity of the inner product, the expression in (3.45) yields

$$\|\mathbf{x}\|^2 = (\mathbf{x}, \mathbf{x}) = \left(\sum_i (\mathbf{x}, \mathbf{e}_i)\mathbf{e}_i, \sum_j (\mathbf{x}, \mathbf{e}_j)\mathbf{e}_j\right) = \sum_i \sum_j ((\mathbf{x}, \mathbf{e}_i)\mathbf{e}_i, (\mathbf{x}, \mathbf{e}_j)\mathbf{e}_j)$$

$$= \sum_i ((\mathbf{x}, \mathbf{e}_i)\mathbf{e}_i, (\mathbf{x}, \mathbf{e}_i)\mathbf{e}_i) = \sum_i (\mathbf{x}, \mathbf{e}_i)^2,$$

from which statement (3.45) follows at once.

Definition 3.24 An orthonormal and complete set of elements $\{e_i\}$ in a Hilbert space H is called an *orthonormal basis* of Hilbert space. The numbers (x, e_i) are called the *Fourier coefficients* of x, the expression $x = \sum (x, e_i)e_i$ is called the *Fourier expansion* of x, and the Eq. (3.45) is called *Parseval's equation.*

These terms come from the classical theory of the Fourier series.

Theorem 3.25 *Every nonzero Hilbert space contains a basis.*

We have thus demonstrated that the Hilbert space is a natural generalization of Euclidean space. It has almost the same properties as Euclidean space, but the Hilbert space elements are formed by much more complicated mathematical objects than simple geometrical points or vectors. This result opens a way to work with these complex objects in the same manner as we work with the geometrical points. For example, we can treat data as the elements of this space. Also, we can treat the models as elements of some Hilbert space. Therefore, we can easily introduce the distance between two different models and two different data sets. For example, we can measure the accuracy of fitting predicted data to observed data by using the distance between corresponding data sets. In other words, we can use all the power and simplicity of the geometrical structure of the Hilbert space to solve the inverse problems. I will consider below an example of solving a simple approximation problem in Hilbert space using the geometrical properties of the space.

3.5.2 Approximation Problem in Hilbert Space

In this section, I will illustrate how one can use the geometrical properties of the Hilbert space to solve an approximation problem. Suppose that L is an n dimensional subspace of a Hilbert space H $(L \subset H)$ and L is spanned by a linearly independent set of n vectors $\{\mathbf{d}_1, \mathbf{d}_2, \ldots\ldots, \mathbf{d}_n\}$. The problem is to determine for any $\mathbf{d}_0 \in H$ the vector $\mathbf{d} \in L$ closest to $\mathbf{d}_0$ (Fig. 3.5).

To solve this problem, let us consider the norm of difference $\|\mathbf{d}_0 - \mathbf{d}\|$. Any vector $\mathbf{d} \in L$ can be represented in the form of a linear combination of basis vectors:

$$\mathbf{d} = \alpha_1 \mathbf{d}_1 + \alpha_2 \mathbf{d}_2 + \alpha_3 \mathbf{d}_3 + \cdots + \alpha_n \mathbf{d}_n.$$

Thus, we have the minimization problem

$$\|\mathbf{d}_0 - \mathbf{d}\| = \|\mathbf{d}_0 - (\alpha_1 \mathbf{d}_1 + \alpha_2 \mathbf{d}_2 + \alpha_3 \mathbf{d}_3 + \cdots + \alpha_n \mathbf{d}_n)\| = \min, \tag{3.47}$$

which can be written, using inner product notation, in the following form:

$$\|\mathbf{d}_0 - \mathbf{d}\|^2 = \left\| \mathbf{d}_0 - \sum_{i=1}^{n} \alpha_i \mathbf{d}_i \right\|^2$$

$$= \left(\mathbf{d}_0 - \sum_{i=1}^{n} \alpha_i \mathbf{d}_i,\ \mathbf{d}_0 - \sum_{i=1}^{n} \alpha_i \mathbf{d}_i \right) = \min .$$

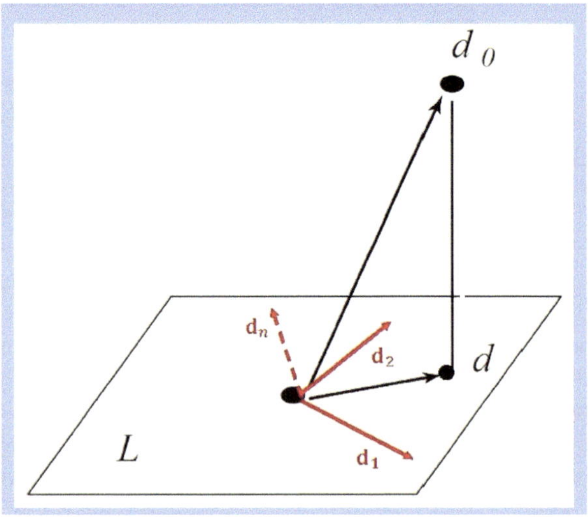

Fig. 3.5 Approximation problem in Hilbert space

Let us calculate the derivatives of the $\|\mathbf{d}_0 - \mathbf{d}\|^2$ with respect to α_j which must vanish at an extremum point:

$$\frac{\partial \|\mathbf{d}_0 - \mathbf{d}\|^2}{\partial \alpha_j} = 2\left(\mathbf{d}_0 - \sum_{i=1}^{n} \alpha_i \mathbf{d}_i,\ \mathbf{d}_j\right) = 0.$$

From the last equation, we have the system of linear equations for the unknown coefficients α_i:

$$\sum_{i=1}^{n} \alpha_i (\mathbf{d}_i, \mathbf{d}_j) = (\mathbf{d}_0, \mathbf{d}_j).$$

We may write the system more compactly as follows:

$$\sum_{i=1}^{n} \Gamma_{ji} \alpha_i = (\mathbf{d}_0, \mathbf{d}_j), \tag{3.48}$$

where the symmetric matrix $\left[\Gamma_{ji}\right] = \left[(\mathbf{d}_i, \mathbf{d}_j)\right]$ is called the *Gram matrix*.

It can be demonstrated that the linear independence of the elements $\mathbf{d}_i$ guarantees that matrix $\left[\Gamma_{ji}\right]$ is nonsingular, which means that the solution to (3.48) $\{\alpha_i,\ i = 1, 2, \ldots n\}$ always exists for any $\mathbf{d}_0$ and is unique.

Note that, we can assume that $\mathbf{d}_0$ are observed data, and $\{\mathbf{d}_1, \mathbf{d}_2, \ldots\ldots, \mathbf{d}_n\}$ is the set of known theoretical data, which would correspond to some inverse problem solution. In this case, the minimization problem (3.47) is equivalent to the problem of observed data approximation by the given theoretical data set. We will discuss the different formulations of this problem in the next sections.

3.5.3 Complex Hilbert Space

We can introduce a *complex Hilbert space* with the scalar coefficients being both real and complex numbers. The vector operations in the complex Hilbert space are based on similar axioms, (3.34)–(3.38), to those for real Hilbert space, but with one significant modification. The point is that the axioms (3.34)–(3.38) cannot be satisfied simultaneously in a complex space. In fact, from (3.34) and (3.36), it follows that

$$(\alpha \mathbf{f}, \alpha \mathbf{f}) = \alpha^2 (\mathbf{f}, \mathbf{f}).$$

In particular, if $\alpha = i$, we have

$$(i\mathbf{f}, i\mathbf{f}) = -(\mathbf{f}, \mathbf{f}).$$

From the last formula we see that if $(i\mathbf{f}, i\mathbf{f}) > 0$, then $(\mathbf{f}, \mathbf{f}) < 0$, and vice versa, which contradicts axiom (3.37). Therefore, we have to introduce a different definition for the inner product of two vectors in the complex space. It is defined as a complex-valued functional, $(\mathbf{f}, \mathbf{g})$, with the properties

$$(\mathbf{f}, \mathbf{g}) = (\mathbf{g}, \mathbf{f})^* \text{ (complex symmetry)}, \tag{3.49}$$

where the asterisk * means complex conjugate,

$$(\mathbf{f} + \mathbf{g}, \mathbf{h}) = (\mathbf{f}, \mathbf{h}) + (\mathbf{g}, \mathbf{h}) \text{ (linearity)}, \tag{3.50}$$

$$(\alpha\mathbf{f}, \mathbf{g}) = \alpha(\mathbf{f}, \mathbf{g}) \text{ (linearity)}. \tag{3.51}$$

This functional has to be also positive definite, i.e.,

$$(\mathbf{f}, \mathbf{f}) > 0, \tag{3.52}$$

and

$$(\mathbf{f}, \mathbf{f}) = 0 \text{ if and only if } \mathbf{f} = \mathbf{0}. \tag{3.53}$$

Thus, we have corrected the first axiom of the real Hilbert space without changing the other axioms. Note that from (3.49) and (3.51) it follows that

$$(\mathbf{f}, \alpha\mathbf{g}) = \alpha^*(\mathbf{f}, \mathbf{g}). \tag{3.54}$$

3.5.4 Properties of the Gram Matrix

In this section, we consider the properties of the Gram matrix of a set of elements (vectors) $f^{(1)}, f^{(2)}, \ldots, f^{(n-1)}, f^{(n)}$, from a complex Hilbert space.

The *Gram matrix* is formed by the inner products between elements $f^{(1)}, f^{(2)}, \ldots, f^{(n-1)}, f^{(n)}$ (which could represent some functions, for example), as follows:

$$\mathbf{G}_n(f^{(1)}, f^{(2)}, \ldots, f^{(n-1)}, f^{(n)})$$

$$= \begin{bmatrix} \left(f^{(1)}, f^{(1)}\right) & \ldots & \left(f^{(1)}, f^{(n-1)}\right) & \left(f^{(1)}, f^{(n)}\right) \\ \left(f^{(2)}, f^{(1)}\right) & \ldots & \left(f^{(2)}, f^{(n-1)}\right) & \left(f^{(2)}, f^{(n)}\right) \\ \vdots & \ddots & \vdots & \vdots \\ \left(f^{(n-1)}, f^{(1)}\right) & \ldots & \left(f^{(n-1)}, f^{(n-1)}\right) & \left(f^{(n-1)}, f^{(n)}\right) \\ \left(f^{(n)}, f^{(1)}\right) & \ldots & \left(f^{(n)}, f^{(n-1)}\right) & \left(f^{(n)}, f^{(n)}\right) \end{bmatrix}. \tag{3.55}$$

We introduce a matrix-column, $\mathbf{F}_n(\mathbf{r})$ formed by elements $f^{(i)}$, $i = 1, 2, \ldots n$:

$$\mathbf{F}_n = \begin{bmatrix} f^{(1)} \\ f^{(2)} \\ \vdots \\ f^{(n-1)} \\ f^{(n)} \end{bmatrix}. \tag{3.56}$$

According to formula (3.55), Gram matrix can be represented in the following form:

$$\mathbf{G}_n(f^{(1)}, f^{(2)}, \ldots, f^{(n-1)}, f^{(n)}) =$$

$$= \left(\begin{bmatrix} f^{(1)} \\ f^{(2)} \\ \vdots \\ f^{(n-1)} \\ f^{(n)} \end{bmatrix}, \left[f^{(1)}, f^{(2)}, \ldots, f^{(n-1)}, f^{(n)} \right] \right) = \left(\mathbf{F}_n, \mathbf{F}_n^T \right), \tag{3.57}$$

where symbol $\left(\mathbf{F}_n, \mathbf{F}_n^T\right)$ denotes the matrix multiplication involving inner product between the corresponding elements of the matrix-column and matrix-row, $\mathbf{F}$ and $\mathbf{F}^T$, as defined in formula (3.55).

From the property (3.49) it follows that the Gram matrix is a Hermitian matrix, that is equal to its own conjugate transpose (Golub and Van Loan 2013):

$$\mathbf{G}_n = \left(\mathbf{F}_n, \mathbf{F}_n^T\right) = \left(\mathbf{F}_n, \mathbf{F}_n^T\right)^* = \mathbf{G}_n^*. \tag{3.58}$$

3.5.4.1 Hermitian Matrices

We now review some basic properties of the Hermitian matrices.

The complex matrix $\mathbf{H}$ is called a *Hermitian matrix*, if it is equal to its own conjugate transpose:

$$\mathbf{H} = \mathbf{H}^*.$$

Definition 3.26 The complex Hermitian $[N \times N]$ matrix, $\mathbf{H}$, is called *positive semidefinite* if for any nonzero vector $\mathbf{x}$ from complex Euclidean space $\mathbf{x} \in E_n^C$:

$$\mathbf{x}^* \mathbf{H} \mathbf{x} \geq 0, \tag{3.59}$$

where $\mathbf{x}^*$ is vector-row formed by the complex conjugate components of vector-column $\mathbf{x}$:

$$\mathbf{x}^* = \left[x_1^*, x_2^*, \ldots\ldots, x_{n-1}^*, x_n^*\right].$$

Definition 3.27 The Hermitian matrix $\mathbf{H}$ is called *positive definite* if the following inequality holds for any nonzero vector $\mathbf{x}$:

$$\mathbf{x}^*\mathbf{H}\mathbf{x} > \mathbf{0}. \tag{3.60}$$

For the complex Hermitian matrix, $\mathbf{H}$, spectral representation (2.43) takes the form:

$$\mathbf{H} = \mathbf{U}\Phi\mathbf{U}^*, \tag{3.61}$$

where $\mathbf{U}$ is the unitary matrix and Φ is real diagonal matrix formed by the eigenvalues $\varphi_1, \varphi_2, \ldots\ldots \varphi_N$, of matrix $\mathbf{H}$.

Note that unitary matrices are the complex analog of real orthogonal matrices. The inverse unitary matrix equals it conjugate transpose:

$$\mathbf{U}^{-1} = \mathbf{U}^*, \text{ and } |\det \mathbf{U}| = 1. \tag{3.62}$$

From Eq. (3.61) it follows at once that the determinant of the Hermitian matrix can be calculated as follows:

$$\det \mathbf{H} = \det\left(\mathbf{U}\Phi\mathbf{U}^*\right) = \det(\mathbf{U})\det(\Phi)\det\left(\mathbf{U}^*\right) = \det(\Phi) = \varphi_1\varphi_2\ldots\ldots\varphi_N, \tag{3.63}$$

where φ_i, $i = 1, 2, \ldots\ldots N$ are the eigenvalues of matrix $\mathbf{H}$.

Finally, it can be shown that a Hermitian (or symmetric) matrix is positive definite if and only if all its eigenvalues are real positive numbers. Indeed, let φ_i is an eigenvalue of matrix $\mathbf{H}$ with a nonzero eigenvector, $\mathbf{x}^{(i)}$:

$$\mathbf{H}\mathbf{x}^{(i)} = \varphi_i\mathbf{x}^{(i)}, \; i = 1, 2, \ldots N. \tag{3.64}$$

Then condition (3.60) of matrix, $\mathbf{H}$, being positive definite can be written as follows:

$$\mathbf{x}^{(i)*}\mathbf{H}\mathbf{x}^{(i)} = \mathbf{x}^{(i)*}\varphi_i\mathbf{x}^{(i)} = \varphi_i\left\|\mathbf{x}^{(i)}\right\|^2 > 0, \; i = 1, 2, \ldots N. \tag{3.65}$$

This condition holds if and only if

$$\varphi_i > 0, \; i = 1, 2, \ldots N. \tag{3.66}$$

From the last formula and Eq. (3.63), one can obtain at once that the determinant of the Hermitian matrix is positive,

$$\det \mathbf{H} > 0, \tag{3.67}$$

if and only if matrix **H** is positive definite.

Thus, we can formulate the following theorem.

Theorem 3.28 *The following conditions are equivalent:*
(a) Hermitian matrix is positive definite

$$\mathbf{x}^*\mathbf{H}\mathbf{x} > 0 \;\; \textit{for all} \;\; \mathbf{x} \neq 0;$$

(b) All eigenvalues of **H** *satisfy the inequality* $\varphi_i > 0$;
(c) Determinant of **H** *is positive,*

$$\det \mathbf{H} > 0.$$

A similar theorem holds for positive semidefinite matrices.

Theorem 3.29 *The following conditions are equivalent:*
(a) Hermitian matrix is positive semidefinite

$$\mathbf{x}^*\mathbf{H}\mathbf{x} \geq 0 \;\; \textit{for all} \;\; \mathbf{x} \neq 0.$$

(b) All eigenvalues of **H** *satisfy* $\varphi_i \geq 0$.
(c) Determinant of **H** *is non-negative,*

$$\det \mathbf{H} \geq 0.$$

3.5.4.2 Gram Matrix as Hermitian Positive Semidefinite Matrix

We can prove that the Gram matrix, $\mathbf{G}_n$, of any system of elements (functions) in Hilbert space is Hermitian positive semidefinite magtrix. Let us calculate the product of vector-row $\mathbf{x}^*$ and matrix $\mathbf{G}_n$:

$$\mathbf{x}^*\mathbf{G}_n =$$

$$= \left[x_1^*, x_2^*, \ldots\ldots, x_{n-1}^*, x_n^*\right] \begin{bmatrix} \left(f^{(1)}, f^{(1)}\right) & \ldots & \left(f^{(1)}, f^{(n-1)}\right) & \left(f^{(1)}, f^{(n)}\right) \\ \left(f^{(2)}, f^{(1)}\right) & \ldots & \left(f^{(2)}, f^{(n-1)}\right) & \left(f^{(2)}, f^{(n)}\right) \\ \ldots & \ldots & \ldots & \ldots \\ \left(f^{(n-1)}, f^{(1)}\right) & \ldots & \left(f^{(n-1)}, f^{(n-1)}\right) & \left(f^{(n-1)}, f^{(n)}\right) \\ \left(f^{(n)}, f^{(1)}\right) & \ldots & \left(f^{(n)}, f^{(n-1)}\right) & \left(f^{(n)}, f^{(n)}\right) \end{bmatrix}$$

$$= \left[\sum_{i=1,2,\ldots,n} x_i^*\left(f^{(i)}, f^{(1)}\right) \ldots\ldots \sum_{i=1,2,\ldots,n} x_i^*\left(f^{(i)}, f^{(n-1)}\right) \;\; \sum_{i=1,2,\ldots,n} x_i^*\left(f^{(i)}, f^{(n)}\right)\right], \tag{3.68}$$

where x_i^* $(i = 1, 2, \ldots, n)$ are the complex conjugate scalar components of nonzero vector $\mathbf{x}$.

We can now calculate the product of expression (3.68) with vector-column $\mathbf{x}$:

$$\mathbf{x}^*\mathbf{G}_n\mathbf{x} =$$

$$= \left[\sum_{i=1,2,\ldots,n} x_i^* \left(f^{(i)}, f^{(1)}\right) \ldots\ldots \sum_{i=1,2,\ldots,n} x_i^* \left(f^{(i)}, f^{(n-1)}\right) \sum_{i=1,2,\ldots,n} x_i^* \left(f^{(i)}, f^{(n)}\right) \right] \begin{bmatrix} x_1 \\ \cdots \\ x_{n-1} \\ x_n \end{bmatrix}$$

$$= \sum_{i=1,2,\ldots,n} \sum_{j=1,2,\ldots,n} x_i^* \left(f^{(i)}, f^{(j)}\right) x_j = \left(\sum_{i=1,2,\ldots,n} x_i^* f^{(i)}, \sum_{j=1,2,\ldots,n} x_j^* f^{(j)} \right)$$

$$= \left\| \sum_{i=1,2,\ldots,n} x_i^* f^{(i)} \right\|^2 \geq 0. \tag{3.69}$$

Note that in the last formula we took into account the property (3.54) of the inner product in the complex Hilbert space:

$$x_i^* \left(f^{(i)}, f^{(j)}\right) x_j = \left(x_i^* f^{(i)}, f^{(j)} x_j^*\right).$$

We can see from formula (3.69) that Gram matrix of any system of elements (functions) in Hilbert space is Hermitian positive semidefinite function. Therefore, according to Theorem 3.29, the determinant of this matrix is non-negative:

$$\det \mathbf{G}_n \geq 0. \tag{3.70}$$

Formula (3.70) is called *Gram inequality* (Everitt 1958; Barth 1999).

Let us assume that elements of the Hilbert space $f^{(1)}, f^{(2)}, \ldots, f^{(n-1)}, f^{(n)}$ form a linear independent set of functions in Hilbert space. This means that any linear combination of these elements (e.g., functions) with the coefficients equal to scalar components of nonzero vector $\mathbf{x}^*$, cannot be equal to zero:

$$\sum_{i=1,2,\ldots,n} x_i^* f^{(i)} \neq 0. \tag{3.71}$$

Therefore, according to (3.69), we have

$$\mathbf{x}^*\mathbf{G}_n\mathbf{x} = \left\| \sum_{i=1,2,\ldots,n} x_i^* f^{(i)} \right\|^2 > 0. \tag{3.72}$$

Equation (3.72) shows that, in this case, Gram matrix, $\mathbf{G}_n$, is positive definite Hermitian matrix.

From the last result we can formulate the following Theorem.

Theorem 3.30 *A necessary and sufficient condition for the elements* $f^{(1)}, f^{(2)}, \ldots, f^{(n-1)}, f^{(n)}$ *to be linearly independent is that the corresponding Gram matrix is positive definite according to (3.72).*

At the same time, according to Theorem 3.28 the Hermitian matrix is positive definite if and only if its determinant is positive:

$$\det \mathbf{G}_n > 0. \tag{3.73}$$

Thus, we can formulate the following results.

Corollary 3.31 *A necessary and sufficient condition for the elements* $f^{(1)}, f^{(2)}, \ldots, f^{(n)}$ *to be linearly independent is that the determinant of the corresponding Gram matrix is positive.*

One can also prove the following statement.

Corollary 3.32 *A necessary and sufficient condition for the elements,* $f^{(1)}, f^{(2)}, \ldots, f^{(n)}$*, to be linearly dependent, is that the determinant of their Gram matrix is equal to zero:*

$$\mathbf{G}_n(f^{(1)}, f^{(2)}, \ldots, f^{(n)}) = 0. \tag{3.74}$$

This means that there exist the nonzero vector $\widetilde{\mathbf{x}}$ with the property

$$\sum_{i=1,2,\ldots,n} \widetilde{x}_i f^{(i)} = 0. \tag{3.75}$$

The proof of the above corollary is straightforward.

(a) If elements $f^{(1)}, f^{(2)}, \ldots, f^{(n)}$ are linearly dependent, then according (3.69) and (3.75) we have

$$\widetilde{\mathbf{x}}^* \mathbf{G}_n \widetilde{\mathbf{x}} = 0. \tag{3.76}$$

Equation (3.76) means that matrix $\mathbf{G}_n$ is not positive definite. In this case, according to Theorem 3.28 its determinant cannot be positive, and, therefore, it is equal to zero, because, according to (3.70), the determinant of the Gram matrix is always non-negative.

(b) If the determinant is equal to zero, the Gram matrix is not positive definite. Therefore, according to Theorem 3.28, elements $f^{(1)}, f^{(2)}, \ldots, f^{(n)}$ are linearly dependent. In other words, if $\det \mathbf{G}_n = 0$, there always exists the nonzero vector $\widetilde{\mathbf{x}}$ with the property (3.75).

The properties of the Gram matrix formulated above provide the basis for using the Gram matrix in the multiphysics inversion (see Chap. 12).

3.6 Examples of Linear Vector Spaces

The Euclidean space formed by the vectors with real components is the simplest example of linear vector space. There are several other fundamental mathematical spaces that are widely used in applications. This section presents several important examples of these spaces. We begin the discussion with the real Euclidean space formed by real vectors.

3.6.1 Euclidean Space Formed by Real Vectors

The simplest example of a linear space is Euclidean space E_N (or R_n), which is a natural generalization of three dimensional (3D) physical space to n dimensions. We have introduced Euclidean space in the beginning of this chapter already. For convenience, let us summarize the main properties of this space. It can be described as a set of all possible vectors of order n:

$$\mathbf{a} = (a_1, a_2, a_3, \ldots\ldots a_n),$$

where the scalars $a_1, a_2, a_3, \ldots\ldots a_n$ are real numbers.

By analogy with the length of the vector in 3D physical space, a *norm of the vector* $\|\mathbf{a}\|$ is defined as follows:

$$\|\mathbf{a}\| = \sqrt{a_1^2 + a_2^2 + a_3^2 + \cdots\cdots\cdots + a_n^2}. \tag{3.77}$$

It is easy to check that the norm introduced above satisfies the conditions

$$\|\mathbf{a}\| > 0 \text{ if } \mathbf{a} \neq 0, \ \ \|\mathbf{a}\| = 0 \text{ if } \mathbf{a} = 0, \tag{3.78}$$

$$\|\lambda \mathbf{a}\| = |\lambda| \, \|\mathbf{a}\|, \tag{3.79}$$

$$\|\mathbf{a} + \mathbf{b}\| \leq \|\mathbf{a}\| + \|\mathbf{b}\|, \tag{3.80}$$

where the last inequality is called *triangle inequality*

We have also introduced an operation on two vectors, called the *inner (dot) product,* as follows:

$$\mathbf{a} \cdot \mathbf{b} = \sum_{i=1}^{n} a_i b_i. \tag{3.81}$$

The norm (3.77) is equal to the square root of the dot product of the vector with itself:

$$\|\mathbf{a}\| = \sqrt{\mathbf{a} \cdot \mathbf{a}}. \tag{3.82}$$

3.6.2 Complex Euclidean Space

We can introduce a complex Euclidean space, where the scalar components of the vectors, the scalars $a_1, a_2, a_3, \ldots\ldots a_n$, are the complex numbers. However, in this case we have to modify definitions (3.77) and (3.81) for the norm of the vector and the inner product in order to satisfy conditions (3.78)–(3.80). The norm of a vector in the complex Euclidean space is introduced as follows:

$$\|\mathbf{a}\| = \sqrt{|a|_1^2 + |a|_2^2 + |a|_3^2 + \cdots\cdots\cdots + |a|_n^2}. \tag{3.83}$$

The inner (dot) product of two vectors is introduced as a complex value, determined by the following formula

$$\mathbf{a} \cdot \mathbf{b} = \sum_{i=1}^{n} a_i b_i^*, \tag{3.84}$$

where asterisk * means complex conjugate.

Obviously, a norm can still be determined as a square root of the dot square of the vector:

$$\|\mathbf{a}\| = \sqrt{\mathbf{a} \cdot \mathbf{a}} = \sqrt{a_1 a_1^* + a_2 a_2^* + a_3 a_3^* + \cdots\cdots\cdots + a_n a_n^*}. \tag{3.85}$$

Note that in the complex Euclidean space the inner product operation is not symmetrical:

$$\mathbf{a} \cdot \mathbf{b} = (\mathbf{b} \cdot \mathbf{a})^* . \tag{3.86}$$

It also follows from Eq. (3.86), that

$$\mathbf{a} \cdot \alpha\mathbf{b} = (\alpha\mathbf{b} \cdot \mathbf{a})^* = \alpha^* (\mathbf{b} \cdot \mathbf{a})^* = \alpha^* \mathbf{a} \cdot \mathbf{b}. \tag{3.87}$$

3.6.3 Typical Mathematical Function Spaces

3.6.3.1 C_0 [a, b] Space

The next example can be constructed using a set of real functions, continuously differentiable to order n on the real interval $[a, b]$. Obviously, the sum of two differentiable functions is another differentiable function, and the multiplication of the function by a constant scalar is a differentiable function as well. Therefore, this set of functions forms some linear space, denoted by C_n $[a, b]$. However, it is not a normed space yet, because we did not introduce a norm of the function.

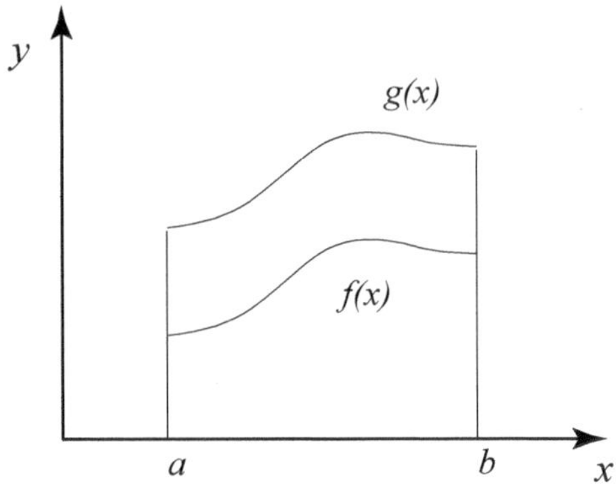

Fig. 3.6 The plot of one function, $f(x)$, is shifted vertically with respect to the plot of the other function, $g(x)$. The size of this shift is equal to the "distance" between two functions in the space with the uniform norm

There are many different ways to introduce a norm in a function space. The simplest one is as follows:

$$\|f\|_{\infty} = \max_{a \leq x \leq b} |f(x)|. \tag{3.88}$$

It is easy to verify that the norm introduced in (3.88) satisfies all required conditions for the norm of the normed space, and I leave this proof to the reader as an exercise. This norm is called the *uniform norm*. The linear space of continuous functions, $C_0[a, b]$, equipped with the uniform norm, forms a normed space, denoted by $C[a, b]$. The distance between two functions, $f(x)$ and $g(x)$, in the space $C[a, b]$ with the uniform norm is equal to

$$\mu(f(x), g(x)) = \max_{a \leq x \leq b} |f(x) - g(x)|. \tag{3.89}$$

Figure 3.6 gives an illustration of this distance as applied to functions $f(x)$ and $g(x)$. In this case, one can see that the distance corresponds to the shift between the plots of these two functions. However, one can notice that the distance between two functions in this norm is determined by the extremum of the difference $[f(x) - g(x)]$.

In other words, even if these two functions go very close to each other along the interval $[a, b]$, but are different only in a few points (as shown in Fig. 3.7), these differences will determine the distance between these two functions in the metric of the space $C[a, b]$.

3.6.3.2 L_p and Sobolev Spaces

In practical applications, having a metric that reflects the average discrepancy between two functions is much more convenient. This metric can be introduced by a so-called L_1 norm:

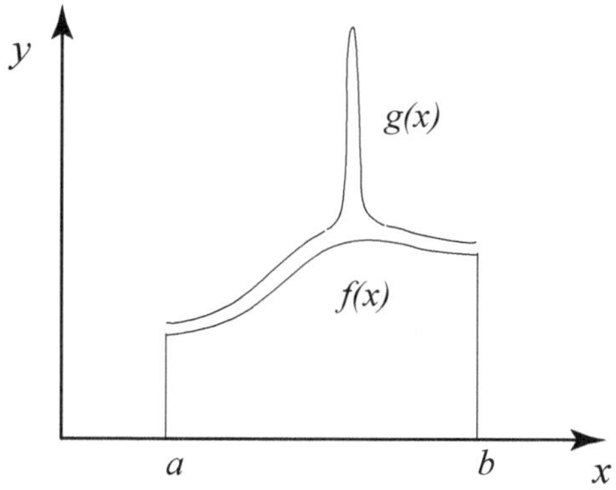

Fig. 3.7 The plots of two functions follow each other very closely along an entire interval $[a, b]$, with the exception of a few outliers. These two functions are considered to be very different in the function space with a uniform norm. However, in the function space with an L_1 norm these two functions are close to each other

$$\|f\|_{L_1} = \int_a^b |f(x)|\,dx. \tag{3.90}$$

In this case, the distance between two functions is given by the following formula:

$$\mu(f(x), g(x)) = \int_a^b |f(x) - g(x)|\,dx. \tag{3.91}$$

Thus, now two functions will be close to each other if the integral of their difference is small enough. The presence of one or two outliers will not affect the result significantly (see Fig. 3.7). The linear space of continuous functions on the real interval $[a, b]$ equipped with the L_1 norm, is called the $L_1[a, b]$ space. This is a linear normed space but not a Hilbert space because it has no inner product operation.

L_1 space is a special case of L_p spaces, when $p = 1$. The metric in L_p spaces is introduced according to the following formula:

$$\|f\|_{L_p} = \left(\int_a^b |f(x)|^p\,dx\right)^{1/p}, \quad 0 < p < \infty. \tag{3.92}$$

Space L_p is a linear normed space, with norm satisfying the *Minkowski inequality*:

$$\|f + g\|_{L_p} \leq \|f\|_{L_p} + \|g\|_{L_p}. \tag{3.93}$$

Minkowski inequality is an analog to the triangle inequality of the metric space.

Space L_2 plays a special role in the function space theory because one can introduce the Hilbert metric in this space. Indeed, in space L_2 we define the norm as follows:

$$\|f\|_{L_2} = \sqrt{\int_a^b f^2(x)\,dx}.$$

This norm is called an L_2 norm. It follows that the distance between two functions will be measured as

$$\mu(f(x), g(x)) = \|f(x) - g(x)\|_{L_2} = \sqrt{\int_a^b [f(x) - g(x)]^2 \, dx}. \tag{3.94}$$

The advantage of this norm is that one can derive it from the inner product of two functions, defined as follows:

$$(f(x), g(x)) = \int_a^b f(x)g(x)dx. \tag{3.95}$$

Thus, a linear normed space $L_2[a, b]$ is a Hilbert space, and therefore, possesses all the properties of the Hilbert space discussed above.

One can introduce an inner product between two functions, $f(x)$ and $g(x)$, continuously differentiable to order n on the real interval $[a, b]$, using a different formula:

$$(f(x), g(x)) = \int_a^b \sum_{k=0}^{n} q_k^2(x) \frac{d^k f(x)}{dx^k} \frac{d^k g(x)}{dx^k} dx, \tag{3.96}$$

where $q_0(x), q_1(x), \ldots, q_n(x)$ are given real functions, ($q_n(x)$ is not identically equal to zero). The corresponding Hilbert space is called a *Sobolev space*, W_2^n. The metric in the Sobolev space W_2^n (the distance between two functions) is determined according to the following formula

$$\mu_{W_2^n}(f(x),g(x)) = \left\{ \int_a^b \sum_{k=0}^{p} q_k^2(x) \left\{ \frac{d^k [f(x)-g(x)]}{dx^k} \right\}^2 dx \right\}^{\frac{1}{2}}. \tag{3.97}$$

Thus, two functions in Sobolev space will now be close to each other if the integral of their difference, and all their derivatives up to the order n, are small enough. In other words, in Sobolev space, not only the functions $f(x)$ and $g(x)$ themselves but also all their derivatives (to order n) should be close to each other. Therefore, the Sobolev metric imposes more control on the function behavior than the conventional L_2 metric.

Another example of the Hilbert space is a space $L_2^C[a, b]$ formed by the sets of complex functions, integrable on the real interval $[a, b]$ and equipped with the inner product

$$(f(x), g(x)) = \int_a^b f(x)g^*(x)dx. \tag{3.98}$$

It is easy to check that expression (3.98) satisfies all axioms, (3.49)–(3.53), for the complex Hilbert space.

Table 3.1 summarizes the examples of linear vector spaces.

Table 3.1 Examples of linear vector spaces

Symbol	Description	Name/comment
E_N	The set of real vectors of order N: $(a_1, a_2, ...a_N)$	Euclidean space
$C_n[a, b]$	The set of functions, continuously differentiable to order n on the real interval $[a, b]$	Not a normed space
$C[a, b]$	$C_0[a, b]$ equipped with the uniform norm $\|f\|_\infty = \max_{a \le x \le b} \|f\|$	Normed space
$L_1[a, b]$	$C_0[a, b]$ equipped with the L_1 norm $\|f\|_{L_1} = \int_a^b \|f\| dx$	Normed space
$L_2[a, b]$	Set of real functions integrable on the real interval $[a, b]$, equipped with the inner product $(f, g)_{L_2} = \int_a^b fgdx$	Hilbert Space
$W_2^n[a, b]$	$C_n[a, b]$ equipped with the norm $\|f\|_{W_2^n}^2 = \left\{\int_a^b \sum_{k=0}^n q_k^2 \left[\frac{d^k f}{dx^k}\right]^2 dx\right\}^{\frac{1}{2}}$	Sobolev space
$L_2^C[a, b]$	Set of complex functions integrable on the real interval $[a, b]$, equipped with the inner product $(f, g)_{L_2^C} = \int_a^b fg^* dx$	Complex Hilbert space

3.7 Gramian Spaces and Their Properties

3.7.1 Inner Product in Gramian Space

We assume that there is a set of complex integrable functions, $f^{(i)}(\mathbf{r})$ $(i = 1, 2, 3, \ldots, n)$, of a radius-vector $\mathbf{r} = (x, y, z)$ defined within some volume V of a 3D space. We can consider these functions as the elements of a complex Hilbert space $L_2^C[V]$ with a L_2 norm, defined by the corresponding inner product:

$$(f, g) = \int_V f(\mathbf{r})\, g^*(\mathbf{r})\, dv, \quad \|f\|^2 = (f, f), \tag{3.99}$$

where asterisk "*" denotes the complex conjugate value.

Let us consider two arbitrary functions from this Hilbert space, $p(\mathbf{r})$ and $q(\mathbf{r}) \in L_2^C[V]$. We can introduce a new inner product operation, $(p, q)_{G^{(n)}}$, between two functions, p and q, as the determinant of the following matrix:

$$(p, q)_{G^{(n)}} =$$

$$= \begin{vmatrix} (f^{(1)}, f^{(1)}) & (f^{(1)}, f^{(2)}) & \dots & (f^{(1)}, f^{(n-1)}) & (f^{(1)}, q) \\ (f^{(2)}, f^{(1)}) & (f^{(2)}, f^{(2)}) & \dots & (f^{(2)}, f^{(n-1)}) & (f^{(2)}, q) \\ \dots & \dots & \dots & \dots & \dots \\ (f^{(n-1)}, f^{(1)}) & (f^{(n-1)}, f^{(2)}) & \dots & (f^{(n-1)}, f^{(n-1)}) & (f^{(n-1)}, q) \\ (p, f^{(1)}) & (p, f^{(2)}) & \dots & (p, f^{(n-1)}) & (p, q) \end{vmatrix}. \tag{3.100}$$

It is easy to check that all the properties of the inner product hold:

$$(p, q)_{G^{(n)}} = (q, p)^*_{G^{(n)}}, \tag{3.101}$$

$$\left(\alpha_1 p^{(1)} + \alpha_2 p^{(2)}, q\right)_{G^{(n)}} = \alpha_1 \left(p^{(1)}, q\right)_{G^{(n)}} + \alpha_2 \left(p^{(2)}, q\right)_{G^{(n)}}, \tag{3.102}$$

and

$$(p, p)_{G^{(n)}} \geq 0. \tag{3.103}$$

The last property (3.103) follows from the fact that the norm square of a function, $\|p\|^2_{G^{(n)}}$, is equal to the determinant, $G(f^{(1)}, f^{(2)}, \dots, f^{(n-1)}, p)$, of the Gram matrix of a set of functions, $(f^{(1)}, f^{(2)}, \dots, f^{(n-1)}, p,)$, which is called a Gramian:

$$\|p\|^2_{G^{(n)}} = (p, p)_{G^{(n)}} = G(f^{(1)}, f^{(2)}, \dots, f^{(n-1)}, p)$$

$$= \begin{vmatrix} (f^{(1)}, f^{(1)}) & (f^{(1)}, f^{(2)}) & \dots & (f^{(1)}, f^{(n-1)}) & (f^{(1)}, p) \\ (f^{(2)}, f^{(1)}) & (f^{(2)}, f^{(2)}) & \dots & (f^{(2)}, f^{(n-1)}) & (f^{(2)}, p) \\ \dots & \dots & \dots & \dots & \dots \\ (f^{(n-1)}, f^{(1)}) & (f^{(n-1)}, f^{(2)}) & \dots & (f^{(n-1)}, f^{(n-1)}) & (f^{(n-1)}, p) \\ (p, f^{(1)}) & (p, f^{(2)}) & \dots & (p, f^{(n-1)}) & (p, p) \end{vmatrix}. \tag{3.104}$$

We have demonstrated above in Sect. 3.5.4 that Gramian satisfies to Gram inequality:

$$G(f^{(1)}, f^{(2)}, \dots, f^{(n-1)}, p) \geq 0. \tag{3.105}$$

Note that equality holds in (3.105) if and only if the system of functions $\left(f^{(1)}, f^{(2)}, \dots, f^{(n-1)}, p\right)$ is linearly dependent.

We will call the Hilbert space formed by the integrable functions, defined within some volume V of a 3D space, with the inner product operation introduced by formula (3.100), a *Gramian space*, $G^{(n)}$. The set of complex integrable functions, $f^{(i)}(\mathbf{r})$ $(i = 1, 2, 3, \ldots, n)$, is called a *Gramian core* set.

3.7.2 Properties of the Norm in Gramian Space

The main property of the Gramian space is that the norm of the function p in the Gramian space provides a measure of correlation between this function and functions $f^{(1)}, f^{(2)}, \ldots, f^{(n-1)}$ from the corresponding Gramian core set.

One can also introduce the Gramian space $G^{(j)}$, where inner product is defined by an expression similar to (3.100) with the only difference that functions p and q are located within the row and column with number j, respectively:

$$(p, q)_{G^{(j)}} =$$
$$= \begin{vmatrix} \left(f^{(1)}, f^{(1)}\right) & \left(f^{(1)}, f^{(2)}\right) & \ldots & \left(f^{(1)}, q\right) & \ldots \left(f^{(1)}, f^{(n)}\right) \\ \ldots & \ldots & \ldots & \ldots & \ldots\ldots\ldots \\ \left(p, f^{(1)}\right) & \left(p, f^{(2)}\right) & \ldots & (p, q) & \ldots \left(p, f^{(n)}\right) \\ \ldots & \ldots & \ldots & \ldots & \ldots \\ \left(f^{(n)}, f^{(1)}\right) & \left(f^{(n)}, f^{(2)}\right) & \ldots & \left(f^{(n)}, q\right) \ldots & \left(f^{(n)}, f^{(n)}\right) \end{vmatrix}. \tag{3.106}$$

In the Gramian space $G^{(j)}$, the norm square of a function, $\|p\|^2_{G^{(j)}}$, is equal to the Gramian of a set of functions, $(f^{(1)}, f^{(2)}, \ldots, f^{(j-1)}, p, f^{(j+1)}, \ldots f^{(n)})$:

$$\|p\|^2_{G^{(j)}} = (p, p)_{G^{(j)}} = G(f^{(1)}, f^{(2)}, \ldots, f^{(j-1)}, p, f^{(j+1)}, \ldots f^{(n)}). \tag{3.107}$$

Therefore, the norm of the function in the Gramian space $G^{(j)}$ provides a measure of linear dependence between this function and all other functions from the Gramian core set, with the exception of function $f^{(j)}$: $f^{(1)}, f^{(2)}, \ldots\ldots f^{(j-1)}, f^{(j+1)}, \ldots, f^{(n)}$.

Note that the Gramian norm has the following property:

$$\left\|f^{(i)}\right\|^2_{G^{(i)}} = \left\|f^{(j)}\right\|^2_{G^{(j)}}, \quad \text{for } i = 1, 2, \ldots, n; \ j = 1, 2, \ldots, n. \tag{3.108}$$

The last formula demonstrates that all the functions, $f^{(1)}, f^{(2)}, \ldots\ldots, f^{(n)}$, have the same norm in the corresponding Gramian spaces $G^{(j)}$, $j = 1, 2, \ldots, n$.

Gramian spaces serve as an effective mathematical instrument for solving the problems of joint inversion of multimodal physical data.

3.8 Operators in Mathematical Spaces

3.8.1 Operators and Their Properties

We can treat the elements of mathematical spaces as geometrical points and consider different transformations of these points from one to another. These transformations can be described by corresponding rules, which are called *operators*. We now give a more strict definition of an operator.

Definition 3.33 Let X and Y be metric spaces and D some subdomain of $X : D \subset X$. If for any $\mathbf{x} \in D$, we can assign according to a certain rule some element $\mathbf{y} \in Y$ we say that the *operator* A is given on D with the values in Y:

$$\mathbf{y} = A(\mathbf{x}), \ \mathbf{x} \in D \subset X, \ \mathbf{y} \in Y.$$

Figure 3.8 gives an illustration of operator A acting from space X into space Y.

Thus, we can see that the operator is a natural generalization of the function for an abstract mathematical space.

We know that one of the fundamental properties of a function is whether it is continuous or discontinuous. This property can be applied to operators as well.

Definition 3.34 Let X and Y be metric spaces with metrics $\mu_1(\mathbf{x}', \mathbf{x}'')$ and $\mu_2(\mathbf{y}', \mathbf{y}'')$, and let A be an operator, transforming elements of X into Y. A is said to be *continuous* at a point $\mathbf{x}_0$ in X if for each real number $\varepsilon > 0$ there exists a real number $\delta > 0$ such that, for any two elements $\mathbf{x}, \mathbf{x}_0 \in X$, satisfying the condition $\mu_1(\mathbf{x}, \mathbf{x}_0) < \delta$, the distance between the results of their transformation by operator A is less than $\varepsilon : \ \mu_2(A(\mathbf{x}), A(\mathbf{x}_0)) < \varepsilon$.

It is important to consider the sequence of elements $\mathbf{x}_1, \mathbf{x}_2, \mathbf{x}_3, \ldots \mathbf{x}_n, \ldots \in X$ in the metric space. This sequence is said to converge to element $\mathbf{x}_0$ if $\mu(\mathbf{x}_n, \mathbf{x}_0) \to 0$ as $n \to \infty$.

Theorem 3.35 *Let X and Y be metric spaces and A an operator of X into Y. Then A is continuous at* $\mathbf{x}_0$ *if and only if the convergence of some sequence of the elements from the metric space* $\{\mathbf{x}_n\}$ *to element* $\mathbf{x}_0$ $(\mathbf{x}_n \to \mathbf{x}_0)$ *results in the convergence of the transformed sequences of the elements* $\{A(\mathbf{x}_n)\}$ *to the element* $A(\mathbf{x}_0) : A(\mathbf{x}_n) \to A(\mathbf{x}_0)$.

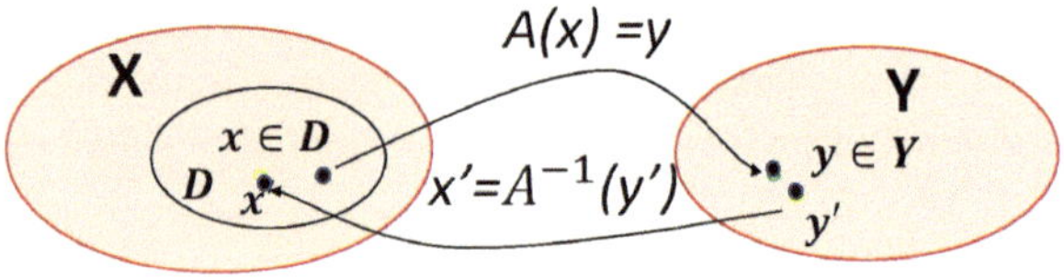

Fig. 3.8 Introducing forward and inverse operators in metric spaces

3.8.2 Linear Operators

Let X and Y be normed spaces with the same system of scalars. There is a very important class of operators in the normed linear spaces which are called *linear operators*.

Definition 3.36 An operator $\mathbf{y} = L(\mathbf{x})$ is called *linear* if for any $\mathbf{x}_i \in X$ and any scalars α_i:

$$L(\alpha_1\mathbf{x}_1 + \alpha_2\mathbf{x}_2 + \cdots\cdots + \alpha_n\mathbf{x}_n) = \alpha_1 L(\mathbf{x}_1) + \alpha_2 L(\mathbf{x}_2) + \cdots\cdots + \alpha_n L(\mathbf{x}_n). \tag{3.109}$$

Definition 3.37 An operator $\mathbf{y} = L(\mathbf{x})$ is called *bounded* if there exists a real number M with the property that

$$\|L(\mathbf{x})\| \leq M\|\mathbf{x}\| \tag{3.110}$$

for every $\mathbf{x} \in X$.

It is easy to prove the following theorem:

Theorem 3.38 *A linear operator L is continuous if and only if it is bounded.*

Definition 3.39 The smallest constant M for which condition (3.110) holds for any $\mathbf{x} \in X$ is called the *norm* of the operator:

$$\|L\| = \min\{M : M \geq 0,\ \|L(\mathbf{x})\| \leq M\|\mathbf{x}\|,\ \text{for any } \mathbf{x} \in X\}.$$

From the last formula we see at once that

$$\|L(\mathbf{x})\| \leq \|L\|\ \|\mathbf{x}\|$$

for all $\mathbf{x} \in X$.

3.8.3 Inverse Operators

Let us consider the following equation

$$A(\mathbf{x}) = \mathbf{y}. \tag{3.111}$$

If the solution of Eq. (3.111) is unique, then we can assign to any $\mathbf{y}'$, for which Eq. (3.111) is solvable, the corresponding value $\mathbf{x}'$ (see Fig. 3.8). Thus we can determine some operator A^{-1}:

$$\mathbf{x} = A^{-1}y,$$

which we call the *inverse operator*.

Theorem 3.40 *The inverse operator A^{-1} for a given linear operator A exists and is linear if and only if the equation $A\mathbf{x} = \mathbf{0}$ holds only for $\mathbf{x} = \mathbf{0}$.*

Theorem 3.41 *The inverse operator A^{-1} for a given linear operator A exists, and is linear and bounded if and only if there exists a real number $m > 0$ with the property that*

$$\|A(\mathbf{x})\| \geq m\|\mathbf{x}\|,$$

for every $\mathbf{x} \in X$.

In this case

$$\|A^{-1}\| \leq \frac{1}{m}.$$

3.9 Functionals in Mathematical Spaces

3.9.1 Linear Functionals

In the previous section, we discussed the operators that transform elements (e.g., vectors) from a mathematical space into other elements from another space. There is a special class of operators that play a critical role in the theory and applications. This class contains an operator transforming vectors from an arbitrary metric space into real numbers, which can be treated as the elements of one-dimensional Euclidean space E_1. The operators from this class are called *functionals*.

We will now give a more rigorous definition of functionals.

Let X be a metric or vector space, $D \subset X$, and E_1 be a one-dimensional Euclidean space (a set of real numbers). Then, we introduce the following definition.

Definition 3.42 If for any $\mathbf{x} \in D$ we can assign according to a certain rule some real number $y \in E_1$, we say that the *functional* f is given on D.

A functional is a special case of an operator when $Y = E_1$. Figure 3.9 illustrates the action of a functional in the vector space.

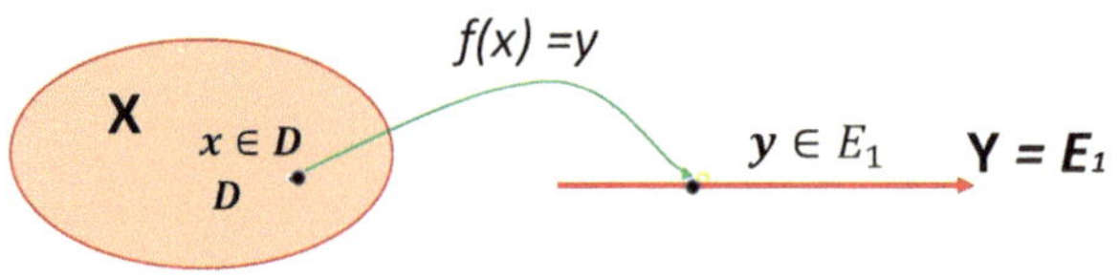

Fig. 3.9 Functional, $f(x)$, transforms vectors from an arbitrary vector space X into the elements of one-dimensional Euclidean space E_1

A linear functional is a special case of a linear operator. Therefore, the following condition should hold for the linear functional defined in the linear vector space:

$$f(\alpha_1\mathbf{x}_1 + \alpha_2\mathbf{x}_2 + \cdots\cdots + \alpha_n\mathbf{x}_n) = \alpha_1 f(\mathbf{x}_1) + \alpha_2 f(\mathbf{x}_2) + \cdots\cdots + \alpha_n f(\mathbf{x}_n). \tag{3.112}$$

In other words, one can open brackets while considering an application of the linear functional to a linear combination of the vectors from a linear vector space.

In a simplified way, we can summarize the actions of conventional functions, functionals, and operators as follows:

(1) functions transform scalars into scalars;
(2) functionals transform vectors into scalars;
(3) operators transform vectors into vectors.

3.9.2 Riesz Representation Theorem

There are many ways to introduce a functional in Hilbert space, H. The simplest example is the norm of a vector:

$$\varphi(\mathbf{x}) = \|\mathbf{x}\|, \ \mathbf{x} \in H. \tag{3.113}$$

Functional $\varphi((\mathbf{x})$ defined by formula (3.113) is not a linear functional, however, because

$$\varphi(\mathbf{x}+\mathbf{y}) = \|\mathbf{x}+\mathbf{y}\| \neq \|\mathbf{x}\| + \|\mathbf{y}\|. \tag{3.114}$$

Expression (3.114) shows that the requirement (3.112) for the linear functional does not hold in this case.

Let us now introduce some examples of linear functionals in Hilbert space. We consider a fixed element $\mathbf{l}$ of a Hilbert space H. Then we can introduce a linear functional, $f(\mathbf{x})$, as a projection (inner product) of arbitrary vector $\mathbf{x}$ on the fixed vector $\mathbf{l}$:

$$f(\mathbf{x}) = (\mathbf{l}, \mathbf{x}), \tag{3.115}$$

for any $\mathbf{x} \in H$.

It is easy to prove that the functional defined by expression (3.115) is indeed a linear one, considering the linear property of the inner product operation:

$$f(\alpha\mathbf{x}+\beta\mathbf{y}) = (\mathbf{l},\alpha\mathbf{x}+\beta\mathbf{y}) = \alpha(\mathbf{l}, \mathbf{x}) + \beta(\mathbf{l}, \mathbf{y}) = \alpha f(\mathbf{x}) + \beta f(\mathbf{y}).$$

Moreover, $f(\mathbf{x})$ is a bounded linear functional:

$$\mid f(\mathbf{x}) \mid < M\|\mathbf{x}\|,$$

where $M \leq \infty$.

This follows from Schwarz's inequality:

$$| f(\mathbf{x}) |=| (\mathbf{l}, \mathbf{x}) |\leq \|\mathbf{l}\| \|\mathbf{x}\|.$$

We can prove now that any linear functional in a Hilbert space can be represented in the form of (3.115).

Theorem 3.43 (Riesz representation theorem) *Every bounded linear functional* $f(\mathbf{x})$ *in a Hilbert space can be represented as* $(\mathbf{l}, \mathbf{x})$ *and* $\mathbf{l}$ *is uniquely determined by* f.

Proof Consider a basis $\{\mathbf{e}_1, \mathbf{e}_2, \mathbf{e}_3, \ldots\ldots \mathbf{e}_n, \ldots\}$ of the Hilbert space. We know that for any $\mathbf{x} \in X$,

$$\mathbf{x} = \sum_i (\mathbf{x}, \mathbf{e}_i)\mathbf{e}_i.$$

Thus, due to the linearity of the functional, we can write

$$f(\mathbf{x}) = f\left(\sum_i \left(\mathbf{x}, \mathbf{e}_i\right)\mathbf{e}_i\right) = \sum_i (\mathbf{x}, \mathbf{e}_i) f(\mathbf{e}_i). \tag{3.116}$$

Suppose that

$$f(\mathbf{e}_i) = \mathbf{l}_i\ . \tag{3.117}$$

We can introduce a vector $\mathbf{l}$ with the scalar components l_i:

$$\mathbf{l} = \sum_i l_i \mathbf{e}_i. \tag{3.118}$$

Then Eq. (3.116) can be written as follows:

$$f(\mathbf{x}) = \sum_i (\mathbf{x}, \mathbf{e}_i) l_i = (\mathbf{x}, \sum_{\mathbf{i}} l_i \mathbf{e}_i) = (\mathbf{x}, \mathbf{l}), \tag{3.119}$$

from which the first statement of the theorem follows.

Suppose now that there is another vector $\mathbf{l}^{(1)}$ such that

$$f(\mathbf{x}) = (\mathbf{x}, \mathbf{l}^{(1)}). \tag{3.120}$$

Then

$$f(\mathbf{e}_i) = (\mathbf{e}_i, \mathbf{l}^{(1)}) = l_i^{(1)}.$$

At the same time, according to formula (3.119), we have:

$$f(\mathbf{e}_i) = l_i.$$

From the last two equations, it follows that,

$$l_i = l_i^{(1)}, \quad \text{and } \mathbf{l} = \mathbf{l}^1.$$

Thus, we have proved that vector $\mathbf{l}$ is uniquely determined by f.

3.9.3 Norm of the Functional

The functional norm can be introduced by analogy with the norm of the operator, based on the following definitions.

Definition 3.44 The functional $y = f(\mathbf{x})$ is called *bounded* if there exists a real number M with the property that

$$| f(\mathbf{x}) | \leq M \|\mathbf{x}\|, \tag{3.121}$$

for every $\mathbf{x} \in X$.

Consider the Schwarz inequality (3.39)

$$\|f(\mathbf{a})\| = | \mathbf{a} \cdot \mathbf{l} | \leq \|\mathbf{a}\| \, \|\mathbf{l}\|. \tag{3.122}$$

Therefore, one can always find a large enough constant M with the following property:

$$\|f(\mathbf{a})\| \leq M\|\mathbf{a}\|. \tag{3.123}$$

Definition 3.45 The norm of the functional f is the minimum value of all possible M that satisfy the inequality (3.123): $\|f\| = \min\{M > 0, \ |f(\mathbf{a})| \leq M\|\mathbf{a}\|\}$, $\|f\| \leq M$.

On the other hand, we have

$$\|f(\mathbf{l})\| = \mathbf{l} \cdot \mathbf{l} = \|\mathbf{l}\| \, \|\mathbf{l}\| \leq M \, \|\mathbf{l}\|.$$

So the minimum value of M that satisfies (3.123), is the norm of the constant vector $\mathbf{l}$:

$$M \geq \|\mathbf{l}\|.$$

Therefore, we have established that the norm of the functional is equal to the norm of the vector $\mathbf{l}$ given by its representation (3.115):

$$\|f\| = \|\mathbf{l}\|.$$

3.9.4 Functional Representation of the Data and An Inverse Problem

Assume that we have physical measurements in a fixed number of observation points $d_j,\ j = 1, 2, \ldots n;\ d_j \in E_1$. These measurements depend on parameters of the corresponding models and therefore can be treated as the functionals,

$$d_j = f_j(\mathbf{m}), \tag{3.124}$$

$$j = 1, 2, \ldots n;\ d_j \in E_1,\ \mathbf{m} \in M,$$

where M is a Hilbert space of model parameters, and $f_j(\mathbf{m})$ are linear functionals, defined on M.

According to the Riesz representation theorem, there exist vectors $\mathbf{l}^{(j)}$ (elements of the space M) such that

$$d_j = (\mathbf{m}, \mathbf{l}^{(j)}), \tag{3.125}$$

$$j = 1, 2, \ldots n;\ ,\ \mathbf{l}^{(j)} \in M.$$

Vectors $\mathbf{l}^{(j)}$ are called "the data kernels" (Parker 1994).

Suppose that we know the data kernels $\mathbf{l}^{(j)}$. The problem is to determine model $\mathbf{m}$, which fits the observed data. In other words, we have to find the solution of the system of Eq. (3.125).

To solve this problem, we assume that $\{\mathbf{l}^{(j)},\ j = 1, 2, \ldots n;\}$ is a system of linear independent vectors, which forms the subspace $L \subset M$. If the dimension of M is greater than L, the element $\mathbf{m}$ is not unequally defined by (3.125). So we can find the solution of (3.125), which possesses the additional properties, for example, the smallest norm.

First of all let us formulate the Decomposition Theorem

Theorem 3.46 (The Decomposition Theorem) *For a given complete subspace $L \subset M$, any element $\mathbf{m} \in M$ can be written as the sum of a part in L and a part in $L^{\perp}$:*

$$\mathbf{m} = \mathbf{l} + \mathbf{h}, \tag{3.126}$$

where $L^{\perp}$ is the orthogonal complement of L, such that if $\mathbf{l} \in L,\ \mathbf{h} \in L^{\perp}$, then $(\mathbf{l}, \mathbf{h}) = 0$ (Fig. 3.10).

By definition:

$$(\mathbf{l}^{(j)}, \mathbf{h}) = 0. \tag{3.127}$$

Substituting (3.126) into (3.125), we have

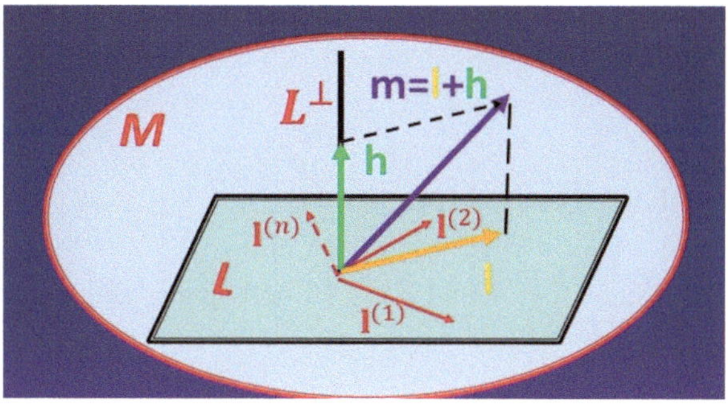

Fig. 3.10 Decomposition theorem

$$d_j = (\mathbf{l}, \mathbf{l}^{(j)}), \tag{3.128}$$

$$j = 1, 2, \ldots n; \; \mathbf{l}^{(j)} \in M,$$

so that only element $\mathbf{l}$ is to be determined to fit Eq. (3.125). An element $\mathbf{h}$ can be chosen from the other conditions, for example, from the condition that the norm of $\mathbf{m}$ is minimum.

Let us calculate this norm:

$$\|\mathbf{m}\|^2 = (\mathbf{l} + \mathbf{h}, \mathbf{l} + \mathbf{h}) = \|\mathbf{l}\|^2 + 2(\mathbf{l}, \mathbf{h}) + \|\mathbf{h}\|^2 = \|\mathbf{l}\|^2 + \|\mathbf{h}\|^2. \tag{3.129}$$

From (3.129) it follows that $\|\mathbf{m}\| = \min$ if

$$\mathbf{h} = 0. \tag{3.130}$$

Thus we have the following solution for $\mathbf{m}$:

$$\mathbf{m} = \mathbf{l} = \sum_{i=1}^{n} \beta_i \mathbf{l}^{(i)}, \tag{3.131}$$

where β_i $(i = 1, 2, \ldots n)$ are unknown coefficients which have to be determined from the observed data.

By substituting (3.131) into (3.125), we have

$$d_j = \sum_{i=1}^{n} \beta_i (\mathbf{l}^{(i)}, \mathbf{l}^{(j)}) = \sum_{i=1}^{n} \Gamma_{ji} \beta_i, \tag{3.132}$$

where

$$\Gamma_{ji} = (\mathbf{l}^{(i)}, \mathbf{l}^{(j)})$$

is the corresponding Gram matrix, which is nonsingular because the vectors $\mathbf{l}^{(j)}$ are assumed to be linear independent.

Thus coefficients β_i $(i = 1, 2, \ldots n)$ are found as the solution of the system of Eq. (3.132).

3.10 Adjoint Operators

We assume that X and Y are Hilbert spaces and A is a linear operator from X to Y:

$$\mathbf{y} = A\mathbf{x}. \tag{3.133}$$

Theorem 3.47 *For any linear operator A on X and any $\mathbf{y} \in Y$, there exists a unique element $\mathbf{x}^\star \in X$ such that for all $\mathbf{x} \in X$*

$$(A\mathbf{x}, \mathbf{y})_Y = (\mathbf{x}, \mathbf{x}^\star)_X, \tag{3.134}$$

where

$$\|\mathbf{x}^\star\|_X \le \|A\| \|\mathbf{y}\|_Y. \tag{3.135}$$

Proof If element $\mathbf{y}$ is fixed, we can consider $(A\mathbf{x}, \mathbf{y})_Y$ as a linear functional with respect to $\mathbf{x}$:

$$(A\mathbf{x}, \mathbf{y})_Y = f(\mathbf{x}). \tag{3.136}$$

According to the Riesz representation theorem, any linear functional can be represented as follows:

$$f(\mathbf{x}) = (\mathbf{x}, \mathbf{x}^\star), \tag{3.137}$$

where $\mathbf{x}^\star \in X$ exists and is unique and

$$\|f\| = \|\mathbf{x}^\star\|_X. \tag{3.138}$$

On the other hand, according to (3.136) and the Schwarz inequality, we have

$$\|f(\mathbf{x})\| \le \|A\mathbf{x}\| \|\mathbf{y}\| \le \|A\| \|\mathbf{x}\|_X \|\mathbf{y}\|_Y. \tag{3.139}$$

Dividing the left-hand and the right-hand sides of (3.139) by $\|\mathbf{x}\|$, we have

$$\|f(\tilde{\mathbf{x}})\| \le \|A\| \|\mathbf{y}\|_Y, \tag{3.140}$$

where $\tilde{\mathbf{x}} = \mathbf{x}/\|\mathbf{x}\|$ and $f(\tilde{\mathbf{x}}) = f(\mathbf{x})/\|\mathbf{x}\|$ by linearity.

From the last inequality we find the following:

$$\|f\| = \sup\{f(\tilde{\mathbf{x}}), \|\tilde{\mathbf{x}}\| = 1\} \le \|A\| \|\mathbf{y}\|_Y. \tag{3.141}$$

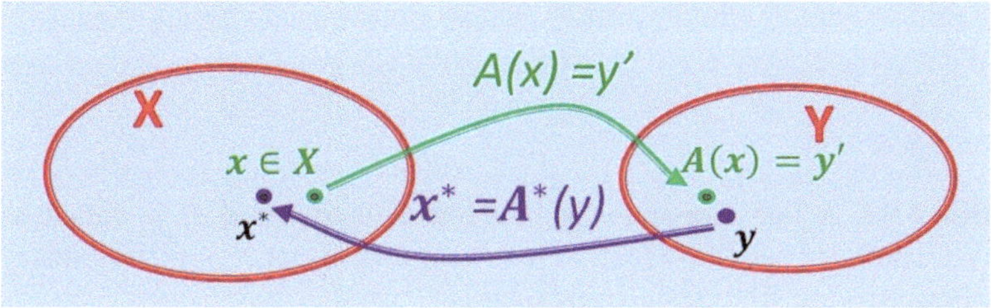

Fig. 3.11 Introducing an adjoint operator

By comparison of (3.138) and (3.141), we see that

$$\|\mathbf{x}^\star\|_{\mathbf{X}} \leq \|A\| \|\mathbf{y}\|_Y. \tag{3.142}$$

From (3.137) and (3.142) the statement of the Theorem 3.47 follows at once.
Figure 3.11 illustrates an idea of the adjoint operator.

Definition 3.48 On the basis of Theorem 3.47 we can determine the operator $A^\star$ which maps an element $\mathbf{y} \in Y$ into the proper element $\mathbf{x}^\star \in X$, according to formula (3.134):

$$\mathbf{x}^\star = A^\star \mathbf{y}. \tag{3.143}$$

The operator $A^\star$ is called the *adjoint* operator of A:

$$(A\mathbf{x}, \mathbf{y}) = (\mathbf{x}, A^\star \mathbf{y}). \tag{3.144}$$

Theorem 3.49 *The adjoint operator is a linear operator and*

$$\|A^\star\| = \|A\|. \tag{3.145}$$

Definition 3.50 A linear operator A in a Hilbert space H is called *self-adjoint* (or *symmetric*) if

$$A = A^\star. \tag{3.146}$$

Thus for a self-adjoint operator we have

$$(A\mathbf{x}, \mathbf{z}) = (\mathbf{x}, A\mathbf{z}). \tag{3.147}$$

Definition 3.51 A symmetric operator is said to be positive on some subset $S \subset H$ if for all $\mathbf{x} \in S$

$$(A\mathbf{x}, \mathbf{x}) \geq 0 \tag{3.148}$$

and

$$(A\mathbf{x}, \mathbf{x}) = 0,$$

if and only if $\mathbf{x} = 0$.

Definition 3.52 A linear operator A in a real Hilbert space H is called positive definite in some subset $S \subset H$, if we can find a constant $\gamma > 0$ such that, for all $\mathbf{x} \in S$, the following relationship holds:

$$(A\mathbf{x}, \mathbf{x}) \geq \gamma (\mathbf{x}, \mathbf{x}) = \gamma \|\mathbf{x}\|^2 . \tag{3.149}$$

The last definition can be extended to the case of the complex Hilbert space H.

Definition 3.53 A linear operator A in a complex Hilbert space H is called an absolutely positive definite (APD) operator in some subset $S \subset H$, if we can find a constant $\gamma > 0$ such that, for all $\mathbf{x} \in S$, the following relationship holds:

$$|(A\mathbf{x}, \mathbf{x})| \geq \gamma (\mathbf{x}, \mathbf{x}) = \gamma \|\mathbf{x}\|^2 . \tag{3.150}$$

3.11 Concepts from Variational Calculus

3.11.1 Differentiation of Operators and Functionals

Assume that X and Y are two Banach spaces (complete normed linear spaces) and A is some operator from X to Y.

Definition 3.54 The operator A is called *differentiable* at some point $\mathbf{x} \in X$ if there exists a linear bounded operator F_x, acting from X to Y, such that

$$A(\mathbf{x}+\delta\mathbf{x}) - A(\mathbf{x}) = F_x(\delta\mathbf{x}) + o(\|\delta\mathbf{x}\|), \tag{3.151}$$

where

$$\frac{o(\|\delta\mathbf{x}\|)}{\|\delta\mathbf{x}\|} \to 0,$$

when $\|\delta\mathbf{x}\| \to 0$.

The operator F_x is called the *Fréchet derivative* of A at $\mathbf{x}$ and is written as follows:

$$F_x = A'(\mathbf{x}). \tag{3.152}$$

The expression $F_x(\delta\mathbf{x})$ is called the *Fréchet differential* of $A(\mathbf{x})$ at $\mathbf{x}$ and is written as

$$F_x(\delta\mathbf{x}) = \delta A(\mathbf{x}, \delta\mathbf{x}). \tag{3.153}$$

In the particular case when we have a linear operator B, its derivative is equal to the operator B itself:

$$B'(\mathbf{x}) = B.$$

Suppose now that X is a Banach space and $f(\mathbf{x})$ is a functional in it.

Definition 3.55 If there exists such linear functional F_x^f that in some point $\mathbf{x} \in X$,

$$f(\mathbf{x}+\delta\mathbf{x}) - f(\mathbf{x}) = F_x^f(\delta\mathbf{x}) + o(\|\delta\mathbf{x}\|), \tag{3.154}$$

where

$$\frac{o(\|\delta\mathbf{x}\|)}{\|\delta\mathbf{x}\|} \to 0, \text{ when } \|\delta\mathbf{x}\| \to 0,$$

the functional $f(\mathbf{x})$ is called *differentiable* at the point x.

The F_x^f is called the *Fréchet derivative* of $f(\mathbf{x})$ at $\mathbf{x}$ and is written as follows:

$$F_x^f = f'(\mathbf{x}). \tag{3.155}$$

The expression $F_x^f(\delta\mathbf{x})$ is called the *Fréchet differential* of $f(\mathbf{x})$ at $\mathbf{x}$ and is written as

$$F_x^f(\delta\mathbf{x}) = df(\mathbf{x},\delta\mathbf{x}). \tag{3.156}$$

Example 3.56 Let us consider the functional $f(\mathbf{x})$ defined on the Hilbert space X:

$$f(\mathbf{x}) = \|\mathbf{x}\|^2$$

Then

$$\|\mathbf{x}+\delta\mathbf{x}\|^2 - \|\mathbf{x}\|^2 = 2(\mathbf{x},\delta\mathbf{x}) + \|\delta\mathbf{x}\|^2,$$

from which we have at once

$$F_x^f(\delta\mathbf{x}) = df(\mathbf{x},\delta\mathbf{x}) = 2(\mathbf{x},\delta\mathbf{x}). \tag{3.157}$$

3.11.2 Variational Operator

In the calculus of variations it is a common practice to use δA or $\delta\mathbf{x}$ to denote a variation of A or $\mathbf{x}$:

$$\delta A\mathbf{x} = A(\mathbf{x} + \delta\mathbf{x}) - A(\mathbf{x}) \approx F_x(\delta\mathbf{x}) = \delta A(\mathbf{x}, \delta\mathbf{x}). \tag{3.158}$$

The operator δ is called the *variational operator* and $\delta A(\mathbf{x}, \delta\mathbf{x})$ is called the *first variation* of A.

Note that, in expression (3.151), we can take into account the second order term with respect to $\|\delta x\|$:

$$A(\mathbf{x} + \delta\mathbf{x}) - A(\mathbf{x}) = F_x(\delta\mathbf{x}) + \frac{1}{2} F_x^{(2)}(\delta\mathbf{x}) + o(\|\delta\mathbf{x}\|^2). \tag{3.159}$$

Operator $F_x^{(2)}$ is the operator of the second variation of operator A. It has the meaning of the second order derivative of the original operator A.

Similar to expression (3.159) for an operator, we can write for a functional

$$f(\mathbf{x} + \delta\mathbf{x}) - f(\mathbf{x}) = F_x^f(\delta\mathbf{x}) + \frac{1}{2} H_x^f(\delta\mathbf{x}) + o(\|\delta\mathbf{x}\|^2), \tag{3.160}$$

where H_x^f is the so-called Hessian operator, or second variation (second derivative) of the functional f.

Example 3.57 Let us consider the functional $g(\mathbf{x})$ defined on the Hilbert space X:

$$g(\mathbf{x}) = \|A\mathbf{x} - \mathbf{y}_0\|^2, \tag{3.161}$$

where A is some operator from X to Hilbert space Y: $\mathbf{y} = A\mathbf{x}$. Then

$$\begin{aligned} g(\mathbf{x} + \delta\mathbf{x}) - g(\mathbf{x}) &= (A(\mathbf{x} + \delta\mathbf{x}) - \mathbf{y}_0,\ A(\mathbf{x} + \delta\mathbf{x}) - \mathbf{y}_0) \\ &\quad - \left(A\mathbf{x} - \mathbf{y}_0,\ A\mathbf{x} - \mathbf{y}_0\right) \\ &= \left(\left[A\mathbf{x} + F_x(\delta\mathbf{x}) + \frac{1}{2} F_x^{(2)}(\delta\mathbf{x}) + o(\|\delta\mathbf{x}\|^2) - \mathbf{y}_0\right],\right. \\ &\quad \left.\left[A\mathbf{x} + F_x(\delta\mathbf{x}) + \frac{1}{2} F_x^{(2)}(\delta\mathbf{x}) + o(\|\delta\mathbf{x}\|^2) - \mathbf{y}_0\right]\right) \\ &\quad - \left(A\mathbf{x} - \mathbf{y}_0,\ A\mathbf{x} - \mathbf{y}_0\right) = 2\left(A\mathbf{x} - \mathbf{y}_0,\ F_x(\delta\mathbf{x})\right) \\ &\quad + \left(F_x(\delta\mathbf{x}),\ F_x(\delta\mathbf{x})\right) + \left(A\mathbf{x} - \mathbf{y}_0,\ F_x^{(2)}(\delta\mathbf{x})\right) + o(\|\delta\mathbf{x}\|^2). \end{aligned}$$

From the last formula we obtain the expression for the Fréchet differential of the functional g:

$$F_x^g(\delta\mathbf{x}) = \delta g(\mathbf{x}, \delta\mathbf{x}) = 2(A\mathbf{x} - \mathbf{y}_0,\ F_x(\delta\mathbf{x})), \tag{3.162}$$

and for its Hessian:

$$H_x^g(\delta\mathbf{x}) = 2\left(F_x(\delta\mathbf{x}),\ F_x(\delta\mathbf{x})\right) + 2\left(A\mathbf{x} - \mathbf{y}_0,\ F_x^{(2)}(\delta\mathbf{x})\right). \tag{3.163}$$

Note that, similar to the basic formulae of calculus for conventional functions, we can obtain simple rules and operations of the variational calculus. Actually, the variational operator acts like a differential operator. For example, let us consider the following operators:

$$A(\mathbf{x}),\ B(\mathbf{x}),$$

and the operator of two variables,

$$G(\mathbf{x}, \mathbf{z}).$$

We have

$$\delta(A + B) = \delta A + \delta B, \tag{3.164}$$

$$\delta(AB) = B\delta A + A\delta B, \tag{3.165}$$

$$\delta G = G'_x(\delta\mathbf{x}, \mathbf{z}) + G'_z(\mathbf{x}, \delta\mathbf{z}). \tag{3.166}$$

Using these simple rules and the properties of the inner product in the Hilbert space, one can derive, for example, the expression for the Fréchet derivative of the functional $g(\mathbf{x})$ determined by formula (3.161) using the following calculations:

$$\delta g = \delta \left\| A\mathbf{x} - \mathbf{y}_0 \right\|^2$$

$$= \delta(A\mathbf{x} - \mathbf{y}_0, A\mathbf{x} - \mathbf{y}_0) = (\delta A\mathbf{x}, A\mathbf{x} - \mathbf{y}_0) + (A\mathbf{x} - \mathbf{y}_0, \delta A\mathbf{x})$$

$$\approx (F_x(\delta\mathbf{x}), A\mathbf{x} - \mathbf{y}_0) + (A\mathbf{x} - \mathbf{y}_0, F_x(\delta\mathbf{x})) = 2(A\mathbf{x} - \mathbf{y}_0, F_x(\delta\mathbf{x}))).$$

Thus, the first variation of the functional $g(\mathbf{x})$ is equal to

$$\delta g(\mathbf{x}, \delta\mathbf{x}) = 2(A\mathbf{x} - \mathbf{y}_0, F_x(\delta\mathbf{x}))\ , \tag{3.167}$$

which is the same as Eq. (3.162).

3.11.3 Extremum Functional Problems

Theorem 3.58 *A differentiable functional $f(\mathbf{x})$ has an extremum at some point $\mathbf{x}_0$ only if the first variation of the functional at this point is equal to zero for any variation $\delta\mathbf{x}$ of $\mathbf{x}_0$:*

$$\delta f(\mathbf{x}_0, \delta\mathbf{x}) = 0. \tag{3.168}$$

Proof According to the definition,

$$\delta f(\mathbf{x}_0, \lambda\delta\mathbf{x}) = f(\mathbf{x}_0 + \lambda\delta\mathbf{x}) - f(\mathbf{x}_0) - o(\|\delta\mathbf{x}\|). \tag{3.169}$$

However, the first variation $\delta f(\mathbf{x}_0, \delta\mathbf{x})$ is a linear functional with respect to $\delta\mathbf{x}$; therefore

$$\delta f(\mathbf{x}_0, \lambda\delta\mathbf{x}) = \lambda\delta f(\mathbf{x}_0, \delta\mathbf{x}). \tag{3.170}$$

Substituting (3.170) into (3.169) we have

$$f(\mathbf{x}_0 + \lambda\delta\mathbf{x}) - f(\mathbf{x}_0) = \lambda\delta f(\mathbf{x}_0, \delta\mathbf{x}) + o(\|\delta\mathbf{x}\|). \tag{3.171}$$

We now prove the statement of the theorem by contradiction. Let us assume that,

$$\delta f(\mathbf{x}_0, \delta\mathbf{x}) \neq 0. \tag{3.172}$$

The sign of the right-hand side of (3.171) is governed by the sign of $\delta f(\mathbf{x}_0, \delta\mathbf{x})$ and λ. In this case, according to (3.171), the difference $f(\mathbf{x}_0 + \lambda\delta\mathbf{x}) - f(\mathbf{x}_0)$ can be either positive or negative according to the choice of λ which means that there is no extremum at point $\mathbf{x}_0$. Since assuming (3.171) leads to a contradiction, it is concluded that the statement of the theorem is, in fact, true.

Example 3.59 Let us find the minimum of the functional $g(\mathbf{x})$ determined in Example 3.57:

$$g(\mathbf{x}) = \|A\mathbf{x} - \mathbf{y}_0\|^2. \tag{3.173}$$

According to (3.167) and (3.168), we have $\delta g(\mathbf{x},\delta\mathbf{x}) = 2(A\mathbf{x} - \mathbf{y}_0, F_x(\delta\mathbf{x})) = 0$, for any $\delta\mathbf{x} \in X$. Note that the Fréchet derivative F_x is a linear bounded operator. Therefore we can determine the linear and bounded adjoint operator $F_x^\star$, which satisfies the condition

$$(A\mathbf{x} - \mathbf{y}_0, F_x(\delta\mathbf{x})) = (F_x^\star(A\mathbf{x} - \mathbf{y}_0), \delta\mathbf{x}) = 0. \tag{3.174}$$

Equation (3.174) holds for any $\delta\mathbf{x}$ if and only if

$$F_x^\star(A\mathbf{x} - \mathbf{y}_0) = F_x^\star(A\mathbf{x}) - F_x^\star(\mathbf{y}_0) = 0.$$

Thus we have the following equation for the extremum point x_0:

$$F_x^\star A(\mathbf{x}_0) = F_x^\star(\mathbf{y}_0). \tag{3.175}$$

Example 3.60 Suppose now that operator A is a linear operator. Then its Fréchet derivative F_x is equal to operator A itself:

$$F_x = A. \tag{3.176}$$

Substituting (3.176) into (3.175), we have the following equation for the extremum point $\mathbf{x}_0$:

$$A^\star A(\mathbf{x}_0) = A^\star(\mathbf{y}_0). \tag{3.177}$$

It is important to notice that operator $A^{\star}A$ is a self-adjoint (symmetric) positive operator. By inverting the operator $A^{\star}A$, we finally have

$$\mathbf{x}_0 = (A^{\star}A)^{-1}A^{\star}(\mathbf{y}_0). \tag{3.178}$$

Example 3.61 Let us consider the function $\Phi(k)$ of the complex variable k defined by the norm of difference between two vectors in the complex Hilbert space H:

$$\Phi(k) = \|\mathbf{x}-k\mathbf{y}\|, \ \mathbf{x}, \mathbf{y} \in H. \tag{3.179}$$

We would like to find the minimum of this function. It is clear that the minimum of $\Phi(k)$ coincides with the minimum of its square, $\Phi^2(k)$. The first variation of this function is equal

$$\delta\Phi^2(k) = \delta(\mathbf{x}-k\mathbf{y}, \ \mathbf{x}-k\mathbf{y}) = -(\delta k\mathbf{y}, \ \mathbf{x}-k\mathbf{y}) - (\mathbf{x}-k\mathbf{y}, \ \delta k\mathbf{y})$$

$$= -(\delta k\mathbf{y}, \ \mathbf{x}-k\mathbf{y}) - (\delta k\mathbf{y}, \ \mathbf{x}-k\mathbf{y})^*$$

$$= -\delta k(\mathbf{y}, \ \mathbf{x}-k\mathbf{y}) - \left[\delta k(\mathbf{y}, \ \mathbf{x}-k\mathbf{y})\right]^* = -2\,\mathrm{Re}\left[\delta k(\mathbf{y}, \ \mathbf{x}-k\mathbf{y})\right], \tag{3.180}$$

where we take into account the property of the inner product in the complex Hilbert space,

$$(\mathbf{x}, \mathbf{y}) = (\mathbf{y}, \mathbf{x})^*,$$

and asterisk * denotes the complex conjugate.

The necessary condition for the minimum of function $\Phi^2(k)$ is

$$\delta\Phi^2(k) = -2\,\mathrm{Re}\left[\delta k(\mathbf{y}, \ \mathbf{x}-k\mathbf{y})\right] = 0 \text{ for any } \delta k. \tag{3.181}$$

For example, we can select δk as follows

$$\delta k = (\mathbf{y}, \ \mathbf{x}-k\mathbf{y})^*. \tag{3.182}$$

Substituting (3.182) into (3.181), we obtain:

$$|(\mathbf{y}, \ \mathbf{x}-k\mathbf{y})|^2 = 0,$$

and

$$(\mathbf{y}, \ \mathbf{x}) - (\mathbf{y}, \ k\mathbf{y}) = (\mathbf{x}, \ \mathbf{y})^* - k^*(\mathbf{y}, \ \mathbf{y}) = 0,$$

where we use another property of the inner product in the complex Hilbert space,

$$(\mathbf{x}, k\mathbf{y}) = k^*(\mathbf{x}, \mathbf{y}).$$

Therefore we have the following equation for the minimum point k_0:

$$k_0 = \frac{(\mathbf{x}, \mathbf{y})}{(\mathbf{y}, \mathbf{y})}. \tag{3.183}$$

Substituting (3.183) into (3.179), we find the corresponding minimum of the function $\Phi(k)$:

$$\min \Phi(k) = \Phi(k_0) = \sqrt{(\mathbf{x}-k_0\mathbf{y},\ \mathbf{x}-k_0\mathbf{y})} = \left[(\mathbf{x}, \mathbf{x}) - (k_0\mathbf{y},\ \mathbf{x}) - (\mathbf{x},\ k_0\mathbf{y}) + (k_0\mathbf{y},\ k_0\mathbf{y})\right]^{1/2}$$

$$= \left[(\mathbf{x}, \mathbf{x}) - (\mathbf{x},\ k_0\mathbf{y})^* - (\mathbf{x},\ k_0\mathbf{y}) + k_0^* k_0 (\mathbf{y},\ \mathbf{y})\right]^{1/2}$$

$$= \left[(\mathbf{x}, \mathbf{x}) - 2\,\mathrm{Re}\left[k_0^* (\mathbf{x},\ \mathbf{y})\right] + k_0^* k_0 (\mathbf{y},\ \mathbf{y})\right]^{1/2}$$

$$= \left[(\mathbf{x}, \mathbf{x}) - 2\,\mathrm{Re}\left[\frac{(\mathbf{x}, \mathbf{y})^*}{(\mathbf{y}, \mathbf{y})}(\mathbf{x},\ \mathbf{y})\right] + \frac{(\mathbf{x}, \mathbf{y})^*}{(\mathbf{y}, \mathbf{y})}(\mathbf{x}, \mathbf{y})\right]^{1/2}$$

$$= \left[(\mathbf{x}, \mathbf{x}) - 2\frac{|(\mathbf{x}, \mathbf{y})|^2}{(\mathbf{y}, \mathbf{y})} + \frac{|(\mathbf{x}, \mathbf{y})|^2}{(\mathbf{y}, \mathbf{y})}\right]^{1/2} = \|\mathbf{x}\| \sqrt{1 - \frac{|(\mathbf{x}, \mathbf{y})|^2}{\|\mathbf{x}\|\ \|\mathbf{y}\|}}. \tag{3.184}$$

References and Recommended Reading to This Chapter

Barth N (1999) The Gramian and k-volume in n-space: some classical results in linear algebra. J Young Investig 2

Everitt WN (1958) Some properties of Gram matrices and determinants. Q J Math 9(1):87–98

Golub GH, Van Loan CF (2013) Matrix computations, 4th edn. The Johns Hopkins University Press, Baltimore and London, 753 pp

Kreyszig E (1989) Introductory functional analysis with applications. Wiley, 688 pp

Mitrinović DS, Pečarić JE, Fink AM (1993) Gram's inequality. In: Classical and new inequalities in analysis. Mathematics and its applications (East European Series), vol 61. Springer

Parker RL (1994) Geophysical inverse theory. Princeton University Press, Princeton, NJ, 386 pp

Reddy BD (1998) Introductory functional analysis. Springer, 472 pp

Zhdanov MS (2002) Geophysical inverse theory and regularization problems. Elsevier, 609 pp

Zhdanov MS (2015) Inverse theory and applications in geophysics. Elsevier, 704 pp

Part II
Foundations of Inverse Theory

Chapter 4
Principles of Regularization Theory

Abstract The regularization theory provides the framework for solving the ill-posed inverse problems. In this chapter, we discuss the foundations of the regularization theory, starting with the rigorous mathematical formulation of the well-posed and ill-posed problems and introducing the regularizing operators, stabilizing functionals, and Tikhonov parametric functional. The concepts of smoothing and focusing stabilizing functionals are discussed. The important topic of this chapter is the optimal regularization parameter selection, which can be based on the Tikhonov misfit condition or L-curve method.

Keywords Regularizing operator · Stabilizing functional · Tikhonov parametric functional

The formal solution to the ill-posed inverse problem could result in unstable, unrealistic models. The regularization theory guides how one can overcome this difficulty. The foundations of the regularization theory were developed in numerous publications by Andrei N. Tikhonov, which were reprinted in 1999 as a special book, published by Moscow State University (Tikhonov 1999). In this chapter, I will present a short overview of the basic principles of the Tikhonov regularization theory following Zhdanov (1993). A detailed description of the regularization theory of inverse problem solutions can also be found in Tikhonov and Arsenin (1977) and Zhdanov (2002, 2015).

4.1 Formulation of Well-Posed and Ill-Posed Problems

4.1.1 Formulation of the Inverse Problem in General Mathematical Spaces

In the first chapter, we have introduced an inverse problem as the solution of the following operator equation:

$$\mathbf{d} = A(\mathbf{m}), \tag{4.1}$$

M. S. Zhdanov, *Advanced Methods of Joint Inversion and Fusion of Multiphysics Data*, Advances in Geological Science,
https://doi.org/10.1007/978-981-99-6722-3_4

where $\mathbf{m}$ is some function (or vector) describing the model parameters, and $\mathbf{d}$ is a data set, which can also be characterized as a function of the observation point (in the case of continuous observations), or as a vector (in the case of discrete observations). The solution of the inverse problem consists in determining such a model $\mathbf{m}^{pr}$ (predicted model) that generates predicted data $\mathbf{d}^{pr}$ that fit well with the experimental data $\mathbf{d}$. We have already discussed that we do not want to fit the observed data exactly because they always contain some noise that we should not fit. Therefore, we are looking for some predicted data that will be close enough to the observed data (usually, within the accuracy of our observations). But what does "close enough" mean? How can we measure the closeness of two data sets?

The answer to this question was provided in Chap. 3, where we introduced the mathematical technique to measure the distance between the observed and predicted data. This technique was based on the mathematical theory of vector spaces which provides us with guidance to solve this problem. We have also demonstrated in Chap. 3 that the most powerful example of a mathematical space used in inverse theory is the Hilbert space. Using the basic ideas of the mathematical theory of Hilbert spaces and operators acting in these spaces, we can now present a rigorous formulation of mathematical inverse problems.

Let us assume that we are given two Hilbert spaces, M and D, and an operator A that acts from space M to space D:

$$A(\mathbf{m}) = \mathbf{d}, \ \mathbf{m} \in M, \ \mathbf{d} \in D. \tag{4.2}$$

We will call D a space of data sets and M a space of the model parameters. Operator A is a forward modeling operator that transforms any model $\mathbf{m}$ into the corresponding data $\mathbf{d}$. The inverse problem is formulated as the solution of the operator Eq. (4.1). We have already learned in Chap. 1 that there are two important classes of inverse problems: well-posed and ill-posed problems. We can now provide a rigorous mathematical description of these two classes of inverse problems.

4.1.2 Well-Posed Problems

Following classical principles of regularization theory (Tikhonov and Arsenin 1977; Lavrent'ev et al. 1986) we can give the following definition of the well-posed problem.

Definition 4.1 The problem (4.1) is correctly (or well) posed if the following conditions are satisfied: (i) the solution $\mathbf{m}$ of Eq. (4.1) exists, (ii) the solution $\mathbf{m}$ of Eq. (4.1) is unique, and (iii) the solution $\mathbf{m}$ depends continuously on the left-hand side of Eq. (4.1) $\mathbf{d}$.

In other words, the inverse operator A^{-1} is defined throughout space D and is continuous. The last condition means that minor data variations will result in small changes in the model parameters—this requirement is equivalent to the provision of inverse problem stability.

Thus, the well-posed inverse problem possesses all the properties of the "good" solution discussed in Chap. 1: the solution exists, is unique, and is stable.

If at least one of these conditions fails, the inverse problem becomes ill-posed. The following definition reflects this situation.

Definition 4.2 The problem (4.1) is ill-posed if at least one of the conditions, (i), (ii), or (iii), listed above, fails.

In Chap. 1, we have discussed that most inverse problems are ill-posed because at least one of the conditions listed above fails. However, it may happen that if we narrow the class of models used in inversion, the originally ill-posed inverse problem may become well-posed. Mathematically it means that instead of considering $\mathbf{m}$ from the entire model space M, we can select $\mathbf{m}$ from some subspace of M, consisting of simpler and more suitable models for the given inverse problem. Thus, we arrive at the idea of the correctness set and conditionally well-posed inverse problems.

4.1.3 Conditionally Well-Posed Problems

Suppose we know a priori that the exact solution belongs to a set, C, of the solutions with the property that the inverse operator A^{-1}, defined on the image[1] AC, is continuous.

Definition 4.3 The problem (4.1) is conditionally well-posed (Tikhonov's well-posed) if the following conditions are met: (i) we know a priori that a solution of (4.1) exists and belongs to a specified set $C \subset M$, (ii) the operator A is a one-to-one mapping of C onto $AC \subset D$, and (iii) the operator A^{-1} is continuous on $AC \subset D$.

We call set C *the correctness set*. In contrast to the standard well-posed problem, a conditionally well-posed problem does not require solvability over the entire space. Also the requirement of the continuity of A^{-1} over the entire space D is substituted by the requirement of continuity over the image of C in D. Thus, introducing a correctness set makes even an ill-posed problem well-posed.

Tikhonov and Arsenin (1977) introduced the mathematical principles for selecting the correctness set C. For example, if a finite number of bounded parameters describes the models, they form correctness set C in the Euclidean space of the model parameters. This result can be generalized for any metric space.

First, we introduce the following definition.

Definition 4.4 The subset K of a metric space M is called *compact* if any sequence $\mathbf{m}_l \in K$ of elements in K contains a convergent subsequence $\mathbf{m}_{l_j} \in K$, which converges to an element $\mathbf{m}$ in K.

[1] The domain $AC \subset D$ formed by all vectors obtained as a result of operator A applied to all vectors $\mathbf{m}$ from the set C, $\mathbf{m} \in C$, is called an image of the set C in space D.

For example, it is known that any subset R of Euclidean space E_n is compact if and only if it is bounded:

$$\|x\| \leq c, \ c > 0, \text{ for any } x \in R.$$

It can be demonstrated that any compact subset of the metric space M can be used as a correctness set for an ill-posed inverse problem (4.2) (Tikhonov and Arsenin 1977; Zhdanov 2002). This fundamental result opens the way to constructing a stable solution to the ill-posed inverse problem.

4.1.4 Quasi-solution of the Ill-Posed Problem

The concept of a quasi-solution represents another critical element of the regularization theory. We assume now that the problem (4.1) is conditionally well-posed (Tikhonov's well-posed). Let us assume, also, that the left-hand side of (4.1) is given with some error:

$$\mathbf{d}_\delta = \mathbf{d} + \delta\mathbf{d}, \tag{4.3}$$

where

$$\mu_D(\mathbf{d}_\delta, \mathbf{d}) \leq \delta. \tag{4.4}$$

We now introduce the following definition of a quasi-solution of the ill-posed inverse problem.

Definition 4.5 A quasi-solution of inverse problem (4.1) in the correctness set C is an element $\mathbf{m}_\delta \in C$ which minimizes the distance $\mu_D(A\mathbf{m}, \mathbf{d}_\delta)$, i.e.:

$$\mu_D(A\mathbf{m}_\delta, \mathbf{d}_\delta) = \inf_{m \in C} \mu_D(A\mathbf{m}, \mathbf{d}_\delta), \tag{4.5}$$

where inf φ denotes the greatest lower bound of the variable φ.

Obviously, we can reach the minimum of the $\mu_D(A\mathbf{m}, \mathbf{d}_\delta)$ in C, if the correctness set is a compact. In this case the quasi-solution exists for any data $\mathbf{d}_\delta$.

Figure 4.1 illustrates the definition of a quasi-solution. The element $\mathbf{m} \in \mathbf{M}$ is an exact solution of the inverse problem

$$\mathbf{d} = A(\mathbf{m}). \tag{4.6}$$

Subset AC of the data space D is an image of the correctness set C obtained as a result of the application of operator A. A quasi-solution, $\mathbf{m}_\delta$, is selected from the correctness set C under the condition that its image, $A(\mathbf{m}_\delta)$, is the closest element in the subset AC to the observed noisy data, $\mathbf{d}_\delta$.

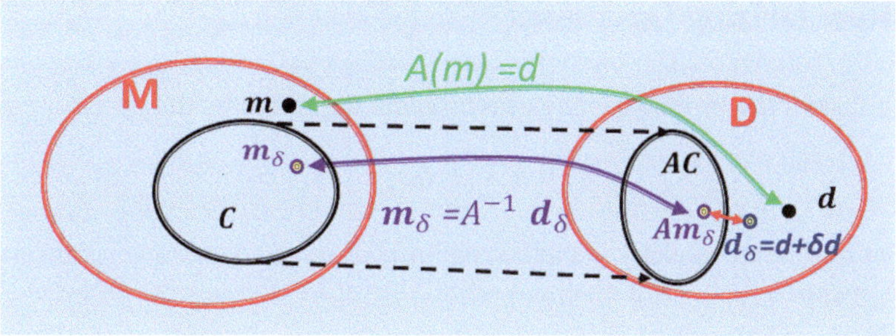

Fig. 4.1 A quasi-solution, $\mathbf{m}_\delta$, is selected from the correctness set C under the condition that its image, $A(\mathbf{m}_\delta)$, is the closest element to the observed noisy data, $\mathbf{d}_\delta$, from the subset AC : $\mu_D(A\mathbf{m}_\delta, \mathbf{d}_\delta) = \inf_{m \in C} \mu_D(A\mathbf{m}, \mathbf{d}_\delta)$

It can be proved also that the quasi-solution is a continuous function of $\mathbf{d}_\delta$. Indeed, let us consider the triangle inequality

$$\mu_D(A\mathbf{m}_\delta, \mathbf{d}) \le \mu_D(A\mathbf{m}_\delta, \mathbf{d}_\delta) + \mu_D(\mathbf{d}_\delta, \mathbf{d}). \tag{4.7}$$

According to the definition of the quasi-solution and condition (4.4), it follows from inequality (4.7) that

$$\mu_D(A\mathbf{m}_\delta, \mathbf{d}) \le 2\delta. \tag{4.8}$$

Based on Tikhonov's definition of the correctness set, we know that operator A^{-1} is a continuous one on the image AC of the correctness set C. Therefore, we conclude from (4.8) that the quasi-solution is a continuous function of $\mathbf{d}_\delta$. Note that this property holds only in the correctness set C. If one selects a solution, $\widetilde{\mathbf{m}}_\delta$, from outside the correctness set, it may be no longer a continuous function of the data (see Fig. 4.1).

The idea of the quasi-solution makes it possible to substitute the inverse problem solution by minimization of the distance $\mu_D(A\mathbf{m}, \mathbf{d}_\delta)$ in some appropriate class of suitable models. We can solve this problem using standard functional minimization methods and, therefore, find the quasi-solution. In this way, we significantly simplify the inverse problem solution. However, this approach is practical only if we know a priori the corresponding class of the models (the correctness set) where we are searching for the solution. It is difficult to describe this class entirely in many situations. Also, we usually prefer not to restrict ourselves to some specific class. In this case, we have to use a more general approach to a stable solution of the inverse problem.

4.2 Regularizing Operators

Let us consider the inverse geophysical problem described by the operator equation

$$\mathbf{d} = A(\mathbf{m}), \tag{4.9}$$

where $\mathbf{m}$ represents model parameters, and $\mathbf{d}$ is observed data. In general cases, the inverse operator A^{-1} is not continuous and, therefore, the inverse problem (4.9) is ill-posed.

The main idea of any regularization algorithm is to consider, instead of one ill-posed inverse problem (4.9), a family of well-posed problems,

$$\mathbf{d} = A_\alpha(\mathbf{m}), \tag{4.10}$$

which approximate the original inverse problem in some sense (Strakhov 1968, 1969a, b; Zhdanov 2002, 2015). The scalar parameter $\alpha > 0$ is called *a regularization parameter.*

Because all problems (4.10) are well-posed, their solutions exist and are unique. Therefore, we can introduce the inverse operators, A_α^{-1}, which transform the data into the solutions of these well-posed problems. We require that these solutions,

$$\mathbf{m}_\alpha = A_\alpha^{-1}(\mathbf{d}), \tag{4.11}$$

asymptotically go to the true solution of the original problem, $\mathbf{m}_t$, when regularization parameter α goes to zero:

$$\mathbf{m}_\alpha = A_\alpha^{-1}(\mathbf{d}) \rightarrow \mathbf{m}_t, \text{ if } \alpha \rightarrow 0. \tag{4.12}$$

Note that inverse operator of the well-posed problem, A_α^{-1}, is by definition a continuous and bounded operator.

Thus, we replace the solution of one ill-posed inverse problem with the solutions of the family of well-posed problems, assuming that these solutions, $\mathbf{m}_\alpha$, asymptotically go to the true solution, as α tends to zero.

In other words, any regularization algorithm is based on the approximation of the noncontinuous inverse operator A^{-1} by the family of continuous inverse operators $A_\alpha^{-1}(\mathbf{d})$ that depend on the regularization parameter α. The regularization must be such that, as α vanishes, the operators in the family should approach the exact inverse operator A^{-1}.

Let us now give a more accurate definition.

Inverse operators $A_\alpha^{-1}(\mathbf{d})$ are called the regularizing operators for Eq. (4.9). The regularizing operators are usually denoted as $R(\mathbf{d}, \alpha)$:

$$R(\mathbf{d}, \alpha) = A_\alpha^{-1}(\mathbf{d}). \tag{4.13}$$

The main property of the regularizing operators is that they deliver a stable but an approximate solution of the ill-posed problem:

$$\mathbf{m}_{\alpha} = R(\mathbf{d}, \alpha) \rightarrow \mathbf{m}_{t},$$

when $\alpha \rightarrow 0$.

We can see that the regularizing operators can be constructed by approximating the ill-posed Eq. (4.9) by the system of well-posed Eq. (4.10), where the corresponding inverse operators A_{α}^{-1} are continuous.

There are many different ways how one can construct a family of regularizing operators. This chapter will discuss the most widely used approach, which was originally introduced by Tikhonov and Arsenin (1977). It is based on the concepts of stabilizing and parametric functionals.

4.3 Stabilizing Functionals

We have discussed in the previous sections of the book that the main factors causing the inverse problem to be ill-posed are related to the nonuniqueness and instability of the solutions. As a result, a formal inversion of the ill-posed problem may produce unrealistic models. The natural way to avoid this situation is to reduce the class of mathematically plausible solutions by imposing constraints. These constraints can be based on some a priori knowledge about the properties of the possible solution. For example, one could assume that the physical properties vary slowly within the target, thus restricting the class of possible solutions to the smooth continuous functions. In other scenarios, we may anticipate the target has sharp physical boundaries. In this case, the class of possible solutions should be restricted to models with sharp boundaries. Our goal is to develop the mathematical theory of the regularized inversion, allowing us to impose these different constraints on the solutions. We achieve this goal by introducing a concept of stabilizing functional.

Using the mathematical language, we can say that a stabilizing functional (or a stabilizer) is used to select from space M of all possible models the subset M_c, which is a correctness set.

Definition 4.6 A nonnegative functional $s(\mathbf{m})$ in some metric space M is called *a stabilizing functional* if, for any real number $c > 0$ from the domain of functional values, the subset M_c of the elements $\mathbf{m} \in M$, for which $s(\mathbf{m}) \leq c$, is compact.

We will give now several examples of stabilizing functionals.

4.3.1 *Minimum Norm and Maximum Smoothness Stabilizing Functionals*

We begin our discussion with the most widely used stabilizing functionals based on the L_2 norm of the function.

Example 4.7 Let us consider a real Hilbert space L_2 formed by functions integrable in the interval $[a, b]$. The metric in space L_2 is determined according to the formula

$$\mu(\mathbf{m}_1, \mathbf{m}_2) = \left\{ \int_a^b [m_1(x) - m_2(x)]^2 \, dx \right\}^{\frac{1}{2}}. \tag{4.14}$$

It can be proved that any ball,

$$b(\mathbf{m}_0, c) = \{\mathbf{m} : \mu(\mathbf{m}, \mathbf{m}_0) \leq c, \ c > 0\},$$

is compact in the Hilbert space. Therefore, we can introduce a stabilizing functional as follows:

$$s(\mathbf{m}) = \mu(\mathbf{m}, \mathbf{m}_0), \tag{4.15}$$

where $\mathbf{m}_0$ is any given model from $M = L_2$. Obviously, the subset M_c of the elements $\mathbf{m} \in M$ for which $s(\mathbf{m}) \leq c$,

$$s(m) = \mu(m, m_0) \leq c, \tag{4.16}$$

is compact.

Example 4.8 Let us consider a Sobolev space (which is at the same time a Hilbert space) W_2^p formed by the functions continuously differentiable to the order n in the interval $[a, b]$. The metric in the space W_2^p is determined according to the formula

$$\mu_{W_2^p}(\mathbf{m}_1, \mathbf{m}_2) = \left\{ \int_a^b \sum_{k=0}^{p} q_k^2(x) \left[\frac{d^k \Delta m(x)}{dx^k} \right]^2 dx \right\}^{\frac{1}{2}},$$

where $\Delta m(x) = m_1(x) - m_2(x)$; and $q_0(x), q_1(x), \ldots., q_p(x)$ are the given real functions ($q_p(x) \neq 0$). We can introduce a stabilizing functional in the space W_2^p as follows:

$$s(\mathbf{m}) = \mu_{W_2^p}(\mathbf{m}, \mathbf{m}_0),$$

where $\mathbf{m}_0$ is any given model from $M = W_2^p$, and the sphere

$$s(\mathbf{m}) = \mu_{W_2^p}(\mathbf{m}, \mathbf{m}_0) \leq c$$

is compact.

We have established above that the main role of the stabilizing functional (a stabilizer) is to select the appropriate class of models for inverse problem solution. The examples listed above show that there are several common choices for a stabilizer. One is based on the least-squares criterion, or, in other words, on the L_2 norm for functions describing model parameters:

$$s_{MN}(\mathbf{m}) = \|\mathbf{m}\|_{L_2}^2 = (\mathbf{m}, \mathbf{m})_{L_2} = \int_V |m(\mathbf{r})|^2 \, dv = \min . \tag{4.17}$$

In the last formula we assume that the function $m(\mathbf{r})$, describing model parameters, is given within a three-dimensional domain V, and $\mathbf{r}$ is a radius-vector of an observation point. The conventional argument in support of the norm (4.17) comes from statistics and is based on an assumption that the least-squares image is the best over the entire ensemble of all possible images. We call $s_{MN}(\mathbf{m})$ a *minimum norm stabilizing functional.*

We can use, also, a quadratic functional s_{MNw}:

$$s_{MNw}(\mathbf{m}) = \|W\mathbf{m}\|_{L_2}^2 = (W\mathbf{m}, W\mathbf{m})_{L_2} = \int_V |w(\mathbf{r})\, m(\mathbf{r})|^2 \, dv = \min, \tag{4.18}$$

where $w(\mathbf{r})$ is an arbitrary weighting function, and W is a linear operator of multiplication of function $m(\mathbf{r})$ by the weighting function $w(\mathbf{r})$.

Another stabilizer uses a minimum norm of difference between a selected model and some a priori model $\mathbf{m}_{apr}$:

$$s_{MNapr(m)} = \left\|\mathbf{m} - \mathbf{m}_{apr}\right\|_{L_2}^2 = \min . \tag{4.19}$$

The next several examples of stabilizing functionals arise from the norm in Sobolev space introduced above. The minimum norm criterion (4.17), as applied to the gradient of the model parameters ∇m, brings us to a *maximum smoothness stabilizing functional*

$$s_{\max sm}(\mathbf{m}) = \|\nabla\mathbf{m}\|_{L_2}^2 = (\nabla\mathbf{m}, \nabla\mathbf{m})_{L_2}$$

$$= \int_V |\nabla m(\mathbf{r})|^2 \, dv = \min . \tag{4.20}$$

In some cases, one can use the minimum norm of the Laplacian of model parameters $\nabla^2 m$,

$$s_{\max sm}(m) = \left\|\nabla^2 m\right\|^2 = \left(\nabla^2 m, \nabla^2 m\right) = \min . \tag{4.21}$$

It has been successfully used in many inversion schemes developed for geophysical data interpretation (see, for example, Constable et al. 1987; Smith and Booker 1991; Zhdanov and Fang 1996). This stabilizer produces smooth models, which in many practical situations fail to describe properly the real blocky geological structures. It also can result in spurious oscillations when m is discontinuous.

4.3.2 L_p-Norm Stabilizing Functionals

It can be demonstrated also that the stabilizing functionals can be constructed based on L_p norm as follows:

$$s_{L_p}(\mathbf{m}) = \|\mathbf{m}\|_{L_p}^p = \int_V |m(\mathbf{r})|^p \, dv, \quad 0 \leq p < \infty. \tag{4.22}$$

Using Minkowski inequality, one can prove that the subset M_c of the elements $\mathbf{m} \in M$ for which $s_{L_p(\mathbf{m})} \leq c$,

$$s_{L_p}(\mathbf{m}) = \mu(m, m_0) \leq c,$$

is compact.

Note that, stabilizing functionals can be also introduced as the L_p norm of the gradient of the model parameters:

$$s_{L_p}(\nabla\mathbf{m}) = \|\nabla\mathbf{m}\|_{L_p}^p = \int_V |\nabla m(\mathbf{r})|^p \, dv. \tag{4.23}$$

For example, Rudin et al. (1992) introduced a total variation (TV) method for reconstruction of noisy, blurred images using a total variation stabilizing functional, which is essentially the L_1 norm of the gradient:

$$s_{TV}(\mathbf{m}) = \|\nabla\mathbf{m}\|_{L_1} = \int_V |\nabla m(\mathbf{r})| \, dv. \tag{4.24}$$

We call $s_{TV}(\mathbf{m})$ a *total variation stabilizing functional*. The TV functional requires that the distribution of model parameters in some domain V be of bounded variation (for definition and background see Giusti 1984). However, this functional is not differentiable at zero. To avoid this difficulty, Acar and Vogel (1994) introduced a modified TV stabilizing functional:

$$s_{\beta TV}(\mathbf{m}) = \int_V \sqrt{|\nabla m(\mathbf{r})|^2 + e^2 dv}, \tag{4.25}$$

where e is a small number.

The advantage of this functional is that it does not require the function m to be continuous, only piecewise smooth (Vogel and Oman 1998). Since the TV norm does not penalize discontinuity in the model parameters, we can remove oscillations while preserving sharp physical property contrasts. At the same time, it imposes a limit on the total variation of m and on the combined arc length of the curves along which m is discontinuous. That is why this functional produces a much better result than maximum smoothness functionals when the blocky structures are imaged.

TV functionals $s_{TV}(\mathbf{m})$ and $s_{\beta TV}(\mathbf{m})$, however, tend to decrease the bounds of variation of the model parameters, as can be seen from (4.24) and (4.25), and in this sense they still try to "smooth" the real image. However, this "smoothness" is much weaker than in the case of traditional stabilizers (4.20) and (4.19).

4.3.3 Focusing Stabilizing Functionals

In many practical applications the solution of inverse problem is described by the blocky functions, having sharp boundaries separating the domains with different smooth distributions of the model parameters. In order to recover these sharp boundaries, Portniaguine and Zhdanov (1999), and Zhdanov (2002) introduced a family of focusing stabilizing functionals which minimize the area where significant variations of model parameters and/or discontinuity occur. These stabilizers are called *minimum support (MS) or minimum gradient support (MGS) functionals*.

For the sake of simplicity we will discuss first *a minimum support (MS) functional*, which provides a model with a minimum area of the distribution of anomalous parameters. The minimum support functional was considered first by Last and Kubik (1983), where the authors suggested seeking a source distribution with the minimum volume (compactness) to explain the anomaly.

We introduce a support of m (denoted spt m) as the combined closed subdomains of V where $m \neq 0$. We call spt m *a model parameter support*. Consider the following functional of the model parameters:

$$s_e(\mathbf{m}) = \int_V \frac{m^2(\mathbf{r})}{m^2(\mathbf{r}) + e^2} dv = \int_{\text{spt } m} \left[1 - \frac{e^2}{m^2(\mathbf{r}) + e^2} \right] dv$$

$$= \text{spt } m - e^2 \int_{\text{spt } m} \frac{1}{m^2(\mathbf{r}) + e^2} dv, \tag{4.26}$$

where $e > 0$ is a small number called *a focusing parameter*.

From the last expression we can see that

$$s_e(\mathbf{m}) \rightarrow \text{spt } m, \text{ if } e \rightarrow 0. \tag{4.27}$$

Thus, $s_e(\mathbf{m})$ can be treated as a functional, proportional (for a small e) to the model parameter support. We can use this functional to introduce a *minimum support stabilizing functional* $s_{MS}(\mathbf{m})$ as follows:

$$s_{MS}(\mathbf{m}) = s_e\left(\mathbf{m} - \mathbf{m}_{apr}\right) = \int_V \frac{\left(m - m_{apr}\right)^2}{\left(m - m_{apr}\right)^2 + e^2} dv. \tag{4.28}$$

To justify this choice we should prove that $s_{MS}(\mathbf{m})$ can actually be considered as a stabilizer according to regularization theory. According to the definition given above, a nonnegative functional $s(\mathbf{m})$ in some Hilbert space M is called a stabilizing functional if, for any real $c > 0$ from the domain of the functional $s(\mathbf{m})$ values, the subset M_c of elements $\mathbf{m} \in M$, for which $s(\mathbf{m}) \leq c$, is compact.

Let us consider the subset M_c of the elements from M, satisfying the condition

$$s_{MS}(\mathbf{m}) \leq c, \tag{4.29}$$

where $s_{MS}(\mathbf{m})$ is a minimum support stabilizing functional determined by Eq. (4.28). It can be proved that s_{MS} is a monotonically increasing function of $\left\| m - m_{apr} \right\|^2$:

$$s_{MS}(\mathbf{m}_1) < s_{MS}(\mathbf{m}_2), \text{ if } \left\| \mathbf{m}_1 - \mathbf{m}_{apr} \right\|_{L_2} < \left\| \mathbf{m}_2 - \mathbf{m}_{apr} \right\|_{L_2}. \tag{4.30}$$

To prove this, let us consider the first variation of the minimum support functional:

$$\delta s_{MS}(\mathbf{m}) = \delta \int_V \frac{\left(m - m_{apr}\right)^2}{\left(m - m_{apr}\right)^2 + e^2} dv$$

$$= \int_V \frac{e^2}{\left[\left(m - m_{apr}\right)^2 + e^2\right]^2} \delta \left(m - m_{apr}\right)^2 dv = \int_V a^2 \delta \left(m - m_{apr}\right)^2 dv,$$

where

$$a^2 = \frac{e^2}{\left[\left(m - m_{apr}\right)^2 + e^2\right]^2}.$$

Using a mean value theorem, we obtain

$$\delta s_{MS}(\mathbf{m}) = \overline{a}^2 \int_V \delta \left(m - m_{apr}\right)^2 dv$$

$$= \overline{a}^2 \delta \int_V \left(m - m_{apr}\right)^2 dv = \overline{a}^2 \delta \left\| \mathbf{m} - \mathbf{m}_{apr} \right\|_{L_2}^2$$

$$= 2\overline{a}^2 \left\| \mathbf{m} - \mathbf{m}_{apr} \right\|_{L_2} \delta \left\| \mathbf{m} - \mathbf{m}_{apr} \right\|_{L_2}, \tag{4.31}$$

where $\overline{a}^2$ is an average value of a^2 in the volume V. Taking into account that $\overline{a}^2 > 0$ and $\left\| \mathbf{m} - \mathbf{m}_{apr} \right\|_{L_2} > 0$, we obtain (4.30) from (4.31).

Thus, from condition (4.29) and (4.30), we see that

$$\left\| \mathbf{m} - \mathbf{m}_{apr} \right\|_{L_2} \leq q, \ \mathbf{m} \in M_c, \tag{4.32}$$

where $q > 0$ is some constant, i.e., M_c forms a ball in the space M with a center at the point $\mathbf{m}_{apr}$. It is well known that the ball is compact in a Hilbert space. Therefore, the functional $s_{MS}(\mathbf{m})$ is a stabilizing functional.

This functional has an important property: it minimizes the total volume with nonzero departure of the model parameters from the given a priori model. Thus, a dispersed and smooth distribution of the parameters with all values different from the a priori model $\mathbf{m}_{apr}$ results in a big penalty function, while a well-focused distribution with a small departure from $\mathbf{m}_{apr}$ will have a small penalty function.

The concept of minimum support functional could be extended by introducing L_p-norm minimum support functionals ($\mathrm{MS}L_p$) as follows:

$$s_{MSL_p}(\mathbf{m}) = s_{eL_p}(\mathbf{m}) = \int_V \frac{|m|^p}{|m|^p + e^p} dv, \ 0 \leq p < \infty. \tag{4.33}$$

Repeating the derivation above for the MS functional, we can show that $\mathrm{MS}L_p$ functional is also proportional (for a small e) to the model parameter support:

$$s_{MSL_p}(\mathbf{m}) = \int_V \frac{|m|^p}{|m|^p + e^p} dv = \int_{\mathrm{spt}\ m} \left[1 - \frac{e^p}{|m|^p + e^p} \right] dv$$

$$= \mathrm{spt}\ m - e^p \int_{\mathrm{spt}\ m} \frac{1}{m^p(\mathbf{r}) + e^p} dv \rightarrow \mathrm{spt}\ m, \ \text{if}\ e \rightarrow 0. \tag{4.34}$$

By changing factor p one can control the degree of focusing the blocky structures.

Another approach to increase the resolution of blocky structures is based on using *a minimum gradient support functional*, which is defined as follows:

$$s_{MGS}(\mathbf{m}) = s_e[\nabla \mathbf{m}] = \int_V \frac{\nabla m \cdot \nabla m}{\nabla m \cdot \nabla m + e^2} dv. \tag{4.35}$$

We denote by spt ∇m the combined closed subdomains of V where $\nabla m \neq 0$. We call spt∇m *a gradient support*. Then, expression (4.35) can be modified:

$$s_{MGS}(\mathbf{m}) = \mathrm{spt}\, \nabla m - e^2 \int_{\mathrm{spt}\, \nabla m} \frac{1}{\nabla m \cdot \nabla m + e^2} dv. \tag{4.36}$$

From the last expression we can see that

$$s_{MGS}(\mathbf{m}) \rightarrow \mathrm{spt}\, \nabla m, \ \text{if}\ e \rightarrow 0. \tag{4.37}$$

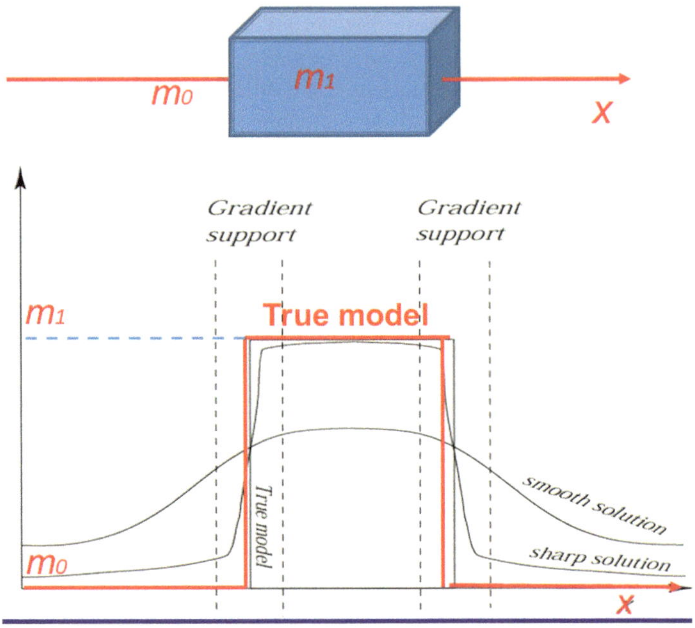

Fig. 4.2 Illustration of the action of the minimum gradient support stabilizing functional. The top section of the figure shows a rectangular body with the physical property value m_1 (e.g., density or conductivity) positioned within the host medium with physical property value m_0. This body is intersected by the horizontal axis x. The bottom section presents the plots of the same physical property along axis x. The bold red line shows the plot of the true physical property change along the axis x. The thin black lines present the same plots of the physical property recovered by inversions produced with the smooth of focused stabilizers. A smooth stabilizer produces a smooth solution to the inverse problem, while the minimum gradient support stabilizer generates a sharp solution close to the true plot of the physical property changes along the axis x

Thus, the functional $s_{MGS}(\mathbf{m})$ can really be treated as a functional proportional (for a small e) to the gradient support. This functional helps to generate a sharp and focused image of the inverse model. Figure 4.2 illustrates this property of the minimum gradient support functional.

The top section of Fig. 4.2 shows a rectangular body with the physical property value m_1 (e.g., density or conductivity) positioned within the host medium with physical property value m_0. This body is intersected by the horizontal axis x. The bottom section of Fig. 4.2 presents the plots of the same physical property along axis x. The plot of the true model looks like a boxcar function with values equal to m_0 outside the rectangular body and equal to m_1 inside the rectangular body. The bold red line shows this plot of the true physical property change along the axis x. The thin black lines present the same plots of the physical property recovered by inversions produced with the smooth of focused stabilizers. The smooth solution shows a smooth increase of the physical property values inside the rectangular body, but it significantly underestimates the actual value of m_1. On the other hand, the sharp solution produced by the inversion with the minimum gradient support stabilizing functional represents the physical properties correctly with the practically true value of m_1 inside the body. This is achieved because the minimum gradient support stabilizer tends to minimize the areas with rapid changes of physical values from m_0 outside the body to m_1 inside the body. We call these areas the "gradient support," and the focusing stabilizer delivers the model with the minimum gradient support, as shown in Fig. 4.2.

Repeating the considerations described above for $s_{MS}(\mathbf{m})$, one can demonstrate that the minimum gradient support functional satisfies the Tikhonov criterion for a stabilizer.

Note that, in principle, we can construct the stabilizers using different monotonic functions of the model parameters or their gradients, for example, exponential functions:

$$s_{\exp m}(\mathbf{m}) = \int_V \exp\left(|m|^2\right) dv, \tag{4.38}$$

or

$$s_{\exp \nabla m}(\mathbf{m}) = \int_V \exp\left(\nabla m \cdot \nabla m\right) dv. \tag{4.39}$$

We just have to check every time that the corresponding functional satisfies all necessary conditions for a stabilizer.

4.3.4 Representation of a Stabilizing Functional in the Form of a Pseudo-quadratic Functional

Note that all stabilizing functionals introduced above can be expressed as pseudo-quadratic functionals of the model parameters:

$$s(\mathbf{m}) = \left(W_e\left(\mathbf{m} - \mathbf{m}_{apr}\right), W_e\left(\mathbf{m} - \mathbf{m}_{apr}\right)\right)_{L_2}$$
$$= \int_V \left|w_e(\mathbf{r})\left(m(\mathbf{r}) - m_{apr}(\mathbf{r})\right)\right|^2 dv, \tag{4.40}$$

where W_e is a linear operator of multiplication of the model parameter function $m(\mathbf{r})$ by function $w_e(\mathbf{r})$, which may depend on m. If operator W_e is independent of $m(\mathbf{r})$, we obtain a quadratic functional, like the minimum norm (4.19) or the maximum smoothness (4.20) stabilizing functionals. In general cases, the function w_e may even be a nonlinear function of m, like the minimum support (4.28) or minimum gradient support (4.35) functionals. In these cases, the functional $s(\mathbf{m})$, determined by formula (4.40), is not quadratic. That is why we call it a "pseudo-quadratic" functional. However, presenting a stabilizing functional in a pseudo-quadratic form simplifies the solution of the regularization problem, and makes it possible to develop a unified approach to regularization with different stabilizers (Zhdanov 2002, 2015).

For example, the maximum smoothness stabilizer is expressed by formula (4.40) if $m_{apr} = 0$ and

$$w_e(\mathbf{r}) = w_e^{\max sm}(\mathbf{r}) = \frac{\nabla m(\mathbf{r})}{\left[m^2(\mathbf{r}) + \beta^2\right]^{1/2}}, \tag{4.41}$$

where, ultimately, we shall let $\beta \to 0$.

In the case of the TV stabilizing functional, $s_{\beta TV}(m)$, we assume $m_{apr} = 0$, and the function $w_e(\mathbf{r})$ in (4.40) is

$$w_e(\mathbf{r}) = w_e^{\beta TV}(\mathbf{r}) = \frac{\left(|\nabla m(\mathbf{r})|^2 + e^2\right)^{1/4}}{\left(m^2(\mathbf{r}) + \beta^2\right)^{1/2}}. \tag{4.42}$$

In the case of the minimum support functional, $s_{MS}(m)$, we have

$$w_e(\mathbf{r}) = w_e^{MS}(\mathbf{r}) = \frac{1}{\left[\left(m(\mathbf{r}) - m_{apr}(\mathbf{r})\right)^2 + e^2\right]^{1/2}}. \tag{4.43}$$

In the case of L_p-norm minimum support functional, $s_{MSL_p}(m)$, we can write

$$w_e(\mathbf{r}) = w_e^{MSLp}(\mathbf{r})$$

$$= \frac{\left(m^{(i)} - m_{apr}^{(i)}\right)^{p/2}}{\left[\left(m(\mathbf{r}) - m_{apr}(\mathbf{r})\right)^p + e^p\right]^{1/2}\left[\left(m^{(i)} - m_{apr}^{(i)}\right)^2 + \beta^2\right]^{1/2}}. \tag{4.44}$$

For the minimum gradient support functional $s_{MGS}(m)$, we assume $m_{apr} = 0$, and find that

$$w_e(\mathbf{r}) = w_e^{MGS}(\mathbf{r}) = \frac{\nabla m(\mathbf{r})}{\left[\nabla m(\mathbf{r}) \cdot \nabla m(\mathbf{r}) + e^2\right]^{1/2}\left[m^2(\mathbf{r}) + \beta^2\right]^{1/2}}. \tag{4.45}$$

And finally, for exponential stabilizers (4.38) and (4.39), we have

$$w_e(\mathbf{r}) = w_e^{\exp m}(\mathbf{r}) = \frac{\exp\left(\frac{1}{2}|m(\mathbf{r})|^2\right)}{\left[m^2(\mathbf{r}) + \beta^2\right]^{1/2}}, \tag{4.46}$$

or

$$w_e(\mathbf{r}) = w_e^{\exp \nabla m}(\mathbf{r}) = \frac{\exp\left[\frac{1}{2}\nabla m(\mathbf{r}) \cdot \nabla m(\mathbf{r})\right]}{\left[m^2(\mathbf{r}) + \beta^2\right]^{1/2}}. \tag{4.47}$$

Similar expressions for $w_e(\mathbf{r})$ can be easily derived for other types of stabilizing functionals.

Thus, we can see that minimization of the stabilizing functionals can impose different conditions on the class of model parameters. For example, one case (e.g., maximum smoothness stabilizer) requires a smooth distribution of the model parameters. Another case (e.g., a minimum gradient support stabilizer) imposes sharpening conditions on the model parameter distribution. As a result, we can select different classes of inverse problem solutions by choosing one or another type of stabilizer. In other words, stabilizing functionals help to use a priori information about the desired properties of inverse problem solutions. This is the central role of stabilizing functionals in regularization theory.

4.3.5 Regularizing Operators Revisited

Let us analyze now more carefully how one can use a stabilizer to select an appropriate class of the models. Assume that the data $\mathbf{d}_\delta$ are observed with some noise $\mathbf{d}_\delta = \mathbf{d}_t + \delta\mathbf{d}$, where $\mathbf{d}_t$ is the true solution of the problem. In other words, we assume that the misfit (distance) between the observed data and true data is less than the given level of the errors, δ, in the observed data,

$$\mu_D(\mathbf{d}_\delta, \mathbf{d}_t) \le \delta, \tag{4.48}$$

where $\delta = \|\delta\mathbf{d}\|$.

In this situation, it is natural to search for an approximate solution in the set Q_δ of the models $\mathbf{m}$, such that

$$\mu_D(A(\mathbf{m}), \mathbf{d}_\delta) \le \delta. \tag{4.49}$$

Thus, $Q_\delta \subset M$ is a set of possible solutions.

The main application of a stabilizer is to select from the set of possible solutions Q_δ the solutions that continuously depend on the data and which possesses a specific property depending on the choice of a stabilizer. Such solutions can be selected by the condition of the minimum of the stabilizing functional:

$$s(\mathbf{m};\ \mathbf{m} \in Q_\delta) = \min. \tag{4.50}$$

We have introduced a stabilizing functional under the condition that it selects a compact subset M_C from a metric space of the model parameters. Therefore, we can say that a stabilizer selects a solution from a set of possible solutions, Q_δ, which at the same time belongs to the correctness set M_C. Figure 4.3 helps to explain this role of the stabilizing functional.

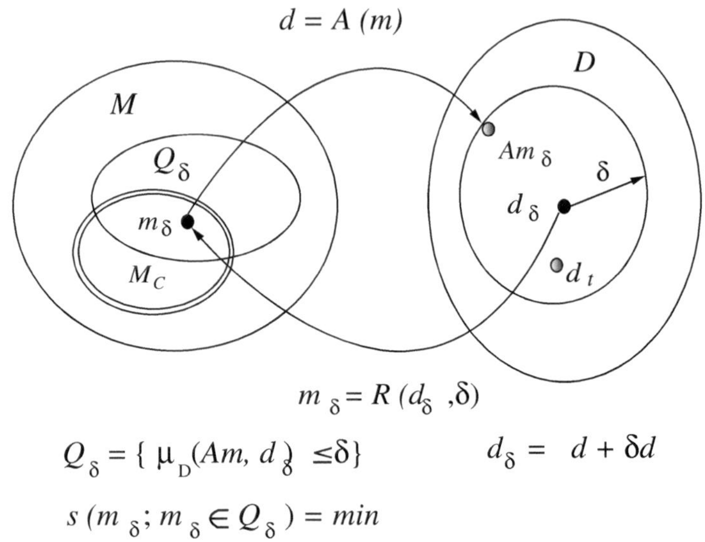

Fig. 4.3 The stabilizing functional selects from a set of the possible solutions, Q_δ, a solution, $\mathbf{m}_\delta$, which at the same time belongs to the correctness set M_C

The existence of the model, minimizing (4.50), was demonstrated by Tikhonov and Arsenin (1977). We will denote this model as $\mathbf{m}_\delta$:

$$s(\mathbf{m}_\delta;\ \mathbf{m}_\delta \in Q_\delta) = \min . \tag{4.51}$$

One can consider a model $\mathbf{m}_\delta$ as the result of an application of the operator $R(\mathbf{d}_\delta, \delta)$ to the observed data $\mathbf{d}_\delta$, depending on the parameter δ:

$$\mathbf{m}_\delta = R(\mathbf{d}_\delta, \delta). \tag{4.52}$$

The following theorem presents an important result of regularization theory:

Theorem 4.9 *The operator $R(\mathbf{d}_\delta, \delta)$, introduced by formula (4.52), is the regularizing operator for the equation (4.9), and $\mathbf{m}_\delta$ can be used as an approximate solution of the inverse problem (note that in this case $\alpha = \delta$, while in general cases $\alpha = \alpha(\delta)$).*

The proof of this theorem can be found in Zhdanov (2015).

Thus, in the framework of the approach we have developed, the problem of the solution of Eq. (4.9) with approximate left-hand part $\mathbf{d}_\delta$ can be reduced to the problem of minimization of the stabilizing functional on the set Q_δ:

$$s(\mathbf{m};\ \mathbf{m} \in Q_\delta) = \min,$$

where

$$Q_\delta = \{\mathbf{m};\ \mu_D(A(\mathbf{m}), \mathbf{d}_\delta) \le \delta\}. \tag{4.53}$$

4.4 Tikhonov Parametric Functional

It has been proved by Tikhonov and Arsenin (1977) that, for a wide class of stabilizing functionals, their minimum is reached on the model $\mathbf{m}_\delta$ such that $\mu_D(A(\mathbf{m}_\delta), \mathbf{d}_\delta) = \delta$. Therefore, we can solve the problem of minimization (4.50) under the condition that

$$\mu_D(A(\mathbf{m}_\delta), \mathbf{d}_\delta) = \delta. \tag{4.54}$$

In other words, one should consider the problem of minimization of the stabilizing functional (4.50), when the model $\mathbf{m}$ is subject to the constraint (4.54). A common way to solve this problem is to introduce an unconstrained *parametric functional* $P^\alpha(\mathbf{m}, \mathbf{d}_\delta),\ \mathbf{m} \in M$, given by the following expression:

$$P^\alpha(\mathbf{m}, \mathbf{d}_\delta) = \mu_D^2(A(\mathbf{m}), \mathbf{d}_\delta) + \alpha s(\mathbf{m}), \tag{4.55}$$

and to solve the problem of minimization of this functional:

$$P^\alpha(\mathbf{m}, \mathbf{d}_\delta) = \min. \tag{4.56}$$

Functional $\mu_D^2(A(\mathbf{m}), d_\delta)$ is often called *a misfit functional*, and is denoted as follows:

$$\mu_D^2(A(\mathbf{m}), \mathbf{d}_\delta) = \varphi(\mathbf{m}). \tag{4.57}$$

Thus, the parametric functional $P^\alpha(\mathbf{m}, \mathbf{d}_\delta)$ is a linear combination of the misfit and stabilizing functionals, and the unknown real parameter α is similar to the Lagrangian multiplier. The regularization parameter α can be determined under the following condition:

$$\mu_D(A(\mathbf{m}_\alpha), \mathbf{d}_\delta) = \delta, \tag{4.58}$$

where $\mathbf{m}_\alpha$ is the element on which $P^\alpha(\mathbf{m}, \mathbf{d}_\delta)$ reaches its minimum. This equality is called a misfit condition.

The functional $P^\alpha(\mathbf{m}, \mathbf{d}_\delta)$ is called *the Tikhonov parametric functional*. The fundamental theorem of regularization theory states that the solution of minimization problem (4.58) always exists and is unique for a wide class of forward modeling operators. Therefore, we can introduce an operator transforming the observed data into the model minimizing the parametric functional, $\mathbf{m}_\alpha$:

$$\mathbf{m}_\alpha = R(\mathbf{d}_\delta, \alpha). \tag{4.59}$$

The fundamental result of the regularization theory is that *this operator*, $R(\mathbf{d}_\delta, \alpha)$, *is a regularizing operator for the problem* (4.9) (Tikhonov and Arsenin 1977).

Therefore, as an approximate solution to the inverse problem (4.9), we take the solution of another problem (4.56) (minimization of the Tikhonov parametric functional $P^\alpha(\mathbf{m}, \mathbf{d}_\delta)$), close to the initial problem for the small values of the data errors δ.

4.5 Definition of the Regularization Parameter

4.5.1 Optimal Regularization Parameter Selection

The regularization parameter α describes the trade-off between the best fitting and most reasonable stabilization. In a case where α is selected to be too small, the minimization of the parametric functional $P^\alpha(\mathbf{m})$ is equivalent to the minimization of the misfit functional; therefore, we have no regularization, which can result in an unstable incorrect solution. When α is too large, the minimization of the parametric functional $P^\alpha(\mathbf{m})$ is equivalent to the minimization of the stabilizing functional $s(\mathbf{m})$, which will force the solution to be closer to the a priori model. Ultimately, we would expect the final model to be exactly like the a priori model, while the observed data are totally ignored in the inversion. Thus, the critical question of the regularization theory is the selection of the optimal regularization parameter α. The basic principles used for determining the regularization parameter α are discussed in Tikhonov and Arsenin (1977). The solution of this problem can be based on the following consideration.

Let us assume that data $\mathbf{d}_\delta$ are observed with some noise, $\mathbf{d}_\delta = \mathbf{d}_t + \delta\mathbf{d}$, where $\mathbf{d}_t$ is the true solution of the problem and the level of the errors in the observed data is equal to δ:

$$\mu_D(\mathbf{d}_\delta, \mathbf{d}_t) \leq \delta. \tag{4.60}$$

Then the regularization parameter can be determined by the misfit condition (4.58).

To justify this approach we will examine more carefully the properties of all three functionals involved in the regularization method: the Tikhonov parametric functional, the stabilizing and misfit functionals.

Let us introduce the following notations:

$$\begin{gathered} p(\alpha) = P^\alpha(\mathbf{m}_\alpha, \mathbf{d}_\delta), \text{ parametric functional,} \\ s(\alpha) = s(\mathbf{m}_\alpha), \text{ stabilizing functional,} \\ i(\alpha) = \mu_D^2(A(\mathbf{m}_\alpha), \mathbf{d}_\delta), \text{ misfit functional.} \end{gathered} \tag{4.61}$$

We examine some properties of the functions $p(\alpha), i(\alpha), s(\alpha)$.

Property 4.1 Functions $p(\alpha), i(\alpha), s(\alpha)$ are monotonic functions: $p(\alpha)$ and $i(\alpha)$ are not decreasing and $s(\alpha)$ is not increasing.

Proof Let $\alpha_1 < \alpha_2$ and

$$p_k = p(\alpha_k) = P^{\alpha_k}(\mathbf{m}_{\alpha_k}, \mathbf{d}_\delta),$$

$$i_k = i(\alpha_k) = \mu_D^2(A(\mathbf{m}_{\alpha_k}), \mathbf{d}_\delta),$$

$$s_k = s(\alpha_k) = s(\mathbf{m}_{\alpha_k}).$$

The following inequality holds:

$$p_2 = i_2 + \alpha_2 s_2 \geq i_2 + \alpha_1 s_2, \tag{4.62}$$

because $\alpha_1 < \alpha_2$.

On the other hand

$$P^{\alpha_1}(\mathbf{m}_{\alpha_2}, \mathbf{d}_\delta) = i_2 + \alpha_1 s_2 \geq i_1 + \alpha_1 s_1 = p_1 = P^{\alpha_1}(\mathbf{m}_{\alpha_1}, \mathbf{d}_\delta), \tag{4.63}$$

because m_{α_1} realizes the minimum p_1 of the functional $P^{\alpha_1}(\mathbf{m}, \mathbf{d}_\delta)$.

Thus from (4.62) and (4.63) we have

$$p_2 \geq p_1 \tag{4.64}$$

for

$$\alpha_2 > \alpha_1,$$

which means that $p(\alpha)$ is a monotonic function of α.

Furthermore,

$$P^{\alpha_2}(\mathbf{m}_{\alpha_1}, \mathbf{d}_\delta) = i_1 + \alpha_2 s_1 \geq i_2 + \alpha_2 s_2 = P^{\alpha_2}(\mathbf{m}_{\alpha_2}, \mathbf{d}_\delta), \tag{4.65}$$

because $\mathbf{m}_{\alpha_2}$ realizes the minimum p_2 of the functional $P^{\alpha_2}(\mathbf{m}, \mathbf{d}_\delta)$.

Subtracting the left-hand side of inequality (4.65) from the right-hand side of inequality (4.63) and the right-hand side of inequality (4.65) from the left-hand side of inequality (4.63), we obtain

$$(\alpha_1 - \alpha_2)s_2 \geq (\alpha_1 - \alpha_2)s_1. \tag{4.66}$$

Since $\alpha_1 < \alpha_2$,

$$s_1 \geq s_2. \tag{4.67}$$

From inequalities (4.63) and (4.67) it follows that

$$i_2 - i_1 \geq \alpha_1(s_1 - s_2)$$

and hence

$$i_2 \geq i_1.$$

Property 4.2 It can be proved that the functions $p(\alpha), i(\alpha), s(\alpha)$ are continuous functions (if the element m_α is unique).

Note, also, that

$$p(\alpha) \to 0 \text{ for } \alpha \to 0,$$

and

$$p(0) = 0. \tag{4.68}$$

From the fact that

$$i(\alpha) + \alpha s(\alpha) = p(\alpha) \to 0, \text{ for } \alpha \to 0,$$

it follows that

$$i(0) = 0. \tag{4.69}$$

Thus we have proved the following theorem.

Theorem 4.10 *If $i(\alpha)$ is a one-to-one function, then, for any positive number $\delta < \delta_0 = \mu_D(A(\mathbf{m}_0), \mathbf{d}_\delta)$ (where $\mathbf{m}_0$ is some a priori model), there exists $\alpha(\delta)$ such that $\mu_D(A(\mathbf{m}_{\alpha(\delta)}), \mathbf{d}_\delta) = \delta$.*

Note that $i(\alpha)$ is a one-to-one function when element $\mathbf{m}_\alpha$ is unique. It happens, for example, when A is a linear operator, D is a Hilbert space, and $s(\mathbf{m})$ is a quadratic functional.

Figure 4.4 helps in understanding of the principle of optimal regularization parameter selection. One can see that because of the monotonic character of the function $i(\alpha)$, there is only one point, α_0, where $i(\alpha_0) = \mu_D^2(A(\mathbf{m}_{\alpha_0}), \mathbf{d}_\delta) = \delta^2$.

Let us consider one simple numerical method for determining the parameter α. Consider, for example, a progression of numbers:

$$\alpha_k = \alpha_1 q^{k-1}, \; k = 1, 2, ..., n; \;\; 0 < q < 1. \tag{4.70}$$

For any number α_k we can find the element $\mathbf{m}_{\alpha_k}$ minimizing $P^{\alpha_k}(\mathbf{m}, \mathbf{d}_\delta)$ and calculate the misfit $\mu_D(A(\mathbf{m}_{\alpha_k}), \mathbf{d}_\delta)$. The optimal value of the parameter α is the number $\alpha_0 = \alpha_{k0}$, for which, with the necessary accuracy, we have the equality

$$\mu_D(A(\mathbf{m}_{\alpha_{k0}}), \mathbf{d}_\delta) = \delta. \tag{4.71}$$

As we noted above, equality (4.71) is called the *misfit condition*.

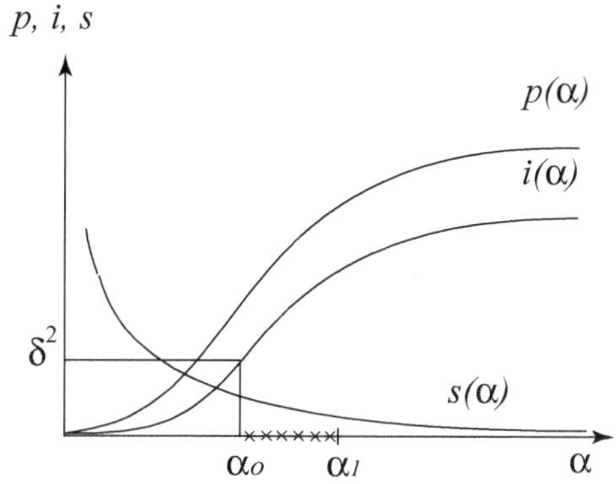

Fig. 4.4 Illustration of the principle of optimal regularization parameter selection

4.5.2 L-Curve Method of Regularization Parameter Selection

L-curve analysis (Hansen 1998) represents a simple graphical tool for qualitative selection of the quasi-optimal regularization parameter.

It is based on plotting for all possible α the curve of the misfit functional, $i(\alpha)$, versus the stabilizing functional, $s(\alpha)$ (where we use notations (4.61)). The L-curve illustrates the trade-off between the best fitting (minimizing a misfit) and most reasonable stabilization (minimizing a stabilizer). In a case where α is selected to be too small, the minimization of the parametric functional $P^{\alpha}(\mathbf{m})$ is equivalent to the minimization of the misfit functional; therefore $i(\alpha)$ decreases, while $s(\alpha)$ increases. When α is too large, the minimization of the parametric functional $P^{\alpha}(\mathbf{m})$ is equivalent to the minimization of the stabilizing functional; therefore $s(\alpha)$ decreases, while $i(\alpha)$ increases. As a result, it turns out that the L-curve, when it is plotted in log-log scale, very often has the characteristic L-shape appearance (Fig. 4.5) that justifies its name (Hansen 1998).

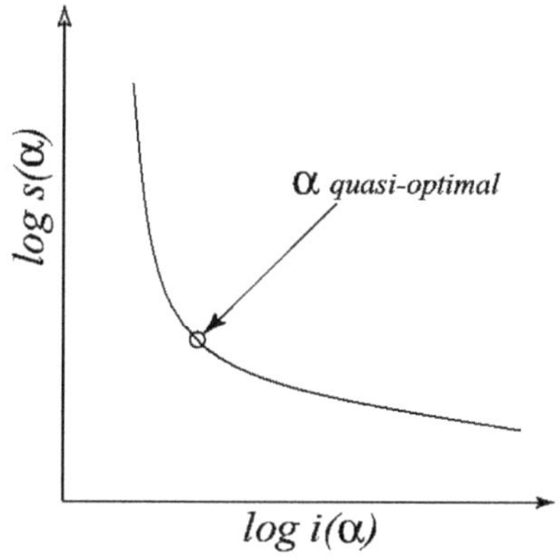

Fig. 4.5 L-curve represents a simple curve for all possible α of the misfit functional, $i(\alpha)$, versus stabilizing functional, $s(\alpha)$, plotted in log-log scale. The approximate corner, separating the vertical and the horizontal branches of this curve, corresponds to the quasi-optimal value of the regularization parameter α

The approximate corner, separating the vertical and the horizontal branches of this curve, corresponds to the quasi-optimal value of the regularization parameter α. However, in practice, the plot of the misfit functional versus the stabilizing functional does not have a distinct corner (as shown in Fig. 4.5), which makes it hard to determine the optimal regularization parameter using the L-curve method. Another limitation of this approach is that it requires solving a large number of minimization problems (4.56) with different values of regularization parameter α to produce a representative L-curve.

References and Recommended Reading to This Chapter

Acar R, Vogel CR (1994) Analysis of total variation penalty methods. Inverse Probl 10:1217–1229

Constable SC, Parker RC, Constable GG (1987) Occam's inversion: a practical algorithm for generating smooth models from EM sounding data. Geophysics 52:289–300

Dmitriev VI, Editor in chief (1990) Computational mathematics and techniques in exploration geophysics (in Russian). Nedra, Moscow, 498 pp

Giusti E (1984) Minimal surfaces and functions of bounded variations. Birkhauser-Verlag, 240 pp

Hansen C (1998) Rank-deficient and discrete ill-posed problems. Numerical aspects of linear inversion. Department of mathematical modeling, Technical University of Denmark, Lyngby, 247 pp

Last BJ, Kubik K (1983) Compact gravity inversion. Geophysics 48:713–721

Lavrent'ev MM, Romanov VG, Shishatskii SP (1986) Ill-posed problems of mathematical physics and analysis. Translations of mathematical monographs, vol 64. American Mathematical Society, Providence, Rhode Island, 290 pp

Portniaguine O, Zhdanov MS (1999) Focusing geophysical inversion images. Geophysics 64(3):874–887

Ramos FM, Campos Velho HF, Carvalho JC, Ferreira NJ (1999) Novel approaches to entropic regularization. Inverse Probl 15:1139–1148

Rudin LI, Osher S, Fatemi E (1992) Nonlinear total variation based noise removal algorithms. Physica D 60:259–268

Smith JT, Booker JR (1991) Rapid inversion of two- and three-dimensional magnetotelluric data. J Geophys Res 96:3905–3922

Strakhov VN (1968) Numerical solution of incorrect problems representable by integral equations of convolution type (in Russian). DAN SSSR 178(2):299

Strakhov VN (1969a) Theory of approximate solution of the linear ill-posed problems in a Hilbert space and its application in applied geophysics, part I (in Russian). Izvestia AN SSSR, Fizika Zemli, No 8, pp 30–53

Strakhov VN (1969b) Theory of approximate solution of the linear ill-posed problems in a Hilbert space and its application in applied geophysics, part II (in Russian). Izvestia AN SSSR, Fizika Zemli, No 9, pp 64–96

Tikhonov AN (1999) Mathematical geophysics (in Russian). Moscow State University, 476 pp

Tikhonov AN, Arsenin VY (1977) Solution of ill-posed problems. V. H. Winston and Sons, 258 pp

Vogel CR, Oman ME (1998) Fast total variation based reconstruction of noisy, blurred images. IEEE Trans Image Process 7:813–824

Wernecke SJ, D'Addario LR (1977) Maximum entropy image reconstruction. IEEE Trans Comput 26:351–364

Zhdanov MS, Fang S (1996) 3-D quasi-linear electromagnetic inversion. Radio Sci 3(4):741–754

Zhdanov MS (1993) Tutorial: regularization in inversion theory. CWP-136, Colorado School of Mines, 47 pp

Zhdanov MS (2002) Geophysical inverse theory and regularization problems. Elsevier, 609 pp

Zhdanov MS (2015) Inverse theory and applications in geophysics. Elsevier, 704 pp

Chapter 5
Linear Inverse Problems

Abstract Linear inverse problems represent the most important and, at the same time, relatively simple type of inverse problems. Even for problems with nonlinear relationships, it is often possible to find an accurate linearized approximation, e.g., Born approximation for electromagnetic or acoustic fields. This chapter considers the general methods of linear discrete inverse problem solutions for both overdetermined and underdetermined systems of linear equations. The classical approach to solving the linear inverse problem is based on the least-squares method. We also discuss the weighted least-squares method, which takes into account the data errors. Tikhonov's regularization method for solving the ill-posed linear inverse problems is considered as well as the classical Levenberg–Marquard method of damped least-squares.

Keywords Linear inverse problem · Least-squares method · Overdetermined system · Underdetermined system · Weighting matrices

Linear problems form the simplest but, at the same time, the most widely used class of inverse problems. The linear relationships between the model parameters and the corresponding data are typical for many physical and imaging problems. For example, the gravity field $\mathbf{g}(\mathbf{r}')$ is related to the density distribution, $\rho(\mathbf{r})$, within a domain D by the linear Newton law:

$$\mathbf{g}(\mathbf{r}') = \gamma \iiint_D \rho(\mathbf{r}) \frac{\mathbf{r} - \mathbf{r}'}{\mid \mathbf{r} - \mathbf{r}' \mid^3} dv, \tag{5.1}$$

where γ is the universal gravitational constant.

The magnetic field $\mathbf{H}(\mathbf{r}')$ is linearly proportional to the magnetization distribution within a domain D according to the Gauss law:

$$\mathbf{H}(\mathbf{r}') = \nabla' \iiint_D \mathbf{I}(\mathbf{r}) \cdot \nabla' \frac{1}{|\mathbf{r} - \mathbf{r}'|} dv, \tag{5.2}$$

where $\mathbf{I}(\mathbf{r})$ is the magnetization.

M. S. Zhdanov, *Advanced Methods of Joint Inversion and Fusion of Multiphysics Data*, Advances in Geological Science,
https://doi.org/10.1007/978-981-99-6722-3_5

The observed blurred images are related to the original images by the linear equation:

$$\mathbf{d} = \mathbf{Bm}, \tag{5.3}$$

where **d** is the degraded (blurred) image, **m** is the original (ideal) image, and **B** is the blurring linear operator of the imaging system.

Even for problems with nonlinear relationships, one can always find an accurate linearized approximation, e.g., Born approximation for electromagnetic or acoustic fields (Zhdanov 2002, 2018).

In addition, in most applications, we have to work with discrete inverse problems where both the model parameters and the data sets are represented by a finite number of data points. This chapter considers the general methods of linear discrete inverse problem solutions following Zhdanov (1993, 2002, 2015) and Menke (2018).

5.1 Least-Squares Inversion of the Linear Discrete Inverse Problem

5.1.1 Linear Discrete Inverse Problem

Let us consider a general inverse geophysical problem described by the following operator equation:

$$\mathbf{d} = A(\mathbf{m}), \tag{5.4}$$

where **m** represents the model parameters, and **d** are observed data. We assume that N measurements are performed in some physical experiments. Then we can treat these values as the components of the N-dimensional vector **d**. Similarly, some model parameters can be represented as the components of a vector **m** of order L:

$$\mathbf{d} = [d_1, d_2, d_3, ..., d_N]^T,$$

$$\mathbf{m} = [m_1, m_2, m_3, ..., m_L]^T,$$

where the superscript T denotes the transpose of the two vectors.

In a case where A is a linear operator, Eq. (5.4) can be rewritten in matrix notation as follows:

$$\mathbf{d} = \mathbf{Am}, \tag{5.5}$$

where **A** is the $N \times L$ matrix of the linear operator A.

Expression (5.5) describes a system of N linear equations for L unknown parameters, $m_1, m_2, m_3, ..., m_L$:

$$d_i = \sum_{j=1}^{L} A_{ij} m_j, \quad i = 1, 2, 3, ..., N. \tag{5.6}$$

Therefore, solving inverse problem (5.4) means solving the system of linear equations (5.6) with respect to parameters $m_1, m_2, m_3, ..., m_L$.

The system (5.6) is called *underdetermined* if $N < L$. The system (5.6) is called *overdetermined* if $N > L$. In applications, we often work with an overdetermined system wherein the number of observations exceeds the number of model parameters. At the same time, in many situations, it may be necessary to work with an underdetermined system. We will examine both types of linear equation systems below.

Thus, we can see that in the case of a linear discrete inverse problem, operator equation (5.4) is reduced to matrix equation (5.5). To solve this equation, we have to use some formulae and rules from matrix algebra (Golub and Van Loan 2013; Zhdanov 2015).

5.1.2 Least-Squares Method

The least-squares method is the most popular mathematical technique of analysis of experimental data. In this section, I will discuss the basic principles of the least-squares method as applied to the solution of linear inverse problems. The interested reader can find more details on this subject in many publications dedicated to least-squares applications in science and engineering (e.g., Wolberg 2006; Hansen et al. 2013).

Let us consider again a system of linear equations determining the relationship between the observed data $\{d_1, d_2, d_3, ..., d_N\}$ and parameters of the model $\{m_1, m_2, m_3, ..., m_L\}$:

$$d_i = \sum_{j=1}^{L} A_{ij} m_j, \quad i = 1, 2, 3, ..., N, \tag{5.7}$$

where $N > L$. In other words, we assume now that Eq. (5.7) describes an overdetermined system. We know the column vector $\mathbf{d} = [d_1, d_2, d_3, ..., d_N]^T$ of the observed data and matrix $\mathbf{A}$ of the linear operator of the forward problem. Our goal is to determine the column vector $\mathbf{m} = [m_1, m_2, m_3, ..., m_L]^T$ of the model parameters. Note that in this chapter, we assume, for simplicity, that all model parameters and data are represented by real numbers.

Let us denote by $\mathbf{d}^p = [d_1^p, d_2^p, d_3^p, ..., d_N^p]^T$ a vector of the predicted data:

$$d_i^p = \sum_{j=1}^{L} A_{ij} m_j, \quad i = 1, 2, 3, ..., N.$$

We can now write

$$\mathbf{r} = \mathbf{d}^p - \mathbf{d},$$

where $\mathbf{r} = [r_1, r_2, r_3, ..., r_N]^T$ is the column vector of the residuals (errors) between the observed, $\mathbf{d}$, and predicted, $\mathbf{d}^p$, data.

However, we have more data $\{d_1, d_2, d_3, ..., d_N\}$ than model parameters $\{m_1, m_2, m_3, ..., m_L\}$, so we cannot fit all the data. The best that we can do is to minimize the misfit between the observed and predicted data, which in the Euclidean metric of data space can be calculated as a sum of the squares of the errors:

$$f(m_1, m_2, m_3, ..., m_L) = \|\mathbf{r}\|^2 = \sum_{i=1}^{N} r_i^2 = \min . \tag{5.8}$$

Functional f is called a *misfit functional*. It can be written in the following form:

$$f(\mathbf{m}) = \|\mathbf{d}^p - \mathbf{d}\|^2 = \|\mathbf{Am} - \mathbf{d}\|^2$$

$$= (\mathbf{Am} - \mathbf{d}, \mathbf{Am} - \mathbf{d}) = \min . \tag{5.9}$$

Note that, using matrix notations, one can rewrite the last equation as follows:

$$f(\mathbf{m}) = (\mathbf{Am} - \mathbf{d})^T(\mathbf{Am} - \mathbf{d}) = \min . \tag{5.10}$$

The column vector $\mathbf{m}_0 = [m_{01}, m_{02}, m_{03}, ..., m_{0L}]^T$, in which the misfit functional reaches its minimum, is called *a pseudo-solution* of system (5.7). If $f(\mathbf{m}_0) = 0$, then $\mathbf{m}_0$ is the conventional solution of system (5.7).

5.1.2.1 Minimization of the Misfit Functional

It is well known that the best way to solve an optimization problem for conventional functions is based on differentiating the functions and equating the resulting derivatives to zero. A similar approach can be applied in principle to functionals. However, we have to use an analog of calculus for functionals and operators called variational calculus. This generalization has been discussed in Chap. 3.

The problem of minimization of the misfit functional (5.10) can be solved using variational calculus. Let us calculate the first variation of $f(\mathbf{m})$:

$$\delta f(\mathbf{m}) = 2(\mathbf{A}\delta\mathbf{m})^T(\mathbf{Am} - \mathbf{d}).$$

The necessary condition for the minimum of the functional $f(\mathbf{m})$ has the following form:

$$\delta f(\mathbf{m}) = 2(\mathbf{A}\delta\mathbf{m})^T(\mathbf{Am} - \mathbf{d}) = 2(\delta\mathbf{m})^T\mathbf{A}^T(\mathbf{Am} - \mathbf{d}) = 0,$$

for any $\delta\mathbf{m}$.

From the last formula, we have the following system of equations:

$$\mathbf{A}^T\mathbf{Am} = \mathbf{A}^T\mathbf{d}. \tag{5.11}$$

System (5.11) is called the *normal system* for (5.7). Matrix $\mathbf{A}^T\mathbf{A}$ is an $L \times L$ square matrix. Thus the solution of the normal system can be given in the following form:

$$\mathbf{m}_0 = (\mathbf{A}^T\mathbf{A})^{-1}\mathbf{A}^T\mathbf{d}, \tag{5.12}$$

where $\mathbf{m}_0$ is called *a pseudo-solution* of original system (5.7).

Note that the normal system can be obtained formally by multiplication of the original system (5.5) by the transposed matrix $\mathbf{A}^T$. However, in general cases, the pseudo-solution $\mathbf{m}_0$ is not equivalent to the solution of the original system because the new system described by Eq. (5.11) is not equivalent to the original system (5.5) if matrix $\mathbf{A}$ is not square. The main characteristic of the pseudo-solution is that it provides the minimum of the misfit functional.

To find a pseudo-solution numerically, we can apply the method of singular value decomposition and obtain the following result:

$$\mathbf{A}^T\mathbf{A} = (\mathbf{UQV}^T)^T\mathbf{UQV}^T$$

$$= \mathbf{VQU}^T\mathbf{UQV}^T = \mathbf{VQ}^2\mathbf{V}^T, \tag{5.13}$$

where $\mathbf{Q}$ is diagonal, and $\mathbf{U}$ and $\mathbf{V}$ are orthogonal matrices representing matrix $\mathbf{A}$:

$$\mathbf{A} = \mathbf{UQV}^T.$$

Thus, system (5.11) can be rewritten as follows:

$$\mathbf{VQ}^2\mathbf{V}^T\mathbf{m}_0 = \mathbf{VQU}^T\mathbf{d}. \tag{5.14}$$

Let us apply the inverse matrix $(\mathbf{VQ}^2\mathbf{V}^T)^{-1}$ to the right-hand and left-hand parts of Eq. (5.14):

$$\mathbf{m}_0 = (\mathbf{VQ}^2\mathbf{V}^T)^{-1}\mathbf{VQU}^T\mathbf{d} = \mathbf{VQ}^{-1}\mathbf{U}^T\mathbf{d} = \mathbf{V}\left[\mathbf{diag}(\frac{1}{Q_i})\right]\mathbf{U}^T\mathbf{d}. \tag{5.15}$$

Expression (5.15) gives directly the pseudo-solution of system (5.7).

Matrix

$$\mathbf{A}^+ = \mathbf{V}\left[\mathbf{diag}(\frac{1}{Q_i})\right]\mathbf{U}^T \tag{5.16}$$

is called the *pseudo-inverse or generalized inverse* matrix for $\mathbf{A}$. The pseudo-inverse matrix is equal to the inverse matrix of a square matrix $\mathbf{A}$:

$$\mathbf{A}^{+} = \mathbf{A}^{-1}.$$

Thus, minimization of the misfit functional opens a way to construct a generalized inverse matrix for any matrix, rectangular or square, with the only limitation being that the elements of the diagonal matrix $\mathbf{Q}$ are not equal to zero: $Q_i \neq 0,\ i = 1, 2, ...L$.

5.1.2.2 Underdetermined System of Linear Equations

Assume that the inverse problem,

$$\mathbf{d} = \mathbf{A}\mathbf{m}, \tag{5.17}$$

is purely underdetermined. This means that $N < L$, and there are no inconsistencies in these equations. It is, therefore, possible to find more than one set of model parameters that precisely fit the observed data. Let us try to select from all possible solutions the one which is the simplest in some sense, for example, it has the smallest Euclidean norm:

$$l(\mathbf{m}) = \|\mathbf{m}\|^2 = \min, \tag{5.18}$$

where

$$\|\mathbf{m}\|^2 = (\mathbf{m}, \mathbf{m}) = \mathbf{m}^T\mathbf{m}. \tag{5.19}$$

Therefore, we have the following problem: find the $\mathbf{m}^{est}$ that minimizes $l(\mathbf{m})$ subject to the following constraint:

$$\mathbf{d} - \mathbf{A}\mathbf{m} = 0. \tag{5.20}$$

We can solve this problem of the conditioned minimum by using the method of Lagrange multipliers:

$$\phi(\mathbf{m}) = \mathbf{m}^T\mathbf{m} + \boldsymbol{\lambda}(\mathbf{d} - \mathbf{A}\mathbf{m})^T = \min, \tag{5.21}$$

where $\boldsymbol{\lambda}$ is a diagonal matrix of the Lagrange multipliers.

Let us calculate the first variation of the functional $\phi(\mathbf{m})$:

$$\delta\phi(\mathbf{m}) = 2\delta\mathbf{m}^T\mathbf{m} - \delta\mathbf{m}^T\mathbf{A}^T\boldsymbol{\lambda} = \delta\mathbf{m}^T(2\mathbf{m} - \mathbf{A}^T\boldsymbol{\lambda}). \tag{5.22}$$

The necessary minimum condition for the functional gives

$$\delta\phi(\mathbf{m}) = \delta\mathbf{m}^T(2\mathbf{m} - \mathbf{A}^T\boldsymbol{\lambda}) = 0, \tag{5.23}$$

for any $\delta\mathbf{m}^T$.

Thus, we have the following expression for the solution $\mathbf{m}^{est}$ of our problem:

$$\mathbf{m}^{est} = \frac{1}{2}\mathbf{A}^T\boldsymbol{\lambda}. \tag{5.24}$$

On the other hand, this solution must satisfy Eq. (5.17):

$$\mathbf{d} = \frac{1}{2}\mathbf{A}\mathbf{A}^T\boldsymbol{\lambda}.$$

The matrix $\mathbf{A}\mathbf{A}^T$ is a square $N \times N$ matrix and, if it is not singular, can be inverted:

$$\boldsymbol{\lambda} = 2(\mathbf{A}\mathbf{A}^T)^{-1}\mathbf{d}. \tag{5.25}$$

By substituting the last equation into (5.24), we have

$$\mathbf{m}^{est} = \mathbf{A}^T(\mathbf{A}\mathbf{A}^T)^{-1}\mathbf{d}. \tag{5.26}$$

Formula (5.26) provides a minimum norm solution of the underdetermined problem.

5.1.3 Weighted Least-Squares Method

Let us introduce some *weighting factors* w_i^2 for estimation of the residuals r_i. The reason for the weighing is that, in practice, some observations are made with more accuracy than others. In this case, one would like the prediction errors r_i of the more accurate observations to have a greater weight than the inaccurate observations. To accomplish this weighting, we define the weighted misfit functional f_w as follows:

$$f_w(m_1, m_2, m_3, ..., m_L) = \|\mathbf{r}\|_w^2 = \sum_{i=1}^{N}(w_i r_i)^2 = \min. \tag{5.27}$$

We can introduce the *weighting operator* W, which is a linear operator acting in the space of data D and having the diagonal matrix $\mathbf{W}$:

$$\mathbf{W} = [\mathbf{diag}(w_i)]. \tag{5.28}$$

Then Eq. (5.27) can be rewritten in the following form:

$$f_w(\mathbf{m}) = \|\mathbf{W}\mathbf{d}^p - \mathbf{W}\mathbf{d}\|^2 = \|\mathbf{W}\mathbf{A}\mathbf{m} - \mathbf{W}\mathbf{d}\|^2$$

$$= (\mathbf{W}\mathbf{A}\mathbf{m} - \mathbf{W}\mathbf{d})^T(\mathbf{W}\mathbf{A}\mathbf{m} - \mathbf{W}\mathbf{d}) = \min. \tag{5.29}$$

The problem of minimization of the weighted misfit functional can be solved by calculating the first variation of this functional and setting it equal to zero:

$$\delta f_w(\mathbf{m}) = 2\,(\delta \mathbf{WAm})^T\,(\mathbf{WAm} - \mathbf{Wd}) = 2(\delta \mathbf{m})^T\,(\mathbf{WA})^T\,(\mathbf{WAm} - \mathbf{Wd}) = 0.$$

Thus, we obtain the following system of equations:

$$(\mathbf{WA})^T\,\mathbf{WAm} = (\mathbf{WA})^T\mathbf{Wd}$$

or

$$\mathbf{A}^T\mathbf{W}^2\mathbf{Am} = \mathbf{A}^T\mathbf{W}^2\mathbf{d}. \tag{5.30}$$

Assuming that matrix $\mathbf{A}^T\mathbf{W}^2\mathbf{A}$ is non-singular, we can write

$$\mathbf{m}_0 = (\mathbf{A}^T\mathbf{W}^2\mathbf{A})^{-1}\mathbf{A}^T\mathbf{W}^2\mathbf{d}. \tag{5.31}$$

Equation (5.31) delivers the solution of the weighted least-squares problem.The method of selecting the proper weights, w_i, will be discussed in the next sections.

The following matrix

$$\mathbf{A}^{-w} = (\mathbf{A}^T\mathbf{W}^2\mathbf{A})^{-1}\mathbf{A}^T\mathbf{W}^2 \tag{5.32}$$

is called the *weighted generalized inverse matrix*.

5.2 Tikhonov Regularization Method

5.2.1 Tikhonov Parametric Functional Revisited

Different modifications of least-squares solutions of the linear inverse problem discussed above have resulted from the direct minimization of the corresponding misfit functionals. However, all these solutions have many limitations and are very sensitive to minor variations of the observed data. An obvious limitation occurs when the inverse matrices $(\mathbf{A}^T\mathbf{A})^{-1}$ or $(\mathbf{A}^T\mathbf{W}^2\mathbf{A})^{-1}$ do not exist. However, even when the inverse matrices exist, they can still be ill-conditioned (become nearly singular). In this case, the solution would be extremely unstable and unrealistic. To overcome these difficulties, we have to apply regularization methods.

Let us consider first the general approach based on the Tikhonov regularization technique (Tikhonov and Arsenin 1977). The corresponding parametric functional can be introduced in the following form:

$$P^{\alpha}(\mathbf{m}, \mathbf{d}) = \|\mathbf{W}_d\mathbf{Am} - \mathbf{W}_d\mathbf{d}\|^2 + \alpha\|\mathbf{W}_m\mathbf{m} - \mathbf{W}_m\mathbf{m}_{apr}\|^2,$$

where $\mathbf{W}_d$ and $\mathbf{W}_m$ are some weighting matrices of data and model (not necessarily diagonal); $\mathbf{m}_{apr}$ is some *a priori* model, and $\|...\|$ denotes the Euclidean norm in the spaces of data and models.

In the majority of practical applications, we assume that $\mathbf{W}_m = \mathbf{I}$, but it also can be chosen arbitrarily (for example, as a matrix of first or second-order finite-difference differentiation to obtain a smooth solution). We will discuss some specific choices of $\mathbf{W}_m$ later.

We can also rewrite functional $P^{\alpha}(\mathbf{m}, \mathbf{d})$ in matrix notations as follows:

$$P^{\alpha}(\mathbf{m}, \mathbf{d}) = (\mathbf{W}_d\mathbf{A}\mathbf{m} - \mathbf{W}_d\mathbf{d})^T(\mathbf{W}_d\mathbf{A}\mathbf{m} - \mathbf{W}_d\mathbf{d})$$

$$+ \alpha(\mathbf{W}_m\mathbf{m} - \mathbf{W}_m\mathbf{m}_{apr})^T(\mathbf{W}_m\mathbf{m} - \mathbf{W}_m\mathbf{m}_{apr}).$$

According to the basic principles of the regularization method, we have to find a quasi-solution of the inverse problem as the model $\mathbf{m}_{\alpha}$ that minimizes the parametric functional

$$P^{\alpha}(\mathbf{m}_{\alpha}, \mathbf{d}) = \min .$$

The regularization parameter α is determined from the misfit condition:

$$\|\mathbf{W}_d\mathbf{A}\mathbf{m}_{\alpha} - \mathbf{W}_d\mathbf{d}\| = \delta,$$

where δ is some *a priori* estimation of the level of "weighted" noise of the data:

$$\|\mathbf{W}_d\delta\mathbf{d}\| = \delta. \tag{5.33}$$

To solve this problem, let us calculate the first variation of $P^{\alpha}(\mathbf{m}, \mathbf{d})$:

$$\delta P^{\alpha}(\mathbf{m}, \mathbf{d}) = 2(\mathbf{W}_d\mathbf{A}\delta\mathbf{m})^T(\mathbf{W}_d\mathbf{A}\mathbf{m} - \mathbf{W}_d\mathbf{d})$$

$$+ 2\alpha(\mathbf{W}_m\delta\mathbf{m})^T(\mathbf{W}_m\mathbf{m} - \mathbf{W}_m\mathbf{m}_{apr}) = 0.$$

The last equation can be rewritten in the following form:

$$\delta\mathbf{m}^T\left[(\mathbf{A}^T\mathbf{W}_d^T\mathbf{W}_d\mathbf{A} + \alpha\mathbf{W}_m^T\mathbf{W}_m)\mathbf{m}\right.$$

$$\left. - \left(\mathbf{A}^T\mathbf{W}_d^T\mathbf{W}_d\mathbf{d} + \alpha\mathbf{W}_m^T\mathbf{W}_m\mathbf{m}_{apr}\right)\right] = 0,$$

from which we obtain at once *a regularized normal equation* for the original inverse problem (5.5),

$$(\mathbf{A}^T\mathbf{W}_d^T\mathbf{W}_d\mathbf{A} + \alpha\mathbf{W}_m^T\mathbf{W}_m)\mathbf{m}_{\alpha} = \mathbf{A}^T\mathbf{W}_d^T\mathbf{W}_d\mathbf{d} + \alpha\mathbf{W}_m^T\mathbf{W}_m\mathbf{m}_{apr}, \tag{5.34}$$

and its regularized solution,

$$\mathbf{m}_\alpha = (\mathbf{A}^T \mathbf{W}_d^T \mathbf{W}_d \mathbf{A} + \alpha \mathbf{W}_m^T \mathbf{W}_m)^{-1} (\mathbf{A}^T \mathbf{W}_d^T \mathbf{W}_d \mathbf{d} + \alpha \mathbf{W}_m^T \mathbf{W}_m \mathbf{m}_{apr}). \quad (5.35)$$

Usually, the weighting matrices $\mathbf{W}_d$ and $\mathbf{W}_m$ are selected to be symmetric (or even diagonal), so Eq. (5.35) can be rewritten as follows:

$$\mathbf{m}_\alpha = (\mathbf{A}^T \mathbf{W}_d^2 \mathbf{A} + \alpha \mathbf{W}_m^2)^{-1} (\mathbf{A}^T \mathbf{W}_d^2 \mathbf{d} + \alpha \mathbf{W}_m^2 \mathbf{m}_{apr}). \quad (5.36)$$

The last expression gives the regularized solution of the generalized least-squares problem.

5.2.2 *Integrated Sensitivity*

Let us analyze the sensitivity of the data to the perturbation of one specific parameter m_k. To solve this problem, we apply the variational operator to both sides of Eq. (5.6):

$$\delta d_i = A_{ik} \delta m_k. \quad (5.37)$$

In the last formula, A_{ik} are the elements of matrix $\mathbf{A}$ of the forward modeling operator, and there is no summation over index k. Therefore, the norm of the perturbed vector of the data can be calculated as follows:

$$\|\delta \mathbf{d}\| = \sqrt{\sum_i (\delta d_i)^2} = \sqrt{\sum_i (A_{ik})^2} \, |\delta m_k| \,. \quad (5.38)$$

We determine the integrated sensitivity of the data to parameter m_k as the following ratio:

$$S_k = \frac{\|\delta \mathbf{d}\|}{|\delta m_k|} = \sqrt{\sum_i (A_{ik})^2}. \quad (5.39)$$

One can see that the integrated sensitivity depends on parameter k. In other words, the sensitivity of the data to the different model parameters varies because the contributions of the different parameters to the observation are also variable.

Definition 1 The diagonal matrix with the diagonal elements equal to $S_k = \|\delta \mathbf{d}\| / \delta m_k$ is called *an integrated sensitivity matrix:*

$$\mathbf{S} = \mathbf{diag}\left(\sqrt{\sum_i (A_{ik})^2}\right) = \mathbf{diag}\left(\mathbf{A}^T \mathbf{A}\right)^{1/2}. \quad (5.40)$$

In other words, it is formed by the norms of the columns of matrix $\mathbf{A}$.

5.2.3 *Definition of the Weighting Matrices for the Model Parameters and Data*

The basic idea of introducing a *model weighting matrix*, $\mathbf{W}_m$, for the model parameters can be described as follows. We identify this matrix as the diagonal integrated sensitivity matrix

$$\mathbf{W}_m = [W_j] = [S_j] = \mathbf{S}. \tag{5.41}$$

Thus, the weights are selected to be equal to the sensitivities:

$$W_j = S_j. \tag{5.42}$$

We can now introduce the weighted model parameters:

$$\mathbf{m}^w = \mathbf{W}_m \mathbf{m}. \tag{5.43}$$

Using these notations, we can rewrite the inverse problem, Eq. (5.5), as follows:

$$\mathbf{d} = \mathbf{A}\mathbf{W}_\mathbf{m}^{-1}\mathbf{W}_\mathbf{m}\mathbf{m} = \mathbf{A}^w\mathbf{m}^w, \tag{5.44}$$

where $\mathbf{A}^w$ is a weighted forward modeling operator,

$$\mathbf{A}^w = \mathbf{A}\mathbf{W}_\mathbf{m}^{-1}. \tag{5.45}$$

Now, we perturb the data with respect to one specific weighted parameter m_k^w:

$$\delta d_i = A_{ik}^w \delta m_k^w,$$

and calculate a new integrated sensitivity S_k^w of the data to the weighted parameter m_k^w as the ratio

$$S_k^w = \frac{\|\delta \mathbf{d}\|}{\delta m_k^w} = \frac{\sqrt{\sum_i \left(A_{ik}^w\right)^2} \delta m_k^w}{\delta m_k^w} = \sqrt{\sum_i \left(A_{ik}^w\right)^2}$$

$$= \sqrt{\sum_i \left(A_{ik} W_k^{-1}\right)^2} = W_k^{-1} \sqrt{\sum_i (A_{ik})^2} = W_k^{-1} S_k = 1. \tag{5.46}$$

Formula (5.46) shows that the new matrix of the integrated sensitivity $\mathbf{S}^w$ is a unit matrix:

$$\mathbf{S}^w = \mathbf{I}.$$

Therefore, data are uniformly sensitive to the new weighted model parameters.

Note that the corresponding weighted stabilizing functional takes the form

$$s_w(\mathbf{m}) = (\mathbf{m} - \mathbf{m}_{apr})^T \mathbf{W}_m^2 (\mathbf{m} - \mathbf{m}_{apr})$$
$$= (\mathbf{m} - \mathbf{m}_{apr})^T \mathbf{S}^2 (\mathbf{m} - \mathbf{m}_{apr}). \tag{5.47}$$

It imposes a stronger penalty on departure from the *a priori* model for those parameters that contribute more significantly to the data.

Thus, the model weighting results in practically equal resolution of the inversion with respect to different parameters of the model.

In a similar way, we can define the diagonal *data weighting matrix*, formed by the norms of the rows of matrix $\mathbf{A}$:

$$\mathbf{W}_d = \mathbf{diag}\left(\sqrt{\sum_k (A_{ik})^2}\right) = \mathbf{diag}\left(\mathbf{A}\mathbf{A}^T\right)^{1/2}. \tag{5.48}$$

These weights make normalized data less dependent on the specific parameters of observations (for example, frequency and distance from the anomalous domain), which improves the resolution of the inverse method.

5.3 Levenberg–Marquardt Method

Let us consider a special case when $\mathbf{W}_d = \mathbf{I}$ and $\mathbf{W}_m = \mathbf{I}$. Then Eq. (5.36) takes the following form:

$$\mathbf{m}_\alpha = (\mathbf{A}^T\mathbf{A} + \alpha\mathbf{I})^{-1}(\mathbf{A}^T\mathbf{d} + \alpha\mathbf{m}_{apr}). \tag{5.49}$$

Assume now that $\mathbf{m}_{apr} = 0$:

$$\mathbf{m}_\alpha = (\mathbf{A}^T\mathbf{A} + \alpha\mathbf{I})^{-1}\mathbf{A}^T\mathbf{d}. \tag{5.50}$$

Solution (5.50) describes the classical Levenberg–Marquardt methodof damped least-squares, where α plays the role of a "damping factor" (Levenberg 1944; Marquardt 1963).

For a better understanding of how the regularization parameter or "damping factor" α works, let us apply a singular value decomposition method to matrix $\mathbf{A}$:

$$\mathbf{A} = \mathbf{U}\mathbf{Q}\mathbf{V}^T.$$

Then we have

$$\mathbf{A}^T\mathbf{A} = \mathbf{V}\mathbf{Q}^2\mathbf{V}^T. \tag{5.51}$$

From the last equation, we obtain at once

$$(\mathbf{A}^T\mathbf{A} + \alpha\mathbf{I})^{-1} = (\mathbf{V}\mathbf{Q}^2\mathbf{V}^T + \alpha\mathbf{V}\mathbf{I}\mathbf{V}^T)^{-1}$$

$$+ (\mathbf{V}\left[\mathbf{diag}(\alpha + Q_i^2)\right]\mathbf{V}^T)^{-1} = \mathbf{V}\left[\mathbf{diag}(\frac{1}{\alpha + Q_i^2})\right]\mathbf{V}^T. \tag{5.52}$$

We can clearly see from Eq. (5.52) how regularization makes the nearly singular matrix well-conditioned because even if $Q_i \to 0$ division by zero does not occur.

References and Recommended Reading to This Chapter

Golub GH, Van Loan CF (2013) Matrix computations, 4th edn. The Johns Hopkins University Press, Baltimore and London, pp 753

Hansen C, Pereyra V, Scherer G (2013) Least squares data fitting with applications. Johns Hopkins University Press, pp 328

Levenberg K (1944) A method for the solution of certain nonlinear problems in least squares. Quart Appl Math 2:164–168

Marquardt DW (1963) An algorithm for least squares estimation of nonlinear parameters. SIAM J. 11:431–441

Menke W (2018) Geophysical data analysis: discrete inverse theory, 4th edn. Academic Press, pp 330

Tikhonov AN, Arsenin VY (1977) Solution of ill-posed problems. Wiley, pp 258

Wolberg J (2006) Data analysis using the method of least squares. Springer, pp 250

Zhdanov MS (1993) Tutorial: regularization in inversion theory. CWP-136, Colorado School of Mines, pp 47

Zhdanov MS (2002) Geophysical inverse theory and regularization problems. Elsevier, pp 609

Zhdanov MS (2015) Inverse theory and applications in geophysics. Elsevier, pp 704

Zhdanov MS (2018) Foundations of geophysical electromagnetic theory and methods. Elsevier, pp 770

Chapter 6
Probabilistic Methods of Inverse Problem Solution

Abstract This chapter considers the methods of solving the linear discrete inverse problems using the probabilistic approach. We review two major techniques—the maximum likelihood and the maximum a posteriori estimation methods. The Bayes estimation method makes it possible to introduce some a priori information about the properties of the solution in the inversion. We demonstrate that the numerical implementation of these methods is similar to the weighted least-squares and Tikhonov's regularization methods, respectively. A summary of the typical stochastic inversion techniques, e.g., Monte Carlo, genetic algorithm (GA), and simulated annealing (SA) methods, is also provided.

Keywords Maximum likelihood method · Bayes estimation · Stochastic methods · Monte Carlo · Genetic algorithm (GA) · Simulated annealing (SA)

In Chap. 5, we considered the methods of solving the linear discrete inverse problems using the deterministic approach based on Tikhonov regularization. However, there exists an alternative approach based on the ideas of the probability theory. Therefore, in this chapter presents several methods for inverse problem solutions using the probabilistic approach following Zhdanov (1993, 2002, 2015).

6.1 Maximum Likelihood Method

As discussed in Chap. 2, the probability distribution can be described by a very complicated function in general cases. However, according to the *central limit theorem*, a large sample of a random variable tends to a very simple distribution, the so-called *Gaussian (or normal) distribution* , as the size of the random sample increases.

The joint distribution of two independent Gaussian variables is just the product of two univariate distributions. When the data forming a vector **d** are correlated (with mean $\langle \mathbf{d} \rangle$ and covariance $\boldsymbol{\sigma} = [\sigma_{ij}]$), the appropriate distribution turns out to be as follows (Menke 2018):

M. S. Zhdanov, *Advanced Methods of Joint Inversion and Fusion of Multiphysics Data*, Advances in Geological Science,
https://doi.org/10.1007/978-981-99-6722-3_6

$$P(\mathbf{d}) = \frac{|\,\boldsymbol{\sigma}\,|^{-\frac{1}{2}}}{(2\pi)^{\frac{N}{2}}} \exp[-\frac{1}{2}(\mathbf{d} - \langle \mathbf{d} \rangle)^T \boldsymbol{\sigma}^{-1} (\mathbf{d} - \langle \mathbf{d} \rangle)]. \tag{6.1}$$

The idea that the model and data are related by an explicit relationship,

$$\mathbf{Am} = \mathbf{d}, \tag{6.2}$$

can now be reinterpreted in the sense that this relationship holds only for the mean data:

$$\mathbf{Am} = \langle \mathbf{d} \rangle. \tag{6.3}$$

Substituting (6.3) into (6.1), we can rewrite the distribution of the data as follows:

$$P(\mathbf{d}) = \frac{|\,\boldsymbol{\sigma}\,|^{-\frac{1}{2}}}{(2\pi)^{\frac{N}{2}}} \exp[-\frac{1}{2}(\mathbf{d} - \mathbf{Am})^T \boldsymbol{\sigma}^{-1} (\mathbf{d} - \mathbf{Am})]$$

$$= \frac{|\,\boldsymbol{\sigma}\,|^{-\frac{1}{2}}}{(2\pi)^{\frac{N}{2}}} \exp[-\frac{1}{2} f_\sigma(\mathbf{m})], \tag{6.4}$$

where

$$f_\sigma(\mathbf{m}) = (\mathbf{d} - \mathbf{Am})^T \boldsymbol{\sigma}^{-1} (\mathbf{d} - \mathbf{Am}).$$

Under this assumption, we can say that the optimum values for the model parameters are those that maximize the probability that the observed data are, in fact, observed. Thus, the method of maximum likelihood is based on maximization of the probability function (6.4)

$$P(\mathbf{d}) = \max. \tag{6.5}$$

Clearly, the maximum of $P(\mathbf{d})$ occurs when the argument of the exponential function has maximum or when $f_\sigma(\mathbf{m})$ has minimum:

$$f_\sigma(\mathbf{m}) = (\mathbf{d} - \mathbf{Am})^T \boldsymbol{\sigma}^{-1} (\mathbf{d} - \mathbf{Am}) = \min. \tag{6.6}$$

Let us calculate the first variation of functional f_σ:

$$\delta f_\sigma(\mathbf{m}) = -(\delta\mathbf{Am})^T \boldsymbol{\sigma}^{-1} (\mathbf{d} - \mathbf{Am}) - (\mathbf{d} - \mathbf{Am})^T \boldsymbol{\sigma}^{-1} (\delta\mathbf{Am}).$$

It can be shown that for symmetrical matrix $\boldsymbol{\sigma}^{-1}$, the following equality holds:

$$\mathbf{a}^T \boldsymbol{\sigma}^{-1} \mathbf{b} = \mathbf{b}^T \boldsymbol{\sigma}^{-1} \mathbf{a},$$

where $\mathbf{a}$ and $\mathbf{b}$ are two arbitrary column vectors. Therefore, we can write the necessary condition for the functional f_σ to have a minimum as follows:

$$\delta f_{\sigma}(\mathbf{m}) = -2(\mathbf{A}\delta\mathbf{m})^T \boldsymbol{\sigma}^{-1}(\mathbf{d} - \mathbf{A}\mathbf{m}) = -2(\delta\mathbf{m})^T \mathbf{A}^T \boldsymbol{\sigma}^{-1}(\mathbf{d} - \mathbf{A}\mathbf{m}) = 0. \tag{6.7}$$

From Eq. (6.7), we obtain at once the following equation:

$$\mathbf{A}^T \boldsymbol{\sigma}^{-1}(\mathbf{d} - \mathbf{A}\mathbf{m}) = 0.$$

The last formula provides the following normal system of equations for the "pseudo-solution" of the minimization problem (6.6):

$$\mathbf{A}^T \boldsymbol{\sigma}^{-1}\mathbf{A}\mathbf{m} = \mathbf{A}^T \boldsymbol{\sigma}^{-1}\mathbf{d}. \tag{6.8}$$

If the matrix $\left(\mathbf{A}^T \boldsymbol{\sigma}^{-1}\mathbf{A}\right)$ is non-singular, then we can multiply both sides of normal system (6.8) by inverse matrix, $(\mathbf{A}^T \boldsymbol{\sigma}^{-1}\mathbf{A})^{-1}$, and write the pseudo-solution of minimization problem (6.6) in the explicit form as follows:

$$\mathbf{m}_0 = (\mathbf{A}^T \boldsymbol{\sigma}^{-1}\mathbf{A})^{-1}\mathbf{A}^T \boldsymbol{\sigma}^{-1}\mathbf{d}. \tag{6.9}$$

Comparing the last formula with the corresponding equation for the weighted least-squares method (5.31), we see that we have obtained exactly the same result if we substitute matrix $\mathbf{W}^2$ for $\boldsymbol{\sigma}^{-1}$:

$$\mathbf{W}^2 = \boldsymbol{\sigma}^{-1}. \tag{6.10}$$

Note that if data happen to be uncorrelated, then the covariance matrix becomes diagonal:

$$\boldsymbol{\sigma} = [\mathbf{diag}(\sigma_i^2)], \tag{6.11}$$

and the elements of the main diagonal are the variances of the data. In this case, the weights are given by the following formula:

$$w_i^2 = \frac{1}{\sigma_i^2}. \tag{6.12}$$

The functional

$$f_w(\mathbf{m}) = \chi^2(\mathbf{m}) = \sum_{i=1}^{N} \left(\frac{r_i}{\sigma_i}\right)^2 = \sum_{i=1}^{N} \left(\frac{d_i - d_i^p}{\sigma_i}\right)^2 \tag{6.13}$$

is called a "chi-square."

In the cases where the measurement errors are normally distributed, the quantity χ^2 is a sum of N squares of normally distributed variables, each normalized to its variance. Thus, by applying the weighted least-squares method, we can select the smaller weights for data with bigger standard deviations (less accurate data) and the bigger weights for data with smaller standard deviations (more certain data). If the data have equal variances, σ_0^2, then the weighting matrix becomes scalar:

$$\mathbf{W}^2 = \boldsymbol{\sigma}^{-1} = \frac{1}{\sigma_0^2}\mathbf{I},$$

and the chi-square functional becomes equal to the conventional misfit functional.

6.2 The Maximum a Posteriori Estimation Method (The Bayes Estimation)

Let us consider the regularization technique from the point of view of probability theory (Tarantola 1987). First of all, we introduce the following (normally distributed) densities of probability:
(1) $P(\mathbf{d}/\mathbf{m})$ is a conditional density of probability of the data $\mathbf{d}$, given the model $\mathbf{m}$. It means that it is the probability density of theoretical data $\mathbf{d}$ to be expected from a given model $\mathbf{m}$.
(2) $P(\mathbf{m}/\mathbf{d})$ is a conditional density of probability of a model $\mathbf{m}$, given the data $\mathbf{d}$. According to the Bayes theorem, the following equation holds:

$$P(\mathbf{m}/\mathbf{d}) = \frac{P(\mathbf{d}/\mathbf{m})P(\mathbf{m})}{P(\mathbf{d})}, \tag{6.14}$$

where $P(\mathbf{d})$ and $P(\mathbf{m})$ are unconditional probability densities for data and model parameters, respectively. It is assumed that

$$\langle \mathbf{m} \rangle = \mathbf{m}_{apr},$$

where $\mathbf{m}_{apr}$ is an a priori constrained expectation of the model, and

$$[\mathrm{cov}(m_i, m_j)] = \boldsymbol{\sigma}_m.$$

Thus, considering normally distributed parameters, we have the following probability distribution of the model, $\mathbf{m}$:

$$P(\mathbf{m}) = \frac{|\,\boldsymbol{\sigma}_m\,|^{-\frac{1}{2}}}{(2\pi)^{\frac{L}{2}}} \exp[-\frac{1}{2}(\mathbf{m} - \mathbf{m}_{apr})^T \boldsymbol{\sigma}_m^{-1}(\mathbf{m} - \mathbf{m}_{apr})]. \tag{6.15}$$

Analogously, it is assumed that,

$$[\mathrm{cov}(d_i, d_j)] = \boldsymbol{\sigma}_d$$

and we can write for the conditional density of probability of the data **d**

$$P(\mathbf{d}/\mathbf{m}) = \frac{|\boldsymbol{\sigma}_d|^{-\frac{1}{2}}}{(2\pi)^{\frac{N}{2}}} \exp[-\frac{1}{2}(\mathbf{d}-\mathbf{Am})^T \boldsymbol{\sigma}_d^{-1}(\mathbf{d}-\mathbf{Am})]. \tag{6.16}$$

The maximum likelihood method can now be used to find the model $\mathbf{m}_0$ which maximizes the conditional probability of a model, $P(\mathbf{m}/\mathbf{d})$:

$$P(\mathbf{m}/\mathbf{d}) = \frac{|\boldsymbol{\sigma}_d|^{-\frac{1}{2}}}{(2\pi)^{\frac{N}{2}}} \exp[-\frac{1}{2}(\mathbf{d}-\mathbf{Am})^T \boldsymbol{\sigma}_d^{-1}(\mathbf{d}-\mathbf{Am})] \times$$

$$\times \frac{|\boldsymbol{\sigma}_m|^{-\frac{1}{2}}}{(2\pi)^{\frac{L}{2}}} \exp[-\frac{1}{2}(\mathbf{m}-\mathbf{m}_{apr})^T \boldsymbol{\sigma}_m^{-1}(\mathbf{m}-\mathbf{m}_{apr})]P^{-1}(\mathbf{d}). \tag{6.17}$$

It is evident that, to maximize $P(\mathbf{m}/\mathbf{d})$, we have to minimize the sum of the expressions in the exponential factors in Eq. (6.17):

$$f_{Bayes} = (\mathbf{d}-\mathbf{Am})^T \boldsymbol{\sigma}_d^{-1}(\mathbf{d}-\mathbf{Am}) + (\mathbf{m}-\mathbf{m}_{apr})^T \boldsymbol{\sigma}_m^{-1}(\mathbf{m}-\mathbf{m}_{apr}). \tag{6.18}$$

Note that the minimization of the first term in the above equation gives the classical maximum likelihood or weighted least-squares method.

Let us calculate the first variation of f_{Bayes}:

$$\delta f_{Bayes} = -2(\mathbf{A}\delta\mathbf{m})^T \boldsymbol{\sigma}_d^{-1}(\mathbf{d}-\mathbf{Am}) + 2(\delta\mathbf{m})^T \boldsymbol{\sigma}_m^{-1}(\mathbf{m}-\mathbf{m}_{apr}) = 0.$$

From the last equation, we have

$$(\delta\mathbf{m})^T[\mathbf{A}^T \boldsymbol{\sigma}_d^{-1}(\mathbf{d}-\mathbf{Am}) - \boldsymbol{\sigma}_m^{-1}(\mathbf{m}-\mathbf{m}_{apr})] = 0.$$

Thus, the normal system of equations for minimization of f_{Bayes} can be written as follows:

$$\mathbf{A}^T \boldsymbol{\sigma}_d^{-1}(\mathbf{d}-\mathbf{Am}) - \boldsymbol{\sigma}_m^{-1}(\mathbf{m}-\mathbf{m}_{apr}) = 0,$$

From the last formula, we have at once the following equation:

$$(\mathbf{A}^T \boldsymbol{\sigma}_d^{-1}\mathbf{A} + \boldsymbol{\sigma}_m^{-1})\mathbf{m} = \mathbf{A}^T \boldsymbol{\sigma}_d^{-1}\mathbf{d} + \boldsymbol{\sigma}_m^{-1}\mathbf{m}_{apr}. \tag{6.19}$$

We can write the solution of Eq. (6.19) in the closed form as follows:

$$\mathbf{m}_0 = (\mathbf{A}^T \boldsymbol{\sigma}_d^{-1}\mathbf{A} + \boldsymbol{\sigma}_m^{-1})^{-1}(\mathbf{A}^T \boldsymbol{\sigma}_d^{-1}\mathbf{d} +_m^{-1} \mathbf{m}_{apr}). \tag{6.20}$$

By comparing Eqs. (6.20) and (5.36), we see that

$$\boldsymbol{\sigma}_m^{-1} = \alpha \mathbf{W}_m^2, \tag{6.21}$$

so $\boldsymbol{\sigma}_m^{-1}$ plays the role of the regularization parameter and the model parameter weights simultaneously.

Let us assume now that we have uncorrelated data with equal variances,

$$\boldsymbol{\sigma}_d = \sigma_d^2 \mathbf{I},$$

and similarly for the a priori covariance of the model,

$$\boldsymbol{\sigma}_m = \sigma_m^2 \mathbf{I}.$$

Then Eq. (6.20) takes the following form:

$$\mathbf{m}_0 = (\mathbf{A}^T \mathbf{A} + k\mathbf{I})^{-1} (\mathbf{A}^T \mathbf{d} + k\mathbf{m}_{apr}), \tag{6.22}$$

where

$$k = \frac{\sigma_d^2}{\sigma_m^2} = \alpha \tag{6.23}$$

plays the role of the regularization parameter.

We can see from formula (6.23) that large values of the variance σ_m^2 of the model parameters correspond to a small regularization parameter α, and vice versa, large values of α correspond to a small variance σ_m^2. This means that, without regularization (α close to zero), the uncertainty in determining the inverse model is great, while with regularization, it becomes smaller. The last formula illustrates once again the close connection between the probabilistic (Tarantola 1987) and deterministic (Tikhonov and Arsenin 1977) approaches to regularization.

6.3 Stochastic Methods of Inversion

We have already discussed in this and previous chapters that there are two different major points of view in addressing the inverse problem:
(a) the algebraic (deterministic) point of view, dating back to the works of Lanczos (1961), Backus and Gilbert (1967), Backus (1970a, b, c), Marquardt (1963, 1970), Tikhonov and Arsenin (1977), etc.;
(b) the probabilistic (stochastic) point of view, formulated in the pioneering papers of Foster (1961), Franklin (1970), Jackson (1972), Tarantola and Valette (1982), Tarantola (1987, 2005), etc.

The stochastic point of view is widely used in literature because it is closely associated with the statistical nature of the noise in the data. At the same time, it has been demonstrated in many publications (e.g., the classical work by Sabatier (1977)) that in many cases, both points of view result in similar computational algorithms (see Sects. 6.1 and 6.2).

The Monte Carlo inversion methods represent a general approach based on the stochastic point of view (Metropolis and Ulam 1949; Metropolis et al. 1953). They are named after the famous Casino in Monaco. There are two major types of Monte Carlo methods. The first one is based on an extensive random search in the space M of the model parameters for a solution, which generates the predicted data from the data space, D, close to the observed data, realizing the global minimum of the corresponding misfit functional $f(\mathbf{m})$ (e.g., Cary and Chapman 1988; Khan et al. 2000; Khan and Mosegaard 2001). This method is suitable for problems with misfit functionals having multiple local minimums, where conventional gradient-type minimization methods may have difficulties getting out from a "deep" local minimum (see Chap. 7). The second type of Monte Carlo method uses an optimization algorithm in order to minimize the number of steps required by the random search methods. The most effective global optimization algorithms have been developed based on known physical or biological rules to evolve to the best solution. For example, the simulated annealing (SA) algorithm (Kirkpatrick et al. 1983; Corana et al. 1987) comes from annealing in metallurgy, a technique involving heating and controlled cooling of a material. It is known from physics that, in order to minimize the final lattice energy, one should apply a very slow cooling process. The SA method uses an analogy between the minimization of lattice energy in the framework of the physical process of annealing and numerical problem of determining the global minimum of a misfit functional, $f(\mathbf{m})$.

The genetic algorithm (GA) (Holland 1975; Goldberg 1989; Michalewicz and Schoenauer 1996; Whitley 1994; Mosegaard and Sambridge 2002) is a heuristic search method that mimics the process of natural evolution. In a pure genetic algorithm, a population of candidate solutions (individuals) for an optimization problem is evolved toward better solutions. Traditionally, the solutions are coded in binary form as strings of 0s and 1s to be mutated and altered. The evolution starts from a population of randomly generated solutions from the search space and proceeds as an iterative process. The population in each iteration is called a *generation*. In each generation, the fitness of every individual is evaluated by an objective functional (e.g., a misfit functional $f(\mathbf{m})$). The individuals who have low misfits are stochastically selected from the current population, and then they are chosen to form a new generation by applying genetic operations (*mutation* and *crossover*). The above steps run iteratively until the inversion process meets the termination conditions.

A detailed overview of the simulated annealing and genetic algorithms can be found, for example, in Zhdanov (2015).

The Monte Carlo methods are considered to be an effective optimization technique for many inverse problems where some general gradient-type methods fail. They can be applied for solving optimization problems with continuous or discrete parameters and with small sample intervals; there is no need to calculate the derivatives; the global minimization problem can be solved for the misfit functional with multiple local minima.

The Monte Carlo methods were first applied to the solutions of earth science problems by Keilis-Borok and Yanovskaya (1967) and Press (1968, 1970a, b). The paper by Sambridge and Mosegaard (2002) provides an excellent review of applications of the Monte Carlo methods to solving geophysical inverse problems.

References and Recommended Reading to This Chapter

Backus GE (1970a) Inference from inadequate and inaccurate data. I Proceedings of the National Academy of Sciences, vol 65, pp 1–7

Backus GE (1970b) Inference from inadequate and inaccurate data, II. Proceedings of the National Academy of Sciences, vol 65, pp 281–287

Backus, GE (1970c) Inference from inadequate and inaccurate data, III: Proceedings of the National Academy of Sciences, vol 67, pp 282–289

Backus GE, Gilbert TI (1967) Numerical applications of a formalism for geophysical inverse problems. Geophys J R Astr Soc 13:247–276

Cary PW, Chapman CH (1988) Automatic 1D waveform inversion of marine seismic refraction data. Geophys J 93:527–46

Corana A, Marchesi M, Martini C, Ridella S (1987) Minimising multimodal functions of continuous variables with the "Simulated Annealing" Algorithm. ACM Trans Math Soft 13:262–280

Foster M (1961) An application of the Wiener-Kolmogorov smoothing theory to matrix inversion. J Soc Ind Appl Math 9:387–392

Franklin JN (1970) Well-posed stochastic extensions of ill-posed linear problems. J Math Anal Appl 31:682–716

Goldberg DE (1989) Genetic Algorithms in search, optimization, and machine learning. Addison-Wesley

Holland JH (1975) Adaptation in natural and artificial systems. Ann Arbor: University of Michigan Press

Jackson DD (1972) Interpretation of inaccurate, insufficient and inconsistent data: Geophys J R Astronom Soc 28:97–110

Keilis-Borok VI, TB Yanovskaya (1967) Inverse problems of seismology. Geophys J 13:223–234

Khan A, Mosegaard K, Rasmussen KL (2000) A new seismic velocity model for the Moon from a Monte Carlo inversion of the Apollo Lunar seismic data. Geophys Res Lett 27:1591–1594

Khan A, Mosegaard K (2001) New information on the deep lunar interior from an inversion of lunar free oscillation periods. Geophys Res Lett 28:1791

Kirkpatrick, S. C., D. Gelatt and M. P. Vecchi, 1983, Optimization by simulated annealing: Science, 220, 671–680 Lanczos C (1961) Linear differential operators. D van Nostrand Co

Marquardt DW (1963) An algorithm for least-squares estimation of nonlinear parameters. J Soc Ind Appl Math 11:431–441

Marquardt DW (1970) Generalized inverses, ridge regression, biased linear estimation, and nonlinear estimation. Technometrics 12:591–612

Menke W (2018) Geophysical data analysis: discrete inverse theory, 4th ed. Academic Press, 330 pp

Metropolis N, Ulam SM (1949) The Monte Carlo method. J Am Stat Assoc **44**:335–341

Metropolis N, Rosenbluth MN, Rosenbluth AW, Teller, Teller E (1953) Equation of state calculations by fast computing machines. J Chem Phys 21:1087–1092

Michalewicz Z, Schoenauer M (1996) Evolutionary algorithms for constrained parameter optimization problems. Evol Comput 4(1):1–32

Mosegaard K, Sambridge M (2002) Monte Carlo analysis of inverse problems. Inverse Probl 18:R29–R54

Press F 1968 Earth models obtained by Monte Carlo inversion. J Geophys Res 73:5223–34

Press F (1970a) Earth models consistent with geophysical data. Phys Earth Planet Inter 3:3–22

Press F (1970b) Regionalized Earth models. J Geophys Res 75:6575–81.

Sabatier PC (1977) On geophysical inverse problems and constraints. J Geophys Res 43:115–137

Sambridge, M, Mosegaard K (2002) Monte Carlo methods in geophysical inverse problems. Rev Geophys 40, 3:1–29

Tarantola A (1987) Inverse problem theory. Elsevier, Amsterdam, Oxford, New York, Tokyo, 613 pp

Tarantola A (2005) Inverse problem theory and methods for model parameter estimation. SIAM, 344 pp

Tarantola A, Valette B (1982) Generalized nonlinear inverse problem solved using the least squares criterion. Rev Geophys Space Phys 20:219–232

Tikhonov AN, Arsenin VY (1977) Solution of ill-posed problems. W H Winston and Sons

Whitley DL (1994) A genetic algorithm tutorial. Stat Comput 4:65–85

Zhdanov MS (1993) Tutorial: regularization in inversion theory. CWP-136, Colorado School of Mines, 47 pp

Zhdanov MS (2002) Geophysical inverse theory and regularization problems. Elsevier, 609 pp

Zhdanov MS (2015) Inverse theory and applications in geophysics. Elsevier, 704 pp

Chapter 7
Gradient-Type Methods of Nonlinear Inversion

Abstract This chapter presents a detailed description of the main methods for solving nonlinear inverse problems based on iterative minimization of the misfit or parametric functionals. The gradient-type methods discussed in this chapter include the steepest descent, the Newton, and the conjugate gradient methods. We also consider the application of the gradient-type methods to solving ill-posed nonlinear discrete inverse problems. The concept of integrated sensitivity is introduced, which plays an important role in the inversion. This chapter also presents a numerical comparison of three main minimization techniques—steepest descent, Newton, and conjugate gradient methods.

Keywords Steepest descent · Newton method · Conjugate gradient method · Integrated sensitivity

In previous chapters, we mainly considered the linear inverse problems. In a general case, however, the relationships between the data and the model parameters are nonlinear. This requires developing more general methods of nonlinear inversion. We found above that the solution of the linear discrete inverse problems can be presented in closed form using matrix inversion and matrix multiplications. However, there is no closed-form solution for nonlinear inverse problems. Most nonlinear inverse methods are based on iterative processes aimed at minimizing the misfit or parametric functionals (in the case of regularized inversion). Over the years, many iterative minimization methods have been developed. The gradient-type methods are most widely used, including the steepest descent, Newton, and conjugate gradient minimization techniques. We will review the basic gradient-type methods in this chapter, following the works by Zhdanov (1993, 2002, and 2015).

7.1 Method of Steepest Descent

Consider again the inverse problem

$$\mathbf{d} = A(\mathbf{m}), \tag{7.1}$$

M. S. Zhdanov, *Advanced Methods of Joint Inversion and Fusion of Multiphysics Data*, Advances in Geological Science,
https://doi.org/10.1007/978-981-99-6722-3_7

where $\mathbf{m} \in M$ is some element (vector) from a real Hilbert space M of the model parameters, $\mathbf{d} \in D$ is an element (vector) from a real Hilbert space D of data sets, and A is a *nonlinear operator.*

To simplify the initial description of the gradient-type methods, we first assume that problem (7.1) is well possed, which means that the solution exists, is unique, and stable. In this case, one can solve inverse problem (7.1) by minimizing the corresponding misfit functional between the observed and predicted data,

$$\phi(\mathbf{m}) = \|A(\mathbf{m}) - \mathbf{d}\|^2 = (A(\mathbf{m}) - \mathbf{d}, A(\mathbf{m}) - \mathbf{d}) = \min, \tag{7.2}$$

where (..., ...) means the inner product of the elements from Hilbert space D.

This minimization problem can be solved directly in the case of the linear operator A of forward modeling. However, in the general case of a nonlinear operator A, the solution can only be found iteratively. There are many different approaches to constructing the iterative process of functional minimization. One of the most widely used techniques for optimization is based on gradient-type methods.

We start our discussion with the most important and easily understandable method of steepest descent. This method is based on the concept of the descent condition, which we discuss below.

7.1.1 Descent Condition

We consider an iterative algorithm of misfit functional minimization. It is reasonable to build this algorithm on the idea that misfit decreases at every iteration $\mathbf{m}_n$. In other words, we impose the following *descent condition :*

$$\phi(\mathbf{m}_{n+1}) < \phi(\mathbf{m}_n) \text{ for all } n \geq 0. \tag{7.3}$$

A method that imposes this condition is called a *descent method.* The question is how to find iterations $\{\mathbf{m}_n\}$ which satisfies the descent condition.

To solve this problem, we calculate the first variation of the misfit functional at point $\mathbf{m}$:

$$\delta\phi(\mathbf{m}) = \delta\mu_D^2(A(\mathbf{m}), \mathbf{d}) = \delta(A(\mathbf{m}) - \mathbf{d}, A(\mathbf{m}) - \mathbf{d}) = 2(\delta A(\mathbf{m}), A(\mathbf{m}) - \mathbf{d}). \tag{7.4}$$

We also assume that operator A is a nonlinear one (in general cases) but that it is a differentiable operator so that

$$\delta A(\mathbf{m}) = F_m \delta\mathbf{m}. \tag{7.5}$$

Here, F_m is a linear operator called the *Fréchet derivative* of A (see Chap. 3).

By substituting Eq. (7.5) into (7.4), we obtain

$$\delta\phi(\mathbf{m}) = \delta\mu_D^2\left(A(\mathbf{m}), \mathbf{d}\right) = 2\left(F_m\delta\mathbf{m},\, A(\mathbf{m}) - \mathbf{d}\right). \tag{7.6}$$

Equation (7.6) can be simplified using the properties of the adjoint of the Fréchet derivative operator, considered in Chap. 3.

The notion of the adjoint operator makes it possible to move a linear operator of a Fréchet derivative, F_m, from the left to the right-hand side of the inner product in Eq. (7.6):

$$\delta\phi(\mathbf{m}) = 2\left(\delta\mathbf{m},\, F_m^\star(A(\mathbf{m}) - \mathbf{d})\right), \tag{7.7}$$

where $F_m^\star$ is the adjoint operator of the Fréchet derivative of A.

In order to satisfy the descent condition (7.3), we select

$$\delta\mathbf{m} = -k\mathbf{l}(\mathbf{m}), \tag{7.8}$$

where k is some positive real number and $\mathbf{l}(\mathbf{m})$ is a direction determined by the following formula:

$$\mathbf{l}(\mathbf{m}) = F_m^\star(A(\mathbf{m}) - \mathbf{d}). \tag{7.9}$$

Certainly, by substituting Eqs. (7.8) and (7.9) into (7.7), we have

$$\delta\phi(\mathbf{m}) = -2k(\mathbf{l}(\mathbf{m}), \mathbf{l}(\mathbf{m})) < 0, \tag{7.10}$$

so $\mathbf{l}(\mathbf{m})$ describes the "direction" of increasing (ascent) of the functional $\phi(\mathbf{m})$, because it is opposite to the descent direction, $\delta\mathbf{m}$.

Thus, we can construct the iterative process as follows:

$$\mathbf{m}_{n+1} = \mathbf{m}_n + \delta\mathbf{m}_n = \mathbf{m}_n - k_n\mathbf{l}(\mathbf{m}_n), \tag{7.11}$$

where k_n is a step length on the n-th iteration.

Iterations (7.11) will satisfy the descent condition (7.3) if the corresponding step lengths, k_n, are appropriately selected. The traditional way of selecting k_n is based on the following condition:

$$\phi\left(\mathbf{m}_{n+1}\right) = \phi\left(\mathbf{m}_n - k_n\mathbf{l}(\mathbf{m}_n)\right) = \Phi(k_n) = \min, \tag{7.12}$$

where the minimum is determined with respect to k_n. The last condition allows us to define a step length along the direction of the steepest descent.

Figure 7.1 shows schematically the plot of the misfit functional value as a function of model parameters $\mathbf{m}$. The vector of the steepest ascent, $\mathbf{l}_n = \mathbf{l}(\mathbf{m}_n)$, shows the direction of "climbing on the hill" along the misfit functional surface. If we are at the point on this surface which corresponds to the nth iteration, we will be moving "downhill" using formula (7.11) along the steepest descent direction to reach the $(n+1)$th iteration.

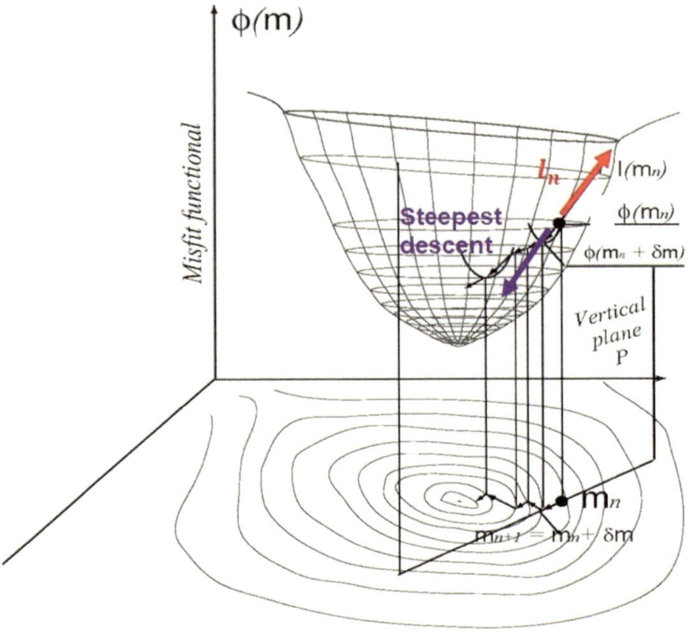

Fig. 7.1 The plot of the misfit functional value as a function of model parameters **m**. The bold red vector of the steepest ascent, $\mathbf{l}_n$, shows the direction of "climbing on the hill" along the misfit functional surface. The intersection between the vertical plane *P* drawn through the direction of the steepest ascent at point $\mathbf{m}_n$ and the misfit functional surface is shown by a solid parabola-type curve. The steepest descent step begins at a point $\phi\,(\mathbf{m}_n)$ and ends at a point $\phi\,(\mathbf{m}_{n+1})$ at the minimum of this curve. The second parabola-type curve is drawn for one of the subsequent iteration points. Repeating the steepest descent iterations, we move along the set of mutually orthogonal segments, as shown by the solid arrows in the space *M* of the model parameters (modified from Zhdanov 2015)

The iterative process (7.11) together with the condition (7.12) gives us a numerical scheme for the steepest descent method applied to misfit functional minimization.

7.1.2 Linear Line Search

We will now discuss the problem of determining the length of the step, k_n. Over the years, numerous line search algorithms have been proposed based on the optimization of the functional $\Phi(k_n)$:

$$\begin{aligned}\Phi(k_n) &= \phi\,(\mathbf{m}_{n+1}) = \phi\,(\mathbf{m}_n - k_n\mathbf{l}(\mathbf{m}_n)) \\ &= (A(\mathbf{m}_n - k_n\mathbf{l}(\mathbf{m}_n)) - \mathbf{d},\ A(\mathbf{m}_n - k_n\mathbf{l}(\mathbf{m}_n)) - \mathbf{d}) = (\mathbf{r}_{n+1},\ \mathbf{r}_{n+1}) = \min,\end{aligned} \tag{7.13}$$

where

$$\mathbf{r}_{n+1} = A(\mathbf{m}_n - k_n \mathbf{l}(\mathbf{m}_n)) - \mathbf{d},$$

is a residual vector at the $(n+1)$th iteration.

We now consider how to find the minimum of the last functional with respect to k_n. Let us calculate the first variation of $\Phi(k_n)$ and equate it to zero at a minimum point:

$$\delta\Phi(k_n) = -2\delta k_n (F_{m_n}\mathbf{l}(\mathbf{m}_n),\ A(\mathbf{m}_n - k_n \mathbf{l}(\mathbf{m}_n)) - \mathbf{d}) = 0. \tag{7.14}$$

The simplest technique of the line search arises if one assumes that in the last equation $k_n \mathbf{l}(\mathbf{m}_n)$ is small enough that one can use a linearized representation of the operator $A(\mathbf{m}_n - k_n \mathbf{l}(\mathbf{m}_n)$:

$$A\left(\mathbf{m}_n - k_n \mathbf{l}(\mathbf{m}_n)\right) \approx A(\mathbf{m}_n) - k_n F_{m_n}\mathbf{l}(\mathbf{m}_n), \tag{7.15}$$

where F_{m_n} is Fréchet derivative operator.

Substituting (7.15) into (7.14), we have

$$\delta\Phi(k_n) = -2\delta k_n (F_{m_n}\mathbf{l}(\mathbf{m}_n),\ A(\mathbf{m}_n) - k_n F_{m_n}\mathbf{l}(\mathbf{m}_n) - \mathbf{d}) = 0. \tag{7.16}$$

Solving the last equation, we finally obtain

$$k_n = \frac{\left(F_{m_n}\mathbf{l}(\mathbf{m}_n),\ A(\mathbf{m}_n) - \mathbf{d}\right)}{\left(F_{m_n}\mathbf{l}(\mathbf{m}_n),\ F_{m_n}\mathbf{l}(\mathbf{m}_n)\right)} = \frac{\left(\mathbf{l}(\mathbf{m}_n),\ F^{\star}_{m_n}(A(\mathbf{m}_n) - \mathbf{d})\right)}{\left\| F_{m_n}\mathbf{l}(\mathbf{m}_n)\right\|^2}$$

$$= \frac{(\mathbf{l}(\mathbf{m}_n),\ \mathbf{l}(\mathbf{m}_n))}{\left\| F_{m_n}\mathbf{l}(\mathbf{m}_n)\right\|^2} = \frac{\|\mathbf{l}(\mathbf{m}_n)\|^2}{\left\| F_{m_n}\mathbf{l}(\mathbf{m}_n)\right\|^2}, \tag{7.17}$$

where $F^*_{m_n}$ is adjoint Fréchet derivative operator.

Formula (7.17) works well if A is close to being a linear operator so that representation (7.16) holds for every iteration step. That is why this technique may be referred to as *a linear line search.*

It can be demonstrated that, if we apply the steepest descent method with the line search, the subsequent gradient directions are mutually orthogonal

$$\left(\mathbf{l}(\mathbf{m}_{n+1}), \mathbf{l}(\mathbf{m}_n)\right) = 0. \tag{7.18}$$

This result comes from the simple fact that, if we minimize a functional along some direction, described by a parametric line, the direction of the steepest ascent must be perpendicular to this line at the minimum point on the line; otherwise, we would still not be reaching the minimum along this line. A formal proof of this result was presented in many textbooks on minimization methods (e.g., Press et al. 1987; Zhdanov 2015).

7.2 The Newton Method

7.2.1 Hessian Operator

The main idea of the Newton method is to try to solve the problem of minimization in one step:

$$\mathbf{m}_1 = \mathbf{m}_0 + \Delta\mathbf{m}. \tag{7.19}$$

Thus, instead of moving downhill along a long path formed by mutually orthogonal directions of the steepest descent, one can try to reach the minimum of the misfit functional along one direction.

To determine this specific direction, $\Delta\mathbf{m}$, let us calculate the misfit functional for this first iteration

$$\begin{aligned}\phi(\mathbf{m}_1) &= (A(\mathbf{m}_1) - \mathbf{d},\ A(\mathbf{m}_1) - \mathbf{d}) \\ &= (A(\mathbf{m}_0 + \Delta\mathbf{m}) - \mathbf{d},\ A(\mathbf{m}_0 + \Delta\mathbf{m}) - \mathbf{d}).\end{aligned} \tag{7.20}$$

The first variation of the misfit functional with respect to $\Delta\mathbf{m}$ is equal to the following:

$$\begin{aligned}\delta_{\Delta\mathbf{m}}\phi(\mathbf{m}_1) &= 2(\delta_{\Delta\mathbf{m}}A(\mathbf{m}_0 + \Delta\mathbf{m}),\ A(\mathbf{m}_0 + \Delta\mathbf{m}) - \mathbf{d}) \\ &= 2(F_{m_0}\delta\Delta\mathbf{m},\ A(\mathbf{m}_0 + \Delta\mathbf{m}) - \mathbf{d}).\end{aligned} \tag{7.21}$$

Using an adjoint operator for the Fréchet derivative, we find

$$\delta_{\Delta\mathbf{m}}\phi(\mathbf{m}_1) = 2(\delta\Delta\mathbf{m},\ F^{\star}_{m_0}[A(\mathbf{m}_0 + \Delta\mathbf{m}) - \mathbf{d}]).$$

Note that, according to Theorem 3.58 of Chap. 3, the first variation of the misfit functional at the minimum must be equal to zero:

$$\delta\phi(\mathbf{m}_1) = 2(\delta\Delta\mathbf{m},\ F^{\star}_{m_0}[A(\mathbf{m}_0 + \Delta\mathbf{m}) - \mathbf{d}]) = 0, \tag{7.22}$$

and Eq. (7.22) must hold for any variation $\delta\Delta\mathbf{m}$. For example, we can select $\delta\Delta\mathbf{m}$ as follows:

$$\delta\Delta\mathbf{m} = F^{\star}_{m_0}[A(\mathbf{m}_0 + \Delta\mathbf{m}) - \mathbf{d}]. \tag{7.23}$$

Substituting Eq. (7.23) back into (7.22), we have:

$$(F^{\star}_{m_0}[A(\mathbf{m}_0 + \Delta\mathbf{m}) - \mathbf{d}],\ F^{\star}_{m_0}[A(\mathbf{m}_0 + \Delta\mathbf{m}) - \mathbf{d}]) = 0.$$

Therefore, the second multiplier in (7.22) is equal to zero as well:

$$F^{*}_{m_0}[A(\mathbf{m}_0 + \Delta\mathbf{m}) - \mathbf{d}] = 0. \tag{7.24}$$

It is difficult to find the exact solution of operator Eq. (7.24). However, one can simplify this problem by linearization of operator $A(\mathbf{m}_0 + \Delta\mathbf{m})$, using a Fréchet derivative operator:

$$A(\mathbf{m}_0 + \Delta\mathbf{m}) \approx A(\mathbf{m}_0) + F_{m_0}\Delta\mathbf{m}. \tag{7.25}$$

Substituting (7.25) into (7.24), we obtain

$$F^{\star}_{m_0}[A(\mathbf{m}_0) + F_{m_0}\Delta\mathbf{m} - \mathbf{d}] = 0.$$

From the last equation, we find immediately that

$$F^{\star}_{m_0}F_{m_0}\Delta\mathbf{m} = -F^{\star}_{m_0}[A(\mathbf{m}_0) - \mathbf{d}]. \tag{7.26}$$

According to (7.9),

$$F^{\star}_{m_0}[A(\mathbf{m}_0) - \mathbf{d}] = \mathbf{l}(\mathbf{m}_0),$$

so, from Eq. (7.26), we have *the normal equation* for the optimum step

$$\frac{1}{2}H_{m_0}\Delta\mathbf{m} = -\mathbf{l}(\mathbf{m}_0), \tag{7.27}$$

where

$$H_{m_0} = 2F^{\star}_{m_0}F_{m_0}. \tag{7.28}$$

Operator H_{m_0} is a quasi-Hessian operator called because it does not take into account the second variation $F^{(2)}_{m_0}$ of operator A (see expression (3.163) of Chap. 3). This operator is linear one. If there exists an inverse quasi-Hessian operator $H^{-1}_{m_0}$, one can solve equation (7.27) as follows:

$$\Delta\mathbf{m} = -\,2H^{-1}_{m_0}\mathbf{l}(\mathbf{m}_0). \tag{7.29}$$

Substituting (7.29) into (7.19), we finally define the update $\mathbf{m}_1$ of the initial model:

$$\mathbf{m}_1 = \mathbf{m}_0 - 2H^{-1}_{m_0}\mathbf{l}(\mathbf{m}_0). \tag{7.30}$$

Note that expression (7.30) produces the final solution of the inverse problem only in the case of a linear forward operator A. In this case, expression (7.30) takes the form

$$\mathbf{m}_1 = \mathbf{m}_0 - \left(F^{\star}_{m_0}F_{m_0}\right)^{-1}\mathbf{l}(\mathbf{m}_0) = \mathbf{m}_0 - \left(A^{\star}A\right)^{-1}\mathbf{l}(\mathbf{m}_0). \tag{7.31}$$

Of course, it is usually not enough to use only one iteration for the solution of a nonlinear inverse problem in the framework of Newton method (because we used the linearized approximation (7.25)). However, we can construct an iterative process based on relationship (7.30):

$$\mathbf{m}_{n+1} = \mathbf{m}_n - 2H_{m_n}^{-1}\mathbf{l}(\mathbf{m}_n).$$

Thus, the algorithm of the Newton method can be summarized as follows:

$$\begin{aligned} \mathbf{r}_n &= A(\mathbf{m}_n) - \mathbf{d}, && (a) \\ \mathbf{l}_n &= \mathbf{l}(\mathbf{m}_n) = F_{m_n}^{\star}\mathbf{r}_n, && (b) \\ \mathbf{m}_{n+1} &= \mathbf{m}_n - 2H_{m_n}^{-1}\mathbf{l}_n. && (c) \end{aligned} \tag{7.32}$$

The iterative process (7.32) is terminated at $n = N$ when the misfit reaches the given level ε_0:

$$\phi(\mathbf{m}_N) = \|\mathbf{r}_N\|^2 \leq \varepsilon_0.$$

7.2.2 The Newton Method with the Line Search

Note that in general cases of an arbitrary nonlinear operator A, algorithm (7.32) may not converge (see for details Fletcher 1995), and, in fact, $\phi(\mathbf{m}_n)$ may not even decrease with the iteration number n. This undesirable possibility can be eliminated by introducing a line search at every step of the Newton method:

$$\mathbf{m}_{n+1} = \mathbf{m}_n - k_n H_{m_n}^{-1}\mathbf{l}_n. \tag{7.33}$$

As in the steepest descent method, the length of the Newton step, k_n, is determined from the condition that

$$\Phi(k_n) = \phi(\mathbf{m}_{n+1}) = \phi\left(\mathbf{m}_n - k_n H_{m_n}^{-1}\mathbf{l}_n\right) = \min. \tag{7.34}$$

Applying a linear line search to the last problem, we obtain the optimum length of the step equal to the following:

$$k_n = \frac{\left(H_{m_n}^{-1}\mathbf{l}(\mathbf{m}_n), \mathbf{l}(\mathbf{m}_n)\right)}{\left\| F_{m_n} H_{m_n}^{-1}\mathbf{l}(\mathbf{m}_n)\right\|^2}. \tag{7.35}$$

Algorithm of the Newton method with the linear line search can be summarized as follows:

$$\begin{aligned} \mathbf{r}_n &= A(\mathbf{m}_n) - \mathbf{d}, && (a) \\ \mathbf{l}_n &= \mathbf{l}(\mathbf{m}_n) = F_{m_n}^{\star}\mathbf{r}_n, \ \ \mathbf{g}_n = F_{m_n} H_{m_n}^{-1}\mathbf{l}_n && (b) \\ k_n &= \left(H_{m_n}^{-1}\mathbf{l}(\mathbf{m}_n), \mathbf{l}(\mathbf{m}_n)\right) / \|\mathbf{g}_n\|^2, && (c) \\ \mathbf{m}_{n+1} &= \mathbf{m}_n - k_n H_{m_n}^{-1}\mathbf{l}_n. && (d) \end{aligned} \tag{7.36}$$

Iterative process (7.36) is terminated at $n = N$ when the misfit reaches the given level ε_0 :

$$\phi(\mathbf{m}_N) = \|\mathbf{r}_N\|^2 \leq \varepsilon_0.$$

The advantage of the Newton method is that it converges much faster than the steepest descent method. However, each iteration of the Newton method is very expensive computationally because it requires the calculation of the inverse Hessian matrix, while the steepest descent iterations involve matrix multiplication only. Therefore, it is desirable to develop an algorithm combining the simplicity of iterations by the steepest descent method with the few iteration steps, similar to Newton's method. I will show in the next section that this can be achieved by the conjugate gradient method.

7.3 The Conjugate Gradient Method

The slow convergence of the steepest descent method is related to the fact that every two subsequent gradient directions are mutually orthogonal according to Eq. (7.18). As a result, the iteration path in the model space resembles a zigzag line. Our goal is to develop an iterative method with better convergence than the steepest descent algorithm. This can be achieved by straightening the zigzag line, which is done by moving along the so-called conjugate gradient directions in the model space, which we will define below.

The conjugate gradient method is based on the same ideas as the steepest descent, and the iteration process is very similar to the last one:

$$\mathbf{m}_{n+1} = \mathbf{m}_n + \Delta\mathbf{m}_n = \mathbf{m}_n - \tilde{k}_n \tilde{\mathbf{l}}(\mathbf{m}_n),$$

where

$$\Delta\mathbf{m}_n = -\tilde{k}_n \tilde{\mathbf{l}}(\mathbf{m}_n). \tag{7.37}$$

However, the "directions" of ascent $\tilde{\mathbf{l}}(\mathbf{m}_n)$ are selected in a different way to achieve faster convergence. In the first step, we use the "direction" of the steepest ascent:

$$\tilde{\mathbf{l}}(\mathbf{m}_0) = \mathbf{l}(\mathbf{m}_0).$$

In the next step, the "direction" of ascent is a linear combination of the steepest ascent on this step and the "direction" of ascent $\tilde{\mathbf{l}}(\mathbf{m}_0)$ on the previous step:

$$\tilde{\mathbf{l}}(\mathbf{m}_1) = \mathbf{l}(\mathbf{m}_1) + \beta_1 \tilde{\mathbf{l}}(\mathbf{m}_0).$$

In the n-th step

$$\tilde{\mathbf{l}}(\mathbf{m}_{n+1}) = \mathbf{l}(\mathbf{m}_{n+1}) + \beta_{n+1} \tilde{\mathbf{l}}(\mathbf{m}_n).$$

The steps $\tilde{k}_n$ are selected, as usual, by a line search to minimize the misfit functional:

$$\phi(\mathbf{m}_{n+1}) = \phi(\mathbf{m}_n - \tilde{k}_n \tilde{\mathbf{l}}(\mathbf{m}_n)) = \Phi(\tilde{k}_n) = \min. \tag{7.38}$$

7.3.1 A Linear Line Search in the Conjugate Gradient Method

To solve problem (7.38), we consider more carefully functional $\Phi(\widetilde{k}_n)$:

$$\Phi(\widetilde{k}_n) = \phi\left(\mathbf{m}_{n+1}\right) = \phi\left(\mathbf{m}_n - \widetilde{k}_n\tilde{\mathbf{l}}(\mathbf{m}_n)\right)$$
$$= (A(\mathbf{m}_n - \widetilde{k}_n\tilde{\mathbf{l}}(\mathbf{m}_n)) - \mathbf{d},\ A(\mathbf{m}_n - \widetilde{k}_n\tilde{\mathbf{l}}(\mathbf{m}_n)) - \mathbf{d}). \tag{7.39}$$

Let us find the minimum of the last functional with respect to $\widetilde{k}_n$. We calculate now the first variation of $\Phi(\widetilde{k}_n)$:

$$\delta\Phi(\widetilde{k}_n) = -2\delta\widetilde{k}_n(F_{m_n}\tilde{\mathbf{l}}(\mathbf{m}_n),\ A(\mathbf{m}_n - \widetilde{k}_n\tilde{\mathbf{l}}(\mathbf{m}_n)) - \mathbf{d}). \tag{7.40}$$

In the last equation, we assume that $\widetilde{k}_n\tilde{\mathbf{l}}(\mathbf{m}_n)$ is small enough that we can use a linearized representation for operator $A(\mathbf{m}_n - \widetilde{k}_n\tilde{\mathbf{l}}(\mathbf{m}_n))$:

$$A(\mathbf{m}_n - \widetilde{k}_n\tilde{\mathbf{l}}(\mathbf{m}_n)) \approx A(\mathbf{m}_n) - \widetilde{k}_n F_{m_n}\tilde{\mathbf{l}}(\mathbf{m}_n). \tag{7.41}$$

Substituting (7.41) into (7.40), we have

$$\delta\Phi(\widetilde{k}_n) = -2\delta\widetilde{k}_n(F_{m_n}\tilde{\mathbf{l}}(\mathbf{m}_n), A(\mathbf{m}_n) - \widetilde{k}_n F_{m_n}\tilde{\mathbf{l}}(\mathbf{m}_n) - \mathbf{d}) = 0. \tag{7.42}$$

Solving the last equation, we finally find

$$\widetilde{k}_n = \frac{(F_{m_n}\tilde{\mathbf{l}}(\mathbf{m}_n),\ A(\mathbf{m}_n) - \mathbf{d})}{(F_{m_n}\tilde{\mathbf{l}}(\mathbf{m}_n),\ F_{m_n}\tilde{\mathbf{l}}(\mathbf{m}_n))} = \frac{(F_{m_n}\tilde{\mathbf{l}}(\mathbf{m}_n),\ A(\mathbf{m}_n) - \mathbf{d})}{\left\| F_{m_n}\tilde{\mathbf{l}}(\mathbf{m}_n)\right\|^2}$$
$$= \frac{(\tilde{\mathbf{l}}(\mathbf{m}_n),\ F^{\star}_{m_n}[A(\mathbf{m}_n) - \mathbf{d}])}{\left\| F_{m_n}\tilde{\mathbf{l}}(\mathbf{m}_n)\right\|^2} = \frac{(\tilde{\mathbf{l}}(\mathbf{m}_n),\ \mathbf{l}\,(\mathbf{m}_n))}{\left\| F_{m_n}\tilde{\mathbf{l}}(\mathbf{m}_n)\right\|^2}. \tag{7.43}$$

7.3.2 Determining the Conjugate Directions

There are different ways of defining coefficients β_n. The basic idea is to make directions $\tilde{\mathbf{l}}(\mathbf{m}_{n+1})$ and $\tilde{\mathbf{l}}(\mathbf{m}_n)$ "conjugate" in some geometrical sense, which would guarantee much faster convergence of the iterations. The conjugate directions are defined as follows:

Definition 7.1 Vectors $\mathbf{u}$ and $\mathbf{v}$ are said to be conjugate if they satisfy the condition

$$(\mathbf{u}, H\mathbf{v}) = 0, \tag{7.44}$$

where H is a linear operator.

Consider first, for simplicity, a linear inverse problem. Suppose that we have moved from point $\mathbf{m}_n$ to point $\mathbf{m}_{n+1}$ in the space of models. The change in the gradient directions can be described by the following formula:

$$\gamma_n = \mathbf{l}(\mathbf{m}_{n+1}) - \mathbf{l}(\mathbf{m}_n)$$

$$= A^\star[A(\mathbf{m}_{n+1}) - \mathbf{d}] - A^\star[A(\mathbf{m}_n) - \mathbf{d}] = A^\star A \Delta\mathbf{m}_n = H_{m_n}\Delta\mathbf{m}_n, \tag{7.45}$$

where

$$H_{m_n} = A^* A.$$

Note that formula (7.45) holds approximately for a nonlinear operator as well:

$$\gamma_n = \mathbf{l}(\mathbf{m}_{n+1}) - \mathbf{l}(\mathbf{m}_n)$$

$$= F^\star_{m_{n+1}}[A(\mathbf{m}_{n+1}) - \mathbf{d}] - F^\star_{m_n}[A(\mathbf{m}_n) - \mathbf{d}]$$

$$\approx F^\star_{m_n} F_{m_n}\Delta\mathbf{m}_n = H_{m_n}\Delta\mathbf{m}_n, \tag{7.46}$$

where $H_{m_n} = F^\star_{m_n} F_{m_n}$ is a *Hessian operator.*

It has been demonstrated above that, if we apply the steepest descent method with the line search, the subsequent gradient directions are mutually orthogonal

$$\left(\mathbf{l}(\mathbf{m}_{n+1}), \mathbf{l}(\mathbf{m}_n)\right) = 0. \tag{7.47}$$

In the framework of the conjugate gradient method, we require that the vectors $\tilde{\mathbf{l}}(\mathbf{m}_n)$ introduced above,

$$\tilde{\mathbf{l}}(\mathbf{m}_n) = \mathbf{l}(\mathbf{m}_n) + \beta_n \tilde{\mathbf{l}}(\mathbf{m}_{n-1}),$$

to be mutually conjugate with operator H equal to Hessian operator at the current iteration. In other words, we require that the following equation holds:

$$\left(\tilde{\mathbf{l}}(\mathbf{m}_n), H_{m_n}\tilde{\mathbf{l}}(\mathbf{m}_{n-1})\right) = 0, \tag{7.48}$$

which is equivalent to the condition

$$\left(\tilde{\mathbf{l}}(\mathbf{m}_n), \gamma_{n-1}\right) = 0. \tag{7.49}$$

Indeed,

$$\left(\tilde{\mathbf{l}}(\mathbf{m}_n), \gamma_{n-1}\right) = \left(\tilde{\mathbf{l}}(\mathbf{m}_n), H_{m_{n-1}} \Delta \mathbf{m}_{n-1}\right) = -\tilde{k}_{n-1} \left(\tilde{\mathbf{l}}(\mathbf{m}_n), H_{m_{n-1}} \tilde{\mathbf{l}}(\mathbf{m}_{n-1})\right) = 0,$$

which we took into account Eq. (7.37). The last formula proves Eq. (7.49).

To simplify our analysis, we will use the notations

$$\mathbf{l}(\mathbf{m}_n) = \mathbf{l}_n, \ \tilde{\mathbf{l}}(\mathbf{m}_n) = \tilde{\mathbf{l}}_n. \tag{7.50}$$

To satisfy condition (7.49), let us calculate

$$\left(\tilde{\mathbf{l}}_n, \gamma_{n-1}\right) = \left(\mathbf{l}_n + \beta_n \tilde{\mathbf{l}}_{n-1}, \mathbf{l}_n - \mathbf{l}_{n-1}\right)$$

$$= (\mathbf{l}_n, \mathbf{l}_n) - \left(\mathbf{l}_n, \mathbf{l}_{n-1}\right) + \beta_n \left(\tilde{\mathbf{l}}_{n-1}, \mathbf{l}_n\right) - \beta_n \left(\tilde{\mathbf{l}}_{n-1}, \mathbf{l}_{n-1}\right)$$

$$= (\mathbf{l}_n, \mathbf{l}_n) - \beta_n \left(\tilde{\mathbf{l}}_{n-1}, \mathbf{l}_{n-1}\right) = 0,$$

because

$$\left(\mathbf{l}_n, \mathbf{l}_{n-1}\right) = 0 \tag{7.51}$$

and

$$\left(\tilde{\mathbf{l}}_{n-1}, \mathbf{l}_n\right) = 0. \tag{7.52}$$

Equation (7.51) follows from (7.47), and Eq. (7.52) holds because, in the previous step, we moved along the search line in the direction $\tilde{\mathbf{l}}_{n-1}$ to the minimum, so the steepest descent direction $\mathbf{l}_n$ at the minimum point will be perpendicular to $\tilde{\mathbf{l}}_{n-1}$.

Also, it can be shown that

$$\left(\tilde{\mathbf{l}}_{n-1}, \mathbf{l}_{n-1}\right) = \left(\mathbf{l}_{n-1} + \beta_{n-1} \tilde{\mathbf{l}}_{n-2}, \mathbf{l}_{n-1}\right)$$

$$= \left(\mathbf{l}_{n-1}, \mathbf{l}_{n-1}\right) + \beta_{n-1} \left(\tilde{\mathbf{l}}_{n-2}, \mathbf{l}_{n-1}\right) = \left(\mathbf{l}_{n-1}, \mathbf{l}_{n-1}\right), \tag{7.53}$$

because

$$\left(\tilde{\mathbf{l}}_{n-2}, \mathbf{l}_{n-1}\right) = 0.$$

Therefore,

$$(\mathbf{l}_n, \mathbf{l}_n) - \beta_n \left(\tilde{\mathbf{l}}_{n-1}, \mathbf{l}_{n-1}\right) = (\mathbf{l}_n, \mathbf{l}_n) - \beta_n \left(\mathbf{l}_{n-1}, \mathbf{l}_{n-1}\right) = 0,$$

and we finally determine β_n :

$$\beta_n = \frac{(\mathbf{l}_n, \mathbf{l}_n)}{\left(\mathbf{l}_{n-1}, \mathbf{l}_{n-1}\right)} = \frac{(\mathbf{l}(\mathbf{m}_n), \mathbf{l}(\mathbf{m}_n))}{\left(\mathbf{l}(\mathbf{m}_{n-1}), \mathbf{l}(\mathbf{m}_{n-1})\right)} = \frac{\|\mathbf{l}(\mathbf{m}_n)\|^2}{\left\|\mathbf{l}(\mathbf{m}_{n-1})\right\|^2}. \tag{7.54}$$

Note that there are several other popular techniques for determining the coefficients β_n, which I do not describe here. Instead, I refer interested readers to Tarantola's (1987) and Fletcher's (1995) books.

The algorithm of the conjugate gradient method described above has been substantiated for a linear inverse problem. For example, in the simplest case, when operator A is a linear operator with a square matrix of order N, it can be proved that the conjugate gradient algorithm will give an exact solution of the inverse problem in N iterations (Fletcher, 1995). This algorithm can also be used, similarly to the Newton method, for the solution of the nonlinear inverse problem. In general nonlinear cases, the number of iterations is not fixed, but the method converges very rapidly.

7.4 Comparison of Three Main Minimization Methods

7.4.1 Three Main Methods of Misfit Functional Minimization

In summary, we can compare three main methods of misfit functional minimization introduced above.

1. The steepest descent method is the simplest minimization technique, with every iteration computed by simple matrix multiplication. However, this method requires a large number of iterations to reach the minimum due to the complex zigzag-like path in the model space and the short length of each iteration step.

2. The Newton method makes it possible to reach the minimum in one or a few large "jumps" from the initial model. However, this is achieved due to the high computational cost of finding the inverse Hessian on every iteration.

3. The conjugate gradient method combines the advantages of both approaches, the steepest descent, and the Newton method. It uses matrix multiplication only for computing each iteration step. At the same time, it also requires fewer iterations than the steepest descent due to the straightened path to the minimum in the model space. This is achieved by moving along the conjugate directions instead of mutually perpendicular directions of the steepest descent method.

We will illustrate the points listed above with a very simple example of the solution to the nonlinear inverse problem.

7.4.2 Numerical Example

We consider the following system of nonlinear equations:

$$\begin{aligned} x^3 + y^2 &= 5, \\ x^2 - y &= -1, \\ -2x + 2y^2 &= 6. \end{aligned} \tag{7.55}$$

This system can be written in operator notations as follows:

$$A(\mathbf{m}) = \mathbf{d}, \tag{7.56}$$

where $\mathbf{m}$ is the vector column containing model parameters, x and y, and $\mathbf{d}$ is the vector column of the right-hand sides of these equations, which we consider the data,

$$\mathbf{m} = \begin{bmatrix} x \\ y \end{bmatrix}, \quad \mathbf{d} = \begin{bmatrix} 5 \\ -1 \\ 6 \end{bmatrix}. \tag{7.57}$$

In Eq. (7.56), A is a non-linear operator acting on two input parameters, x and y, and producing three output data values $\mathbf{d}$ (the predicted data values):

$$A(x, y) = A(\mathbf{m}) = \begin{bmatrix} x^3 + y^2 \\ x^2 - y \\ -2x + 2y^2 \end{bmatrix} = \begin{bmatrix} 5 \\ -1 \\ 6 \end{bmatrix}. \tag{7.58}$$

Our goal is to solve the inverse problem described by operator Eq. (7.56). Because the system of Eqs. (7.55) is overdetermined (there are three data values and only two unknown parameters), the exact solution may not exist. Instead, we are looking for a least-squares solution, which delivers the minimum of the misfit functional $\varphi(x, y)$, defined as the norm square of the residual, $\mathbf{r} = A(\mathbf{m}) - \mathbf{d}$, between the predicted and observed data:

$$\varphi(\mathbf{m}) = \|\mathbf{r}\|^2 = ||A(\mathbf{m}) - \mathbf{d}||^2. \tag{7.59}$$

Using Euclidean norm, we can write the expression for the misfit functional in the following form:

$$\varphi(x, y) = (x^3 + y^2 - 5)^2 + (x^2 - y + 1)^2 + (-2x + 2y^2 - 6)^2. \tag{7.60}$$

Thus, the inverse problem is reduced to the following minimization problem:

$$\varphi(\mathbf{m}) = ||A(\mathbf{m}) - \mathbf{d}||^2 = \min. \tag{7.61}$$

We will now apply all three gradient-type methods introduced above to solve this problem.

7.4.2.1 The Steepest Decent Method

It was shown that the steepest descent method is based on the linearization of the nonlinear operator A in some vicinity of point $\mathbf{m}$:

$$A(\mathbf{m}+\delta\mathbf{m}) \approx A(\mathbf{m}) + F\delta\mathbf{m}, \tag{7.62}$$

where F is the Fréchet derivative operator at point $\mathbf{m}$, and $\delta\mathbf{m}$ is a variation of the model parameters. In our case, the linear Fréchet derivative operator can be represented by a matrix, $\mathbf{F}$, which consists of the partial derivatives of data with respect to the model parameters:

$$\mathbf{F} = \begin{bmatrix} \frac{\partial d_1}{\partial m_1} & \frac{\partial d_1}{\partial m_2} \\ \frac{\partial d_2}{\partial m_1} & \frac{\partial d_2}{\partial m_2} \\ \frac{\partial d_3}{\partial m_1} & \frac{\partial d_3}{\partial m_2} \end{bmatrix} = \begin{bmatrix} \frac{\partial d_1}{\partial x} & \frac{\partial d_1}{\partial y} \\ \frac{\partial d_2}{\partial x} & \frac{\partial d_2}{\partial y} \\ \frac{\partial d_3}{\partial x} & \frac{\partial d_3}{\partial y} \end{bmatrix}. \tag{7.63}$$

Taking into account expression (7.58) for operator A, we calculate the Fréchet derivative matrix as follows:

$$\mathbf{F}(x, y) = \begin{bmatrix} \frac{\partial}{\partial x}(x^3 + y^2) & \frac{\partial}{\partial y}(x^3 + y^2) \\ \frac{\partial}{\partial x}(x^2 - y) & \frac{\partial}{\partial y}(x^2 - y) \\ \frac{\partial}{\partial x}(-2x + 2y^2) & \frac{\partial}{\partial y}(-2x + 2y^2) \end{bmatrix} = \begin{bmatrix} 3x^2 & 2y \\ 2x & -1 \\ -2 & 4y \end{bmatrix}. \tag{7.64}$$

Using matrix notations, we can re-write the iterative process as follows:

$$\mathbf{m}_{n+1} = \mathbf{m}_n + \Delta\mathbf{m}_n, \tag{7.65}$$

where

$$\Delta\mathbf{m}_n = -k_n \mathbf{F}_n^T \mathbf{r}_n, \tag{7.66}$$

$$\mathbf{r}_n = A(\mathbf{m}_n) - \mathbf{d}, \tag{7.67}$$

and $\mathbf{F}_n^T$ is the transposed Fréchet derivative matrix at iteration n.

We begin the iterations with a starting point $\mathbf{m}_0$, compute Fréchet derivative at this point, find the residual vector using (7.67), find optimal model perturbation $\Delta\mathbf{m}_n$ using (7.66), and update model to $\mathbf{m}_1$ using formula (7.65). Then, we repeat the process to update the model to $\mathbf{m}_2$, etc.

Thus, the numerical scheme of the steepest descent method with linear line search can be described by the following formulae:

$$\begin{aligned}
\mathbf{r}_n &= A(\mathbf{m}_n) - \mathbf{d}, && (a)\\
\mathbf{l}_n &= \mathbf{l}(\mathbf{m}_n) = \mathbf{F}_n^T \mathbf{r}_n, && (b)\\
\mathbf{g}_n &= \mathbf{F}_n \mathbf{l}_n, && (c)\\
k_n &= \|\mathbf{l}_n\|^2 / \|\mathbf{g}_n\|^2, && (d)\\
\mathbf{m}_{n+1} &= \mathbf{m}_n - k_n \mathbf{l}_n. && (e)
\end{aligned} \tag{7.68}$$

The iterative process (7.68) is terminated when the misfit reaches the given level ε_0 :

$$\varphi(\mathbf{m}_N) = \|\mathbf{r}_N\|^2 \leq \varepsilon_0, \tag{7.69}$$

or using percent error:

$$PE = \frac{\|\mathbf{r}_n\|}{\|\mathbf{d}\|} \times 100\%. \tag{7.70}$$

7.4.2.2 Newton Method

Newton method aims at solving the inverse problem in one or a very few iterations. It is based on the linearization of the nonlinear operator A in some vicinity of point $\mathbf{m}$:

$$A(\mathbf{m}+\Delta\mathbf{m}) \approx A(\mathbf{m}) + \mathbf{F}\Delta\mathbf{m}, \tag{7.71}$$

where $\mathbf{F}$ is the Fréchet derivative matrix at point $\mathbf{m}$, and $\Delta\mathbf{m}$ is a variation of the model parameters. In our case, Eq. (7.64) represents the Fréchet derivative matrix, $\mathbf{F}$, of the forward modeling operator.

Using matrix notations, we can write Newton's iterative process as follows:

$$\mathbf{m}_{n+1} = \mathbf{m}_n + \Delta\mathbf{m}_n, \tag{7.72}$$

where

$$\Delta\mathbf{m}_n = -\left(\mathbf{F}_n^T \mathbf{F}_n\right)^{-1} \mathbf{F}_n^T \mathbf{r}_n, \tag{7.73}$$

and

$$\mathbf{r}_n = A(\mathbf{m}_n) - \mathbf{d}. \tag{7.74}$$

We begin the iterations with a starting point $\mathbf{m}_0$, compute Fréchet derivative at this point, find the residual vector using (7.74), find optimal model perturbation $\Delta\mathbf{m}_n$ using (7.73), and update model to $\mathbf{m}_1$ using formula (7.72). Then, we repeat the process to update the model to $\mathbf{m}_2$, etc.

To improve the convergence of the Newton method, one should apply the line search on every step of Newton's iterations. The algorithm of the Newton method with the line search can be summarized as follows:

$$\begin{aligned} \mathbf{r}_n &= A(\mathbf{m}_n) - \mathbf{d}, && (a)\\ \mathbf{l}_n &= \mathbf{l}(\mathbf{m}_n) = \mathbf{F}_n^T \mathbf{r}_n, && (b)\\ \mathbf{g}_n &= \mathbf{F}_n \mathbf{H}_n^{-1} \mathbf{l}_n, && (c)\\ k_n &= \left[\left(\mathbf{H}_n^{-1}\mathbf{l}_n\right)\mathbf{l}_n\right] / \|\mathbf{g}_n\|^2, && (d)\\ \mathbf{m}_{n+1} &= \mathbf{m}_n - k_n \mathbf{H}_n^{-1}\mathbf{l}_n, && (e) \end{aligned} \tag{7.75}$$

where $\mathbf{H}_n = \mathbf{F}_n^T \mathbf{F}_n$ is the Hessian matrix.

The iterative process (7.75) is terminated when the misfit reaches the given level ε_0 :

$$\varphi(\mathbf{m}_N) = \|\mathbf{r}_N\|^2 \leq \varepsilon_0. \tag{7.76}$$

7.4.2.3 The Conjugate Gradient Method

The algorithm of the conjugate gradient method introduced in this chapter can be summarized as follows:

$$\begin{aligned} \mathbf{r}_n &= A(\mathbf{m}_n) - \mathbf{d}, && (a)\\ \mathbf{l}_n &= \mathbf{l}(\mathbf{m}_n) = \mathbf{F}_n^T \mathbf{r}_n, && (b)\\ \beta_n &= \|\mathbf{l}_n\|^2 / \|\mathbf{l}_{n-1}\|^2, && (c)\\ \tilde{\mathbf{l}}_n &= \mathbf{l}_n + \beta_n \tilde{\mathbf{l}}_{n-1}, \ \tilde{\mathbf{l}}_0 = \mathbf{l}_0, && (d)\\ \tilde{k}_n &= \left(\tilde{\mathbf{l}}_n^T \mathbf{l}_n\right) / \left\|\mathbf{F}_n \tilde{\mathbf{l}}_n\right\|^2, && (e)\\ \mathbf{m}_{n+1} &= \mathbf{m}_n - \tilde{k}_n \tilde{\mathbf{l}}_n. && (f) \end{aligned} \tag{7.77}$$

The iterative process (7.77) is terminated when the misfit reaches the given level ε_0 (Eq. (7.76)).

We will also use the conjugate gradient method with linear line search and step length checks to ensure a decrease in the misfit functional at every iteration. To this end, we perform conjugate gradient minimization as outlined above, but with an added condition. Since this line search is based on a linear approximation to the nonlinear problem, we may encounter a computed optimal step length that overshoots the minimum and actually will increase the misfit functional at that iteration. The solution is to simply check the misfit of the new model parameters, and if it increased, decrease the step length by some factor, k_{fac}, and repeat the calculations:

$$\mathbf{m}_{n+1} = \mathbf{m}_n - k_{fac} \tilde{k}_n \tilde{\mathbf{l}}_n.$$

The results of solving inverse problem (7.56) by all three gradient-type methods are shown in Figs. 7.2 and 7.3. These figures present the paths of the iterations in the model space and the convergent plots, respectively. Figure 7.2 shows the map of the misfit functional calculated as percent errors according to formula (7.70). The bold black star indicates the position of the starting point of the iteration process, while the red dot shows the minimum with the exact solution equal to

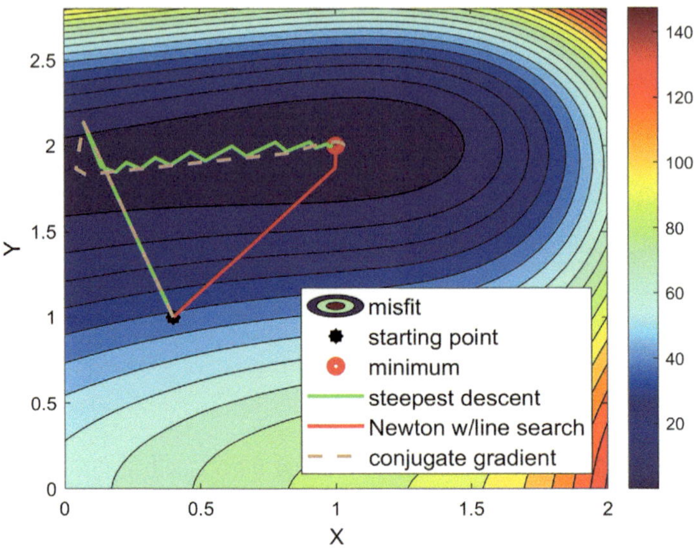

Fig. 7.2 Map of the misfit functional calculated as percent errors according to formula (7.70). The bold black star indicates the position of the starting point of the iteration process, while the red dot is the minimum location. The green line corresponds to the iteration steps of the steepest descent method; the brown dashed line corresponds to the conjugate gradient iterations, and the red line shows the path of Newton's iterations

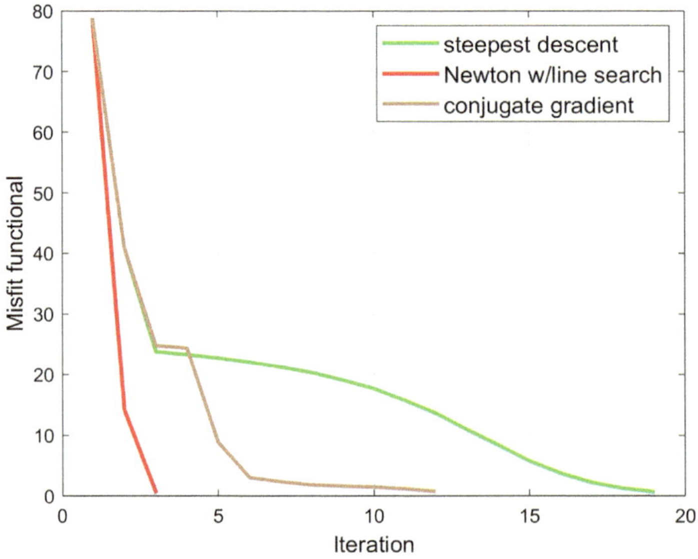

Fig. 7.3 Plots of the misfit functional as percent errors versus the iteration number produced by steepest descent (green line), Newton (red line), and the conjugate gradient (brown line) methods

$$x_{\min} = 1; \; y_{\min} = 2.$$

The green line corresponds to the iteration steps of the steepest descent method; the brown dashed line corresponds to the conjugate gradient iterations, and the red line shows the path of Newton's iterations. We can clearly see the complicated zigzag path of the steepest descent method to the minimum. The initial steps of both the steepest descent and conjugate gradient methods are the same, as seen from the dashed green-brown segment of the iteration paths in this figure. However, after a couple of iterations, the conjugate gradient path is straightened and goes directly to the minimum. The first iteration of the Newton method brings the solution very close to the minimum, and the second iteration completes the job by going straight to the minimum.

The convergence plots also illustrate this behavior of the gradient-type methods. Figure 7.3 shows the plots of the misfit functional calculated as percent errors according to formula (7.70) versus the iteration number produced by steepest descent (green line), Newton (red line), and the conjugate gradient (brown line) methods, respectively. One can see that the steepest descent method requires almost ten times more iterations to reach the minimum than the Newton method. At the same time, the number of the conjugate gradient iterations is somewhat between the Newton and steepest descent method.

Thus, the Newton method is the fastest, usually requiring a few iterations to solve the inverse problem. The downside of the Newton method is the computational cost involved in determining the inverse Hessian matrix $\mathbf{H}_n^{-1}$. Therefore, considering that the cost of every iteration of the conjugate gradient method is practically the same as that of the steepest descent method, this is the method of choice in solving inverse problems.

7.5 The Ill-Posed Nonlinear Discrete Inverse Problem

We now consider again the ill-posed discrete problem, which we have discussed in Sect. 5.2.1. The discrete inverse problem can be written in the form of the following equation:

$$\mathbf{d} = \mathbf{A}(\mathbf{m}), \tag{7.78}$$

where $\mathbf{A}$ is a nonlinear discrete forward operator, $\mathbf{d}$ is the N-dimensional vector of observed data, and $\mathbf{m}$ is the L-dimensional unknown vector of model parameters:

$$\mathbf{d} = [d_1, d_2, d_3, ..., d_N]^T,$$
$$\mathbf{m} = [m_1, m_2, m_3, ..., m_L]^T.$$

For simplicity, we assume within this section that all parameters are real numbers. However, the same technique with a bit of modification can also be applied to complex parameters.

In this and the following sections, we also consider that inverse problem (7.78) is ill-posed. Therefore, we should implement the gradient-type methods in the framework of regularization theory. For a regularized solution of a nonlinear inverse problem, we introduce a parametric functional, similar to the one considered in Sect. 5.2.1:

$$P^{\alpha}(\mathbf{m},\mathbf{d}) = \|\mathbf{W}_d\mathbf{A}(\mathbf{m}) - \mathbf{W}_d\mathbf{d}\|^2 + \alpha\|\mathbf{W}_m\mathbf{m} - \mathbf{W}_m\mathbf{m}_{apr}\|^2$$

$$= (\mathbf{W}_d\mathbf{A}(\mathbf{m}) - \mathbf{W}_d\mathbf{d})^T(\mathbf{W}_d\mathbf{A}(\mathbf{m}) - \mathbf{W}_d\mathbf{d})$$

$$+\,\alpha(\mathbf{W}_m\mathbf{m} - \mathbf{W}_m\mathbf{m}_{apr})^T(\mathbf{W}_m\mathbf{m} - \mathbf{W}_m\mathbf{m}_{apr}), \tag{7.79}$$

where $\mathbf{W}_d$ and $\mathbf{W}_m$ are some weighting matrices of data and model parameters, $\mathbf{m}_{apr}$ is an a priori model, and the upper script "T" denotes transposition, as usual.

The methods of introducing $\mathbf{W}_d$ and $\mathbf{W}_m$ were discussed above in Chap. 5 for linear inverse problems. These methods can be naturally extended to the case of nonlinear problems.

7.5.1 Integrated Sensitivity of the Nonlinear Problem

In order to analyze the sensitivity of the data to the perturbation of one specific model parameter, m_l, we apply the variational operator with respect to this parameter, δ_l to both sides of Eq. (7.78):

$$\delta_l\mathbf{d} = \delta_l\mathbf{A}(\mathbf{m}) = \mathbf{F}\delta_l\mathbf{m},\ l = 1, 2, ..., L, \tag{7.80}$$

where $\mathbf{F}$ is the Fréchet derivative matrix of the nonlinear discrete forward modeling operator, $\mathbf{A}$; $\delta_l\mathbf{m}$ is the vector column with only one nonzero component, δm_l :

$$\delta_l\mathbf{m} = [0, 0, \delta m_l, 0...0]^T. \tag{7.81}$$

The norm square of the perturbed vector of the data can be calculated as follows:

$$\|\delta_l\mathbf{d}\|^2 = (\delta_l\mathbf{d})^T\,\delta_l\mathbf{d} = (\mathbf{F}\delta_l\mathbf{m})^T\,\mathbf{F}\delta_l\mathbf{m} = (\delta_l\mathbf{m})^T\left(\mathbf{F}^T\mathbf{F}\right)\delta_l\mathbf{m}. \tag{7.82}$$

According to definition of vector $\delta_l\mathbf{m}$ (Eq. (7.81)), only diagonal elements of the $[L \times L]$ square matrix $\left(\mathbf{F}^T\mathbf{F}\right)$ are involved in multiplication with $(\delta_l\mathbf{m})^T$ and $\delta_l\mathbf{m}$ in Eq. (7.82). Therefore, we can write Eq. (7.82) in equivalent form:

$$\|\delta_l\mathbf{d}\|^2 = (\delta_l\mathbf{m})^T\,\mathbf{diag}\left(\mathbf{F}^T\mathbf{F}\right)\delta_l\mathbf{m} = (\delta_l\mathbf{m})^T\,\mathbf{S}^2\delta_l\mathbf{m}, \tag{7.83}$$

where

$$\mathbf{S} = \mathbf{diag}\left(\mathbf{F}^T\mathbf{F}\right)^{1/2}. \tag{7.84}$$

A straightforward matrix multiplication shows that

$$(\delta_l \mathbf{m})^T \mathbf{S}^2 \delta_l \mathbf{m} = S_l^2 \left|\delta m_l\right|^2,$$

where S_l is the l-th diagonal element of matrix $\mathbf{S}$:

$$\mathbf{S} = \begin{bmatrix} S_1 & 0 & \dots & 0 \\ 0 & S_2 & \dots & 0 \\ \vdots & \vdots & \ddots & \vdots \\ 0 & 0 & \dots & S_L \end{bmatrix}.$$

Therefore, the norm of the perturbed vector of the data can be calculated as follows:

$$\|\delta_l \mathbf{d}\| = S_l \left|\delta m_l\right|, \tag{7.85}$$

and the diagonal elements of matrix $\mathbf{S}$ represent the integrated sensitivity of the data to the parameter m_l as the ratio

$$S_l = \frac{\|\delta_l \mathbf{d}\|}{|\delta m_k|}. \tag{7.86}$$

As a result of this analysis, we can expand to nonlinear problems the definition of integrated sensitivity originally introduced for linear inverse problems, as follows:

Definition 7.2 The diagonal matrix with the diagonal elements equal to $S_k = \|\delta \mathbf{d}\| / \delta m_k$ is called an integrated sensitivity matrix for nonlinear problem:

$$\mathbf{S} = \mathbf{diag}\left(\sqrt{\sum_i (F_{ik})^2}\right) = \mathbf{diag}\left(\mathbf{F}^T\mathbf{F}\right)^{1/2}. \tag{7.87}$$

In other words, the integrated sensitivity matrix is formed by the norms of the columns of the Fréchet derivative matrix $\mathbf{F}$. Note that, for complex data and model parameters, the expression (7.84) for the diagonal matrix of integrated sensitivity takes the following form:

$$\mathbf{S} = \mathbf{diag}\left(\mathbf{F}^*\mathbf{F}\right)^{1/2}. \tag{7.88}$$

We can recall that the product of the complex conjugate Fréchet derivative matrix by itself is proportional to the Hessian matrix:

$$\mathbf{F}^*\mathbf{F} = \frac{1}{2}\mathbf{H}.$$

Therefore, expression for the integrated sensitivity matrix takes the following form:

$$\mathbf{S} = \frac{1}{2}\mathbf{diag}\,(\mathbf{H})^{1/2}. \tag{7.89}$$

7.5.2 Weighting Matrices of Model Parameters and Data

The basic idea of introducing a weighting matrix, $\mathbf{W}_m$, for the model parameters is as follows. We identify this matrix as the diagonal integrated sensitivity matrix

Thus, the weights are selected to be equal to the sensitivities:

$$\mathbf{W}_m = \mathbf{S}. \tag{7.90}$$

We can now introduce the weighted model parameters:

$$\mathbf{m}^w = \mathbf{W}_m \mathbf{m}. \tag{7.91}$$

Using these notations, we can rewrite the inverse problem Eq. (7.78) as follows:

$$\mathbf{d} = \mathbf{A}\left(\mathbf{W}_\mathbf{m}^{-1}\mathbf{W}_\mathbf{m}\mathbf{m}\right) = \mathbf{A}^w\left(\mathbf{m}^w\right), \tag{7.92}$$

where $\mathbf{A}^w$ is a weighted forward modeling operator.

The variation of the weighted operator is defined as follows:

$$\delta\mathbf{A}^w\left(\mathbf{m}^w\right) = \mathbf{F}^w\delta\mathbf{m}^w = \mathbf{F}^w\mathbf{W}_m\delta\mathbf{m}, \tag{7.93}$$

where $\mathbf{F}^w$ is Fréchet derivative of $\mathbf{A}^w$.

At the same time, according to expression (7.92),

$$\delta\mathbf{A}^w\left(\mathbf{m}^w\right) = \delta\mathbf{A}\left(\mathbf{W}_\mathbf{m}^{-1}\mathbf{W}_\mathbf{m}\mathbf{m}\right) = \mathbf{F}\mathbf{W}_\mathbf{m}^{-1}\mathbf{W}_\mathbf{m}\delta\mathbf{m}, \tag{7.94}$$

where $\mathbf{F}$ is Fréchet derivative of the original operator, $\mathbf{A}$:

$$\delta\mathbf{A}\left(\mathbf{m}\right) = \mathbf{F}\delta\mathbf{m}.$$

Comparing Eqs. (7.93) and (7.94), we obtain at once a simple relationship between the Fréchet derivative matrices of the weighted and original forward modeling operators:

$$\mathbf{F}^w = \mathbf{F}\mathbf{W}_\mathbf{m}^{-1}. \tag{7.95}$$

Now we perturb the data with respect to one specific weighted parameter m_k^w:

$$\delta_k d_i = F_{ik}^w \delta m_k^w,$$

and calculate a new integrated sensitivity S_k^w of the data to the weighted parameter m_k^w as the ratio

$$S_k^w = \frac{\|\delta \mathbf{d}\|}{\delta m_k^w} = \frac{\sqrt{\sum_i \left(F_{ik}^w\right)^2 \delta m_k^w}}{\delta m_k^w} = \sqrt{\sum_i \left(F_{ik}^w\right)^2}$$

$$= \sqrt{\sum_i \left(F_{ik} W_k^{-1}\right)^2} = W_k^{-1} \sqrt{\sum_i (F_{ik})^2} = W_k^{-1} S_k = 1, \tag{7.96}$$

where we took into account Eq. (7.95).

Formula (7.96) shows that the new matrix of the integrated sensitivity $\mathbf{S}^w$ is identity matrix:

$$\mathbf{S}^w = \mathbf{I}.$$

Therefore, data are uniformly sensitive to the new weighted model parameters!

Note that the corresponding weighted stabilizing functional takes the form

$$s_w(\mathbf{m}) = (\mathbf{m} - \mathbf{m}_{apr})^T \mathbf{W}_m^2 (\mathbf{m} - \mathbf{m}_{apr})$$

$$= (\mathbf{m} - \mathbf{m}_{apr})^T \mathbf{S}^2 (\mathbf{m} - \mathbf{m}_{apr}). \tag{7.97}$$

It imposes a stronger penalty on departure from the a priori model for those parameters that contribute more significantly to the data.

Thus, the model weighting results in practically equal resolution of the inversion with respect to different parameters of the model.

In a similar way, we can define the diagonal data weighting matrix, formed by the norms of the rows of the Fréchet derivative matrix $\mathbf{F}$:

$$\mathbf{W}_d = \mathbf{diag}\left(\sqrt{\sum_k (F_{ik})^2}\right) = \mathbf{diag}\left(\mathbf{F}\mathbf{F}^T\right)^{1/2}. \tag{7.98}$$

These weights make normalized data less dependent on the specific parameters of observations (for example, frequency and distance from the anomalous domain), which improves the resolution of the inverse method.

7.6 Steepest Descent Method for Nonlinear Regularized Least-Squares Inversion

7.6.1 *Descent Method of Parametric Functional Minimization*

According to the basic principles of the regularization method, we have to find the model $\mathbf{m}_\alpha$, a quasi-solution of the inverse problem, which minimizes the parametric functional:

$$P^\alpha(\mathbf{m}, \mathbf{d}) = \min .$$

We consider an iterative algorithm for parametric functional minimization. It is reasonable to build this algorithm on the idea that parametric functional decreases at every iteration $\mathbf{m}_n$. In other words, we impose the *descent condition* as follows;

$$P^\alpha(\mathbf{m}_{n+1}) < P^\alpha(\mathbf{m}_n) \text{ for all } n \geq 0. \tag{7.99}$$

A method that imposes this condition is called a *descent method.* The question is how to find iterations $\{\mathbf{m}_n\}$ that satisfy the descent condition. To solve the problem of minimization of the parametric functional using the steepest descent method, let us calculate the first variation of $P^\alpha(\mathbf{m}, \mathbf{d})$, assuming that the operator $\mathbf{A}(\mathbf{m})$ is differentiable, so that

$$\delta\mathbf{A}(\mathbf{m}) = \mathbf{F}_m\delta\mathbf{m}, \tag{7.100}$$

where $\mathbf{F}_m$ is the Fréchet derivative matrix of $\mathbf{A}$.

Thus, we have

$$\delta P^\alpha(\mathbf{m}, \mathbf{d}) = 2(\mathbf{W}_d\mathbf{F}_m\delta\mathbf{m})^T(\mathbf{W}_d\mathbf{A}(\mathbf{m}) - \mathbf{W}_d\mathbf{d}) +$$

$$+2\alpha(\mathbf{W}_m\delta\mathbf{m})^T(\mathbf{W}_m\mathbf{m} - \mathbf{W}_m\mathbf{m}_{apr}),$$

or after some algebra

$$\delta P^\alpha(\mathbf{m}, \mathbf{d}) = 2(\delta\mathbf{m})^T\mathbf{F}_m^T\mathbf{W}_d^2(\mathbf{A}(\mathbf{m}) - \mathbf{d}) + 2\alpha(\delta\mathbf{m})^T\mathbf{W}_m^2(\mathbf{m} - \mathbf{m}_{apr}), \tag{7.101}$$

where we assume, for simplicity, that the matrices $\mathbf{W}_d$ and $\mathbf{W}_m$ are diagonal.

In order to satisfy the descent condition (7.99), we select

$$\delta\mathbf{m} = -k^\alpha\mathbf{l}^\alpha(\mathbf{m}), \tag{7.102}$$

where k^α is some positive real number (length of a step) and $\mathbf{l}^\alpha(\mathbf{m})$ is a column matrix defining the direction of the steepest ascent of the parametric functional:

$$\mathbf{l}^{\alpha}(\mathbf{m}) = \mathbf{F}_m^T \mathbf{W}_d^2 (\mathbf{A}(\mathbf{m}) - \mathbf{d}) + \alpha \mathbf{W}_m^2 (\mathbf{m} - \mathbf{m}_{apr}). \tag{7.103}$$

Certainly, by substituting Eqs. (7.102) and (7.103) into (7.101), we have

$$\delta P^{\alpha}(\mathbf{m}, \mathbf{d}) = -2k^{\alpha}(\mathbf{l}^{\alpha}(\mathbf{m}), \mathbf{l}^{\alpha}(\mathbf{m})) < 0, \tag{7.104}$$

so $\mathbf{l}^{\alpha}(\mathbf{m})$ describes the "direction" of increasing (ascent) of the functional $\phi(\mathbf{m})$, because it is opposite to the descent direction, $\delta\mathbf{m}$.

An iterative process of the method is constructed according to the following formula:

$$\mathbf{m}_{n+1} = \mathbf{m}_n + \delta\mathbf{m} = \mathbf{m}_n - k_n^{\alpha}\mathbf{l}^{\alpha}(\mathbf{m}_n),$$

where coefficient k_n^{α} is defined by a line search according to the condition (Zhdanov 2002):

$$P^{\alpha}(\mathbf{m}_{n+1}) = P^{\alpha}\left(\mathbf{m}_n - k_n^{\alpha}\mathbf{l}^{\alpha}(\mathbf{m}_n)\right) = \Phi^{\alpha}(k_n^{\alpha}) = \min. \tag{7.105}$$

In particular, applying the linear line search, we find that the minimum of the parametric functional is reached if k_n^{α} is determined by the following formula:

$$k_n^{\alpha} = \frac{\|\mathbf{l}^{\alpha}(\mathbf{m}_n)\|^2}{\left\| F_{m_n}\mathbf{l}^{\alpha}(\mathbf{m}_n)\right\|^2 + \alpha \|W\mathbf{l}^{\alpha}(\mathbf{m}_n)\|^2}.$$

7.6.2 *Numerical Schemes of the Regularized Steepest Descent Method*

The algorithm of the regularized steepest descent method can be summarized as follows:

$$\begin{aligned}
\mathbf{r}_n &= A(\mathbf{m}_n) - \mathbf{d}, && (a)\\
\mathbf{l}_n^{\alpha} &= \mathbf{l}^{\alpha}(\mathbf{m_n}) = F_{m_n}^{\star}\mathbf{r_n} + \alpha W^{\star}W(\mathbf{m} - \mathbf{m}_{apr}), && (b)\\
k_n^{\alpha} &= \left\|\mathbf{l}_n^{\alpha}\right\|^2 / \left[\left\| F_{m_n}\mathbf{l}_n^{\alpha}\right\|^2 + \alpha\left\| W\mathbf{l}_n^{\alpha}\right\|^2\right], && (c)\\
\mathbf{m}_{n+1} &= \mathbf{m}_n - k_n^{\alpha}\mathbf{l}_n^{\alpha}. && (d)
\end{aligned} \tag{7.106}$$

The iterative process (7.106) is terminated at $n = N$ when the parametric functional reaches the given level ε_0 :

$$P^{\alpha}(\mathbf{m}_N) \leq \varepsilon_0.$$

7.7 Newton Method of Nonlinear Regularized Least-Squares Inversion

7.7.1 Optimal Newton Step

The main idea of the Newton method is to try to solve the problem of minimization in one step:

$$\mathbf{m}_1 = \mathbf{m}_0 + \Delta\mathbf{m}. \tag{7.107}$$

Thus, instead of moving downhill along a long path formed by mutually orthogonal directions of the steepest descent, one can try to reach the minimum of the misfit functional along one direction. To determine this specific direction, $\Delta\mathbf{m}$, let us write the linearized parametric functional

$$P^{\alpha}(\mathbf{m}_1, \mathbf{d}) = P^{\alpha}(\mathbf{m}_0 + \Delta\mathbf{m}, \mathbf{d})$$

$$= (\mathbf{W}_d\mathbf{A}(\mathbf{m}_0) + \mathbf{W}_d\mathbf{F}_{m_0}\Delta\mathbf{m} - \mathbf{W}_d\mathbf{d})^T(\mathbf{W}_d\mathbf{A}(\mathbf{m}_0) + \mathbf{W}_d\mathbf{F}_{m_0}\Delta\mathbf{m} - \mathbf{W}_d\mathbf{d})$$

$$+ \alpha(\mathbf{W}_m\mathbf{m}_0 + \mathbf{W}_m\Delta\mathbf{m} - \mathbf{W}_m\mathbf{m}_{apr})^T(\mathbf{W}_m\mathbf{m}_0 + \mathbf{W}_m\Delta\mathbf{m} - \mathbf{W}_m\mathbf{m}_{apr}). \tag{7.108}$$

The first variation of the parametric functional is equal to

$$\delta_{\Delta\mathbf{m}}P^{\alpha}(\mathbf{m}_1, \mathbf{d}) = \delta_{\Delta\mathbf{m}}P^{\alpha}(\mathbf{m}_0 + \Delta\mathbf{m}, \mathbf{d})$$

$$= (\delta\Delta\mathbf{m})^T\left[\mathbf{F}_{m_0}^T\mathbf{W}_d^2\left(\mathbf{A}(\mathbf{m}_0) + \mathbf{F}_{m_0}\Delta\mathbf{m} - \mathbf{d}\right) + \alpha\mathbf{W}_m^2(\mathbf{m}_0 + \Delta\mathbf{m} - \mathbf{m}_{apr})\right].$$

It is evident that the necessary condition for the minimum of the parametric functional,

$$\delta P^{\alpha}(\mathbf{m}_1, \mathbf{d}) = 0,$$

is satisfied only if

$$\mathbf{F}_{m_0}^T\mathbf{W}_d^2\left(\mathbf{A}(\mathbf{m}_0) + \mathbf{F}_{m_0}\Delta\mathbf{m} - \mathbf{d}\right) + \alpha\mathbf{W}_m^2(\mathbf{m}_0 + \Delta\mathbf{m} - \mathbf{m}_{apr}) = 0.$$

From the last equation, taking into consideration Eq. (7.103) for the regularized steepest ascent direction, we obtain the following formula for the optimal Newton step:

$$\Delta\mathbf{m} = -2\mathbf{H}_{\alpha,m_0}^{-1}\mathbf{l}^{\alpha}(\mathbf{m}_n),$$

where

$$\mathbf{H}_{\alpha,m_0} = 2\left(\mathbf{F}_{m_0}^T\mathbf{W}_d^2\mathbf{F}_{m_0} + \alpha\mathbf{W}_m^2\right)$$

is the regularized quasi-Hessian matrix.

Thus, the Newton algorithm of the nonlinear regularized least-squares inversion can be expressed by the following formula

$$\mathbf{m}_{n+1} = \mathbf{m}_n - 2\mathbf{H}_{\alpha,m_n}^{-1}\mathbf{l}^{\alpha}(\mathbf{m}_n),$$

where

$$\mathbf{H}_{\alpha,m_n} = 2\left(\mathbf{F}_{m_n}^T\mathbf{W}_d^2\mathbf{F}_{m_n} + \alpha\mathbf{W}_m^2\right).$$

7.7.2 Numerical Scheme of the Newton Method

The algorithm of the Newton method for a discrete inverse problem can be summarized as follows:

$$\begin{array}{rll} \mathbf{r}_n &= \mathbf{A}(\mathbf{m}_n) - \mathbf{d}, & (a) \\ \mathbf{l}_n^{\alpha_n} &= \mathbf{l}^{\alpha_n}(\mathbf{m}_n) = \mathbf{F}_{m_n}^T\mathbf{W}_d^2\mathbf{r}_n + \alpha_n\mathbf{W}_m^2(\mathbf{m}_n - \mathbf{m}_{apr}), & (b) \\ \mathbf{H}_{\alpha_n,m_n} &= 2\left(\mathbf{F}_{m_n}^T\mathbf{W}_d^2\mathbf{F}_{m_n} + \alpha_n\mathbf{W}_m^2\right), & (c) \\ \mathbf{m}_{n+1} &= \mathbf{m}_n - 2\mathbf{H}_{\alpha_n,m_n}^{-1}\mathbf{l}_n^{\alpha_n}(\mathbf{m}_n), & (d) \end{array} \tag{7.109}$$

where α_n are the subsequent values of the regularization parameter, updated on each iteration. This method is called the Newton method with adaptive regularization. The concept of adaptive regularization is discussed below in Sect. 7.8.2.

The iterative process (7.109) is terminated when the misfit reaches the given level ε_0:

$$\phi(\mathbf{m}_N) = \|\mathbf{r}_N\|^2 \leq \varepsilon_0.$$

7.8 Nonlinear Least-Squares Inversion by the Conjugate Gradient Method

7.8.1 Regularized Conjugate Gradient Directions

The conjugate gradient method is based on the same ideas as the steepest descent, and the iteration process is very similar to the last one:

$$\mathbf{m}_{n+1} = \mathbf{m}_n + \delta\mathbf{m}_n = \mathbf{m}_n - \tilde{k}_n^{\alpha}\widetilde{\mathbf{l}}^{\alpha}(\mathbf{m}_n), \tag{7.110}$$

where

$$\delta\mathbf{m}_n = -\tilde{\mathbf{k}}_n^{\alpha}\widetilde{\mathbf{l}}^{\alpha}(\mathbf{m}_n).$$

However, the "directions" of ascent $\widetilde{\mathbf{l}}^{\alpha}(\mathbf{m}_n)$ are selected differently. In the first step, we use the "direction" of the steepest ascent:

$$\widetilde{\mathbf{l}}^{\alpha}(\mathbf{m}_0) = \mathbf{l}^{\alpha}(\mathbf{m}_0).$$

In the next step, the "direction" of ascent is a linear combination of the steepest ascent at this step and the "direction" of ascent $\widetilde{\mathbf{l}}^{\alpha}(\mathbf{m}_0)$ on the previous step:

$$\widetilde{\mathbf{l}}^{\alpha}(\mathbf{m}_1) = \mathbf{l}^{\alpha}(\mathbf{m}_1)+\beta_1\widetilde{\mathbf{l}}^{\alpha}(\mathbf{m}_0).$$

In the n-th step

$$\widetilde{\mathbf{l}}^{\alpha}(\mathbf{m}_{n+1}) = \mathbf{l}^{\alpha}(\mathbf{m}_{n+1}) + \beta^{\alpha}_{n+1}\widetilde{\mathbf{l}}^{\alpha}(\mathbf{m}_n). \tag{7.111}$$

The regularized steepest ascent directions are determined according to the formula for the least-squares method:

$$\mathbf{l}^{\alpha}(\mathbf{m}_n) = \mathbf{F}^T_{m_n}\mathbf{W}^2_d(\mathbf{A}(\mathbf{m}_n) - \mathbf{d}) + \alpha\mathbf{W}^2_{m_n}(\mathbf{m}_n - \mathbf{m}_{apr}). \tag{7.112}$$

Determination of the length of iteration step, a coefficient $\widetilde{k}^{\alpha}_n$, can be based on the linear or parabolic line search:

$$P^{\alpha}(\mathbf{m}_{n+1}) = P^{\alpha}(\mathbf{m}_n - \widetilde{k}^{\alpha}_n\widetilde{\mathbf{l}}^{\alpha}(\mathbf{m}_n)) = f(\widetilde{k}^{\alpha}_n) = \min.$$

Solution of this minimization problem gives the following best estimation for the length of the step using a linear line search:

$$\widetilde{k}^{\alpha}_n = \frac{\widetilde{\mathbf{l}}^{\alpha T}(\mathbf{m}_n)\mathbf{l}^{\alpha}(\mathbf{m}_n)}{\widetilde{\mathbf{l}}^{\alpha T}(\mathbf{m}_n)\left[\left(\mathbf{F}^T_{m_n}\mathbf{W}^2_d\mathbf{F}_{m_n} + \alpha\mathbf{W}^2_m\right)\widetilde{\mathbf{l}}^{\alpha}(\mathbf{m}_n)\right]}$$

$$= \frac{\widetilde{\mathbf{l}}^{\alpha T}(\mathbf{m}_n)\mathbf{l}^{\alpha}(\mathbf{m}_n)}{\left\|\mathbf{W}_d\mathbf{F}_{m_n}\widetilde{\mathbf{l}}^{\alpha}(\mathbf{m}_n)\right\|^2 + \alpha\left\|\mathbf{W}_m\widetilde{\mathbf{l}}^{\alpha}(\mathbf{m}_n)\right\|^2}. \tag{7.113}$$

One can also use a parabolic line search (Fletcher, 1995) to improve the convergence rate of the RCG method.

The CG method requires that the vectors $\widetilde{\mathbf{l}}^{\alpha}(\mathbf{m}_n)$ introduced above will be mutually conjugate. This requirement is fulfilled if the coefficients β_n are determined by the formula similar to Eq. (7.54) of the misfit functional minimization:

$$\beta^{\alpha}_{n+1} = \frac{\|\mathbf{l}^{\alpha}(\mathbf{m}_{n+1})\|^2}{\|\mathbf{l}^{\alpha}(\mathbf{m}_n)\|^2}. \tag{7.114}$$

Thus, vectors $\widetilde{\mathbf{l}}^{\alpha}(\mathbf{m}_n)$ represent the regularized conjugate gradient directions. Using Eqs. (7.110), (7.111), (7.113), and (7.114), we can obtain $\mathbf{m}$ iteratively.

7.8.2 Numerical Scheme of the Regularized Conjugate Gradient Method

In this section, we first consider a method of selecting the parameter α. The regularization parameter α describes a trade-off between the best fitting and most reasonable stabilization. In a case when α is selected to be too small, the minimization of the parametric functional $P^{\alpha}(\mathbf{m})$ is equivalent to the minimization of the misfit functional $\phi(\mathbf{m})$, and therefore we have no regularization, which can result in an unstable incorrect solution. When α is too large, the minimization of the parametric functional $P^{\alpha}(\mathbf{m})$ is equivalent to the minimization of the stabilizing functional $s(\mathbf{m})$, which will force the solution to be closer to the a priori model. Ultimately, we would expect the final model to be exactly like the a priori model, while the observed data are totally ignored in the inversion. Thus, the critical question in the regularized solution of an inverse problem is the selection of the optimal regularization parameter α.

The basic principles used for determining the regularization parameter α were discussed in the previous sections of the book. We introduced in Chap. 4 a simple numerical method to determine parameter α. Consider for example the progression of numbers

$$\alpha_k = \alpha_1 q^{k-1}; \quad k = 1, 2, 3.....; \quad q > 0. \tag{7.115}$$

The first iteration of the steepest descent or any other gradient method is run usually with $\alpha_0 = 0$. The initial value of the regularization parameter, α_1, is determined after the first iteration, $\mathbf{m}_1$, as a ratio:

$$\alpha_1 = \frac{\|\mathbf{W}_d\mathbf{A}(\mathbf{m}_1) - \mathbf{W}_d\mathbf{d}\|^2}{\|\mathbf{W}_m\mathbf{m}_1 - \mathbf{W}_m\mathbf{m}_{apr}\|^2}.$$

In this way, we have an approximate balance between the misfit and stabilizing functional. For any number α_k we can find an element $\mathbf{m}_{\alpha_k}$, minimizing $P^{\alpha_k}(\mathbf{m})$, and calculate the misfit $\| \mathbf{A}\left(\mathbf{m}_{\alpha_k}\right) - \mathbf{d} \|^2$. The optimal value of the parameter α is the number α_{k0}, for which we have

$$\| \mathbf{A}\left(\mathbf{m}_{\alpha_{k0}}\right) - \mathbf{d} \|^2 = \delta, \tag{7.116}$$

where δ is the level of noise in the observed data. The equality (7.116) is *the misfit condition* . This algorithm, as well as the L-curve method (see Chap. 4), has clear practical limitations, because it requires a complete numerical solution of the inverse problem for each value of the regularization parameter α_k.

An alternative approach is based on the simple idea, which we have already discussed above, that the regularization parameter α can be updated in the process of the iterative inversion. For example, one can use the following algorithm for the RCG method:

$$\tilde{\mathbf{l}}^{\alpha_{n+1}}(\mathbf{m}_{n+1}) = \mathbf{l}^{\alpha_{n+1}}(\mathbf{m}_{n+1}) + \beta_{n+1}\tilde{\mathbf{l}}^{\alpha_n}(\mathbf{m}_n),$$

where α_n are the subsequent values of the regularization parameter. This method is called the *adaptive* regularization method. In order to avoid divergence, we begin an iteration from a value of α_1, which can be obtained as a ratio of the misfit functional and the stabilizer for an initial model, then reduce α_n according to formula (7.115) on each subsequent iteration and continuously iterate until the misfit condition (7.116) is reached.

The algorithm of the regularized conjugate gradient method can be summarized as follows:

$$\begin{array}{ll} \mathbf{r}_n = A(\mathbf{m}_n) - \mathbf{d}, & (a) \\ \mathbf{l}_n^{\alpha_n} = \mathbf{l}^{\alpha_n}(\mathbf{m}_n) = \mathbf{F}_{m_n}^T \mathbf{W}_d^2 \mathbf{r}_n + \alpha_n \mathbf{W}_m^2 (\mathbf{m}_n - \mathbf{m}_{apr}), & (b) \\ \beta_n^{\alpha_n} = \left\| \mathbf{l}_n^{\alpha_n} \right\|^2 / \left\| \mathbf{l}_{n-1}^{\alpha_{n-1}} \right\|^2, \quad \tilde{\mathbf{l}}_n^{\alpha_n} = \mathbf{l}_n^{\alpha_n} + \beta_n^{\alpha_n} \tilde{\mathbf{l}}_{n-1}^{\alpha_{n-1}}, \; \tilde{\mathbf{l}}_0^{\alpha_0} = \mathbf{l}_0^{\alpha_0}, & (c) \\ \widetilde{k}_n^{\alpha_n} = \left(\tilde{\mathbf{l}}_n^{\alpha_n T} \mathbf{l}_n^{\alpha_n} \right) / \left\{ \left\| \mathbf{W}_d \mathbf{F}_{m_n} \tilde{\mathbf{l}}_n^{\alpha_n} \right\|^2 + \alpha \left\| \mathbf{W}_m \tilde{\mathbf{l}}_n^{\alpha_n} \right\|^2 \right\}, & (d) \\ \mathbf{m}_{n+1} = \mathbf{m}_n - \widetilde{k}_n^{\alpha_n} \tilde{\mathbf{l}}_n^{\alpha_n}, & (e) \end{array} \tag{7.117}$$

where α_n are the subsequent values of the regularization parameter. The iterative process (7.117) is terminated when the misfit reaches the given level ε_0:

$$\phi(\mathbf{m}_N) = \|\mathbf{r}_N\|^2 \leq \varepsilon_0.$$

This method is called the conjugate gradient method with adaptive regularization.

I should note that the same comparison we made above for the gradient-type methods as applied to misfit functional minimization holds for the case of parametric functional minimization as well. This means that the convergence of the steepest descent method is usually slower than that of the conjugate gradient method because the former requires many short iteration steps to reach the minimum. The Newton method converges rapidly with every iteration making a big "jump" to a minimum; however, every step of the Newton method is computationally very expensive due to the need for the large Hessian matrix inversion. The conjugate gradient method combines the simplicity of iterative steps of the steepest descent technique with a relatively small number of iterations. Therefore, the conjugate gradient method is the most computationally efficient technique for solving large-scale ill-posed inverse problems.

In the conclusion of this chapter, I would like to note that both the gradient-type optimization algorithms and the Monte Carlo-type methods discussed in Chap. 6 have their advantages and disadvantages. The gradient-type methods are characterized by relatively rapid convergence, but they have difficulties in the case of multiple local minima. The Monte Carlo-type methods converge very slowly, but they can find a global minimum even for the functionals with multiple local minima. However, it is possible to consider a hybrid approach to minimization by combining the Monte Carlo-type methods at the initial phase of the iterative inversion to overcome the presence of local minima with the gradient-type methods at the final stage for rapid convergence to the global minimum. The hybrid approach may be useful in many practical applications where the individual techniques are inefficient.

References and Recommended Reading to this Chapter

Fletcher R (1955) Practical methods of optimization. John Willey & Sons, Chichester-New-York, pp. 436

Press WH, Flannery BP, Teukolsky SA, Vettering WT (1987) Numerical recipes, The art of scientific computing, vols. I and II. Cambridge University Press, Cambridge, pp. 1447

Zhdanov MS (1993) Tutorial: regularization in inversion theory. CWP-136, Colorado School of Mines, pp. 47

Zhdanov MS (2002) Geophysical inverse theory and regularization problems. Elsevier, Amsterdam, London, New York, Tokyo, pp. 628

Zhdanov MS (2015) Inverse theory and applications in geophysics. Elsevier, Amsterdam, London, New York, Tokyo, pp. 704

Part III
Joint Inversion of Multiphysics Data

Chapter 8
Joint Inversion Based on Analytical and Statistical Relationships Between Different Physical Properties

Abstract This chapter discusses the general concepts of joint inversion of multiphysics data. The mathematical formulation of the multimodal inverse problem is provided, which serves as a basis for joint inversion methods introduced in the book. We consider the cases of joint inversion when the known functional relationships exist between different model parameters. We also examine the cases of inversion of multiphysics data with different resolution capabilities. The principles of subspace representation of the model parameters and subspace inversion with resampling are also introduced.

Keywords Multiphysics data · Joint inversion · Subspace representation · Resampling

In many applications, researchers collect different types of data representing the same object of investigation. For example, in medical imaging, various imaging techniques, e.g., X-ray, ultrasound, magnetic resonance imaging (MRI), etc., are used to study the internal organs of the human body. In geophysical applications, multiple physical field data, e.g., gravity, magnetic, electromagnetic, seismic, etc., are collected to study the earth's internal structure. In astronomy, optical and radiotelescopes are used to study electromagnetic radiation from the stars and galactic, as well as observations of neutrinos, cosmic rays, or gravitational waves. This list of applications can be expanded to many other fields of science and engineering.

The common feature of all of these applications is that the multiphysics data are used to study the same object of interest while providing information about different physical properties of the target. The joint inversion methods provide a mathematical framework for integrated analysis of multiphysics data which can be applied to all these applications.

As an illustration, we present the general concepts of joint inversion considering geophysical applications following Zhdanov (2015). Different geophysical fields provide information about different physical properties of rock formations. In many cases, this information is mutually complementary, which makes it natural to consider a joint inversion of various geophysical data. There are different approaches to joint inversion (e.g., Dell'Aversana 2013) In a case where the corresponding model

M. S. Zhdanov, *Advanced Methods of Joint Inversion and Fusion of Multiphysics Data*, Advances in Geological Science,
https://doi.org/10.1007/978-981-99-6722-3_8

parameters are identical or mutually correlated, the joint inversion can explore the existence of this correlation (e.g., Jupp and Vozoff 1975; Hoversten et al. 2003, 2006). In a case where the model parameters are not correlated but nevertheless have similar geometrical features, the joint inversion can be based on structure-coupled constraints. This approach has been introduced in several publications (e.g., Fregoso and Gallardo 2009; Gallardo 2007; Gallardo and Meju 2003, 2004, 2007, 2011; Haber and Oldenburg 1997; Haber and Modersitzki 2007; Hu et al. 2009; Meju 2011). It is based on minimizing the value of the cross-gradients between different model parameters. This approach has been widely used for joint inversion of geophysical data (e.g., Colombo and DeStefano 2007; Hu et al. 2009; Jegenet al. 2009; De Stefano et al. 2011; Moorkamp et al. 2011). Note that, in practical applications, the empirical or statistical correlations between different physical properties may exist, but their specific form may be unknown. In addition, there could be both analytical and structural correlations between different attributes of the model parameters. Hence, there is a need for a method of joint inversion, which would not require a priori knowledge about specific empirical or statistical relationships between the different model parameters and/or their attributes.

In the works by Zhdanov et al. (2012a), and Zhdanov (2015), a new approach to the joint inversion of multimodal data using Gramian constraints was introduced. The Gramians are computed as determinants of the corresponding Gram matrices of the multimodal model parameters and/or their different attributes (see Sect. 3.8). The Gramian provides a measure of the correlation between the different model parameters or their attributes. By imposing the additional requirement of the minimum of the Gramian in regularized inversion, we obtain multimodal inverse solutions with enhanced correlations between the different model parameters or their attributes (see Chap. 12).

In this and the following chapters, we will subsequently describe the methods of joint inversion based on analytical and statistical relationships, structural similarity, joint focusing, minimum entropy, and Gramian constraints.

We begin our discussion with the joint inversion based on functional relationships between different model parameters.

8.1 Formulation of the Multimodal Inverse Problem

Considering inverse problems for multiple physical data sets, we can describe these problems by the operator relationships as follows:

$$\mathbf{d}^{(i)} = \mathbf{A}^{(i)}(\mathbf{m}^{(i)}), \quad i = 1, 2, 3, \ldots, N, \tag{8.1}$$

where, in a general case, $\mathbf{A}^{(i)}$ $(i = 1, 2, 3, \ldots, N)$ are nonlinear operators, $\mathbf{d}^{(i)}$ $(i = 1, 2, 3, \ldots, N)$ are different observed data sets (which may have different physical natures), and $\mathbf{m}^{(i)}$ $(i = 1, 2, 3, \ldots, N)$ are the unknown sets of model parameters.

Note that, in a general case, various model parameters may have different physical dimensions (e.g., density is measured in g/cm^3, resistivity is measured in Ohm-m, etc.). Therefore, it is convenient to introduce the dimensionless weighted model parameters, $\widetilde{\mathbf{m}}^{(i)}$, defined as follows:

$$\widetilde{\mathbf{m}}^{(i)} = \mathbf{W}_m^{(i)}\mathbf{m}^{(i)}, \; i = 1, 2, 3, ..., N, \tag{8.2}$$

where $\mathbf{W}_m^{(i)}$ are the corresponding linear operators of the model weighting.

We assume that the dimensionless weighted model parameters are described by integrable functions of a radius-vector $\mathbf{r} = (x, \; y, \; z)$ defined within some volume V of a 3D space. The set of these functions forms a complex Hilbert space of the model parameters, M, with a L_2 norm, defined by the corresponding inner product:

$$\left(\widetilde{\mathbf{m}}^{(i)}, \, \widetilde{\mathbf{m}}^{(j)}\right)_M = \int_V \widetilde{m}^{(i)}(\mathbf{r}) \; \widetilde{m}^{(j)*}(\mathbf{r}) \, dv, \; \left\|\widetilde{\mathbf{m}}^{(i)}\right\|_M^2 = \left(\widetilde{\mathbf{m}}^{(i)}, \, \widetilde{\mathbf{m}}^{(i)}\right)_M, \tag{8.3}$$

where asterisk "*" denotes the complex conjugate value.

Similarly, different data sets, as a rule, have different physical dimensions as well. Therefore, it is convenient to consider dimensionless weighted data, $\widetilde{\mathbf{d}}^{(i)}$, defined as follows:

$$\widetilde{\mathbf{d}}^{(i)} = \mathbf{W}_d^{(i)}\mathbf{d}^{(i)}, \; i = 1, 2, 3, ..., N, \tag{8.4}$$

where $\mathbf{W}_d^{(i)}$ are the corresponding linear operators of the data weighting.

We also assume that the weighted data belong to some complex Hilbert space of the data, D, with the L_2 norm, defined by the corresponding inner product:

$$\left(\widetilde{\mathbf{d}}^{(i)}, \, \widetilde{\mathbf{d}}^{(j)}\right)_D = \int_S \widetilde{d}^{(i)}(\mathbf{r}) \; \widetilde{d}^{(j)*}(\mathbf{r}) \, ds, \; \left\|\widetilde{\mathbf{d}}^{(i)}\right\|_D^2 = \left(\widetilde{\mathbf{d}}^{(i)}, \, \widetilde{\mathbf{d}}^{(i)}\right)_D, \tag{8.5}$$

where S is an observation surface.

The multimodal inverse problem is formulated as a solution of the system of Eq. (8.1) with respect to model parameters $\mathbf{m}^{(i)} \;\; (i = 1, 2, 3, ..., N)$.

8.2 Joint Inversion Based on Functional Relationships Between Different Model Parameters

A priori functional relationships may exist between different model parameters in some cases. For example, let us assume that the relationship between all model parameters can be described by the following constraint equation:

$$\mathbf{C}\left(\widetilde{\mathbf{m}}^{(1)}, \, \widetilde{\mathbf{m}}^{(2)}, ...,\widetilde{\mathbf{m}}^{(n)}\right) = 0, \tag{8.6}$$

where $\mathbf{C}$ is some known operator defined on a set of functions $\widetilde{\mathbf{m}}^{(1)}, \widetilde{\mathbf{m}}^{(2)}, ..., \widetilde{\mathbf{m}}^{(n)}$, from the model space, M, with the values in the model space, M, as well. Note that, in a general case, operator $\mathbf{C}$ is a nonlinear differentiable operator.

For the solution of nonlinear inverse problem (8.1), we introduce the following parametric functional with the constraint stabilizers:

$$P_C^{\alpha}(\widetilde{\mathbf{m}}^{(1)}, \widetilde{\mathbf{m}}^{(2)}, ..., \widetilde{\mathbf{m}}^{(n)}) = \sum_{i=1}^{N} \left\| \widetilde{\mathbf{A}}^{(i)}(\widetilde{\mathbf{m}}^{(i)}) - \widetilde{\mathbf{d}}^{(i)} \right\|_D^2 +$$

$$+ \alpha c_1 \sum_{i=1}^{n} S_{MN,\ MS,\ MGS}^{(i)} + \alpha c_2 S_C(\widetilde{\mathbf{m}}^{(1)}, \widetilde{\mathbf{m}}^{(2)}, ..., \widetilde{\mathbf{m}}^{(n)}), \tag{8.7}$$

where $\widetilde{\mathbf{A}}^{(i)}(\widetilde{\mathbf{m}}^{(i)})$ are the weighted predicted data,

$$\widetilde{\mathbf{A}}^{(i)}(\widetilde{\mathbf{m}}^{(i)}) = \mathbf{W}_d^{(i)} \mathbf{A}^{(i)}(\widetilde{\mathbf{m}}^{(i)});$$

α is the regularization parameter; and c_1 and c_2 are the weighting coefficients determining the weights of the different stabilizers in the parametric functional.

The terms $S_{MN}^{(i)}$, $S_{MS}^{(i)}$, and $S_{MGS}^{(i)}$ are the stabilizing functionals, based on minimum norm, minimum support, and minimum gradient support constraints, respectively (for definitions and properties see Sect. 4.3 above):

$$S_{MN}^{(i)} = \left\| \widetilde{\mathbf{m}}^{(i)} - \widetilde{\mathbf{m}}_{apr}^{(i)} \right\|_M^2 = \int_V \left(\widetilde{m}^{(i)} - \widetilde{m}_{apr}^{(i)} \right)^2 dv,$$

$$S_{MS}^{(i)} = \int_V \frac{\left(\widetilde{m}^{(i)} - \widetilde{m}_{apr}^{(i)} \right)^2}{\left(\widetilde{m}^{(i)} - \widetilde{m}_{apr}^{(i)} \right)^2 + e^2} dv, \tag{8.8}$$

and

$$S_{MGS}^{(i)} = \int_V \frac{\nabla \widetilde{m}^{(i)} \cdot \nabla \widetilde{m}^{(i)}}{\nabla \widetilde{m}^{(i)} \cdot \nabla \widetilde{m}^{(i)} + e^2} dv, \tag{8.9}$$

where e is a *focusing* parameter.

The term S_C is the constraint stabilizing functional, which enforces the parametric relationship (8.6):

$$S_C(\widetilde{\mathbf{m}}^{(1)}, \widetilde{\mathbf{m}}^{(2)}, ..., \widetilde{\mathbf{m}}^{(n)}) = \left\| \mathbf{C}(\widetilde{\mathbf{m}}^{(1)}, \widetilde{\mathbf{m}}^{(2)}, ..., \widetilde{\mathbf{m}}^{(n)}) \right\|_M^2. \tag{8.10}$$

We have demonstrated in Chap. 4 that the regularized solution of the inverse problem (8.1) can be obtained by minimization of the parametric functional (8.7):

$$P_C^{\alpha}(\widetilde{\mathbf{m}}^{(1)}, \widetilde{\mathbf{m}}^{(2)}, ..., \widetilde{\mathbf{m}}^{(n)}) = \min. \tag{8.11}$$

In order to solve the problem of minimization of the parametric functional with the constraint stabilizer, we calculate the first variation

$$\delta P_C^{\alpha}(\widetilde{\mathbf{m}}^{(1)}, \widetilde{\mathbf{m}}^{(2)}, ..., \widetilde{\mathbf{m}}^{(n)})$$

$$= 2\sum_{i=1}^{n} \left(\delta\widetilde{\mathbf{A}}^{(i)}(\widetilde{\mathbf{m}}^{(i)}),\ \widetilde{\mathbf{A}}^{(\mathbf{i})}(\widetilde{\mathbf{m}}^{(\mathbf{i})}) - \widetilde{\mathbf{d}}^{(i)}\right)_D +$$

$$+ 2\alpha\left[c_1\sum_{i=1}^{n}\delta S^{(i)}_{MN,\ MS,\ MGS} + c_2(\delta\mathbf{C}, \mathbf{C})_M\right]. \tag{8.12}$$

Taking into consideration that operators $\widetilde{\mathbf{A}}^{(i)}$ and $\mathbf{C}$ are differentiable, we can write

$$\delta\widetilde{\mathbf{A}}^{(i)}(\widetilde{\mathbf{m}}^{(i)}) = \widetilde{\mathbf{F}}^{(i)}_{\widetilde{m}}\delta\widetilde{\mathbf{m}}^{(i)}, \tag{8.13}$$

and

$$\delta\mathbf{C} = \sum_{i=1}^{n}\mathbf{F}^{C}_{\widetilde{m}^{(i)}}\delta\widetilde{\mathbf{m}}^{(i)}, \tag{8.14}$$

where $\widetilde{\mathbf{F}}^{(i)}_{\widetilde{m}}$ is a linear operator of the Fréchet derivative of $\widetilde{\mathbf{A}}^{(i)}$, and $\mathbf{F}^{C}_{\widetilde{m}^{(i)}}$ are linear operators of the Fréchet derivative of $\mathbf{C}$ with respect to $\widetilde{\mathbf{m}}^{(i)}$.

It can be demonstrated that (Zhdanov 2015)

$$\delta S^{(i)}_{MN,\ MS,\ MGS} = 2\left(\delta\widetilde{\mathbf{m}}^{(i)},\ \mathbf{l}^{(i)}_{MN,\ MS,\ MGS}\right),$$

where vectors $\mathbf{l}^{(i)}_{MN,\ MS,\ MGS}$ are the directions of the steepest ascent for the stabilizing functionals based on minimum norm, minimum support, and minimum gradient support constraints, respectively. These vectors represent the discrete analog of the corresponding gradient direction functions, $l^{(i)}_{MN}$, $l^{(i)}_{MS}$, and $l^{(i)}_{MGS}$:

$$\mathbf{l}^{(i)}_{MN} = \left[l^{(i)}_{MN}\right],\ \mathbf{l}^{(i)}_{MS} = \left[l^{(i)}_{MS}\right],\ \mathbf{l}^{(i)}_{MGS} = \left[l^{(i)}_{MGS}\right], \tag{8.15}$$

where

$$l^{(i)}_{MN} = \left(\widetilde{m}^{(i)} - \widetilde{m}^{(i)}_{apr}\right); \tag{8.16}$$

$$l_{MS}^{(i)} = \frac{e^2\left(\widetilde{m}^{(i)} - \widetilde{m}_{apr}^{(i)}\right)}{\left[\left(\widetilde{m}^{(i)} - \widetilde{m}_{apr}^{(i)}\right)^2 + e^2\right]^2}; \tag{8.17}$$

$$l_{MGS}^{(i)} = \nabla \cdot \frac{e^2 \nabla \widetilde{m}^{(i)}}{\left(\nabla \widetilde{m}^{(i)} \cdot \nabla \widetilde{m}^{(i)} + e^2\right)^2}. \tag{8.18}$$

Substituting expressions (8.13) through (8.18) into formula (8.12), we obtain

$$\delta P^{\alpha}(\widetilde{\mathbf{m}}^{(1)}, \widetilde{\mathbf{m}}^{(2)}, \ldots., \widetilde{\mathbf{m}}^{(n)}) =$$

$$2\sum_{i=1}^{n}\left(\delta\widetilde{\mathbf{m}}^{(i)}, \left[\widetilde{\mathbf{F}}_{\widetilde{m}}^{(i)\star}\left(\widetilde{\mathbf{A}}^{(\mathbf{i})}(\widetilde{\mathbf{m}}^{(\mathbf{i})}) - \widetilde{\mathbf{d}}^{(i)}\right) + \right.\right.$$

$$\left.\left. + \alpha\left(c_1 \mathbf{l}_{MN,\ MS,\ MGS}^{(i)} + c_2 \mathbf{F}_{\widetilde{m}^{(i)}}^{C\star} \mathbf{C}\right)\right]\right)_M, \tag{8.19}$$

where $\widetilde{\mathbf{F}}_{\widetilde{m}}^{(i)\star}$ and $\mathbf{F}_{\widetilde{m}^{(i)}}^{C\star}$ are the adjoint Fréchet derivative operators.

Let us select

$$\delta\widetilde{\mathbf{m}}^{(i)} = -k^{\alpha}\mathbf{l}_C^{\alpha(i)}(\widetilde{\mathbf{m}}^{(1)}, \widetilde{\mathbf{m}}^{(2)}, \ldots., \widetilde{\mathbf{m}}^{(n)}), \tag{8.20}$$

where k^{α} is some positive real number, and $\mathbf{l}_C^{\alpha(i)}(\widetilde{\mathbf{m}}^{(1)}, \widetilde{\mathbf{m}}^{(2)}, \ldots., \widetilde{\mathbf{m}}^{(n)})$ is the direction of the steepest ascent of the functional P_C^{α} as a function of model parameter $\widetilde{\mathbf{m}}^{(i)}$ only,

$$l_C^{\alpha(i)} = \widetilde{\mathbf{F}}_{\widetilde{m}}^{(i)\star}\left(\widetilde{\mathbf{A}}^{(\mathbf{i})}(\widetilde{\mathbf{m}}^{(\mathbf{i})}) - \widetilde{\mathbf{d}}^{(i)}\right) + \alpha\left(c_1 \mathbf{l}_{MN,\ MS,\ MGS}^{(i)} + c_2 \mathbf{F}_{\widetilde{m}^{(i)}}^{C\star}\mathbf{C}\right). \tag{8.21}$$

Then

$$\delta P_C^{\alpha}(\widetilde{\mathbf{m}}^{(1)}, \widetilde{\mathbf{m}}^{(2)}, \ldots., \widetilde{\mathbf{m}}^{(n)}) = -2k^{\alpha}\sum_{i=1}^{n}\left\|\mathbf{l}_C^{\alpha(i)}(\widetilde{\mathbf{m}}^{(1)}, \widetilde{\mathbf{m}}^{(2)}, \ldots., \widetilde{\mathbf{m}}^{(n)})\right\|_M^2. \tag{8.22}$$

The last expression confirms that the selection of the perturbations of the model parameters based on formula (8.20) ensures the decrease of the parametric functional.

We can construct an iterative process for the regularized conjugate gradient (RCG) algorithm of solving minimization problem (8.11), which can be summarized as follows:

$$\begin{aligned} \widetilde{\mathbf{r}}_k &= \widetilde{A}(\widetilde{\mathbf{m}}_k) - \widetilde{\mathbf{d}}, \quad \mathbf{l}_k^\alpha = \mathbf{l}^\alpha(\widetilde{\mathbf{m}}_\mathbf{k}) && (a) \\ \beta_k^\alpha &= \left\|\mathbf{l}_k^\alpha\right\|^2 / \left\|\mathbf{l}_{k-1}^\alpha\right\|^2, \quad \widetilde{\mathbf{l}}_k^\alpha = \mathbf{l}_k^\alpha + \beta_k^\alpha \widetilde{\mathbf{l}}_{k-1}^\alpha, \ \widetilde{\mathbf{l}}_0^\alpha = \mathbf{l}_0^\alpha, && (b) \\ \widetilde{s}_k^\alpha &= \left(\widetilde{\mathbf{l}}_k^\alpha, \mathbf{l}_k^\alpha\right) / \left\{\left\|\widetilde{F}_{m_k}\widetilde{\mathbf{l}}_k^\alpha\right\|^2 + \alpha \left\|W\widetilde{\mathbf{l}}_k^\alpha\right\|^2\right\}, && (c) \\ \widetilde{\mathbf{m}}_{k+1} &= \widetilde{\mathbf{m}}_k - \widetilde{s}_k^\alpha \widetilde{\mathbf{l}}_k^\alpha. && (d) \end{aligned} \tag{8.23}$$

In the last formula, we use the following notations:
$\widetilde{\mathbf{d}}$ is a vector of multiphysics observed data

$$\widetilde{\mathbf{d}} = \left(\widetilde{\mathbf{d}}^{(1)}, \widetilde{\mathbf{d}}^{(2)}, \ldots. \widetilde{\mathbf{d}}^{(n)}\right)^T;$$

$\widetilde{\mathbf{m}}_k$ is a vector of different model parameters computed at iteration number k,

$$\widetilde{\mathbf{m}}_k = \left(\widetilde{\mathbf{m}}_k^{(1)}, \widetilde{\mathbf{m}}_k^{(2)}, \ldots. \widetilde{\mathbf{m}}_k^{(n)}\right)^T;$$

$\widetilde{A}(\widetilde{\mathbf{m}}_k)$ is a vector of the predicted data computed at iteration number k;
$\mathbf{l}_k^\alpha$ is a vector of the direction of the steepest ascent computed at iteration number k,

$$\mathbf{l}_k^\alpha = \left(\mathbf{l}_{Ck}^{\alpha(1)}, \mathbf{l}_{Ck}^{\alpha(2)}, \ldots. \mathbf{l}_{Ck}^{\alpha(n)}\right)^T. \tag{8.24}$$

The expressions for the steepest ascent directions are shown above in formula (8.21).

The iterative process (8.23) is terminated when the misfit reaches the required level:

$$\varphi(\widetilde{\mathbf{m}}_{k+1}) = \|\widetilde{\mathbf{r}}_{k+1}\|_D^2 = \delta_d. \tag{8.25}$$

8.3 Inversion of Multiphysics Data with Different Resolution Capabilities

8.3.1 Different Resolution Capabilities

Various observation methods used in applications have different resolution capabilities. For example, in geophysical applications, seismic data are usually collected with a dense survey and have better resolution than the data produced by gravity, magnetic, or electromagnetic field observations. As a result, seismic data inversion can provide a detailed model of the subsurface distribution of seismic wave velocity. In contrast, the density, magnetization, or conductivity models inverted from the potential field or electromagnetic data do not have the same resolution. Brute force joint inversion of high-resolution (e.g., seismic) data and low-resolution (e.g., potential field or electromagnetic) data may produce fictitious (unreasonably fine) structures for model parameters corresponding to methods with low-resolution capabilities. Accounting for the resolution difference in a framework of a unified inversion method represents a very challenging problem of joint inversion.

In order to address the issue of different resolution capabilities, one should decouple the forward modeling meshes for different physical properties (Tu and Zhdanov 2021). For example, seismic waves have a shorter wavelength compared to low-frequency electromagnetic waves used in geophysics. As a result, the seismic method usually has a better resolution than the electromagnetic measurements. Therefore, one can use a much finer grid to discretize the velocity distribution in the subsurface than the conductivity grid for accurate forward modeling. At the same time, applying the fine grid used for seismic velocity to electromagnetic modeling may result in an oversampled conductivity model. It may increase the computation time of forward modeling and also introduce artificial details of conductivity models unresolvable by electromagnetic data. Thus, one has to use different forward modeling grids for efficient computations and for honoring the different resolution capabilities of distinct data.

At the same time, in order to jointly invert multiphysics data, the corresponding physical property models should be represented at the same scale, especially if the structural similarity between different models is considered. In this case, one should match only the long-wavelength structures of the high-resolution model (e.g., seismic velocity) with the low-resolution model (e.g., conductivity or density), as the fine features of the former model are beyond the resolution of the latter model. These shared discretization models can be produced by the proper parameterization of the different physical properties. Mathematically, this task can be formulated as a *subspace representation* of the model parameters. Subspace representation of different model parameters can be obtained as a solution of approximation and interpolation problems.

8.3.2 Approximation Problem

Let us assume that the vector of model parameter $\mathbf{m}^{(i)}$ corresponds to the values of continuous function, $m^{(i)}(x, y, z)$, defined on the nodes of some regular or irregular grid within some modeling domain D. Function $m^{(i)}(x, y, z)$ can be treated as an element of the Hilbert space $L_2(D)$ formed by a set of functions integrable on the domain D, equipped with the inner product (see Chap. 3):

$$(f, g)_{L_2} = \iiint_D f(x, y, z)\, g(x, y, z)\, dv, \tag{8.26}$$

where $f, g \in L_2(D)$.

We have learned in Chap. 3 that, in a Hilbert space, there always exists a set of functions called basis functions, $e_k(x, y, z)$, $k = 1, 2, ...$, and every function in the Hilbert space can be represented as a linear combination of the basis functions, e.g.,

$$m^{(i)}(x, y, z) = \sum_k \alpha_k e_k(x, y, z), \tag{8.27}$$

where α_k, $k = 1, 2, ...$, are the scalar coefficients of representation (8.27).

In a general case, the number of basis functions in the Hilbert space is counted but unlimited. In practical applications, we always work with a limited number of the basis functions, which result in the following approximate representation of the model function $m^{(i)}(x, y, z)$:

$$m^{(i)}(x, y, z) \approx \sum_{k=1}^{K} \alpha_k e_k(x, y, z). \tag{8.28}$$

In this case, the coefficients of approximation (8.27) can be found using the solution of the approximation problem in a Hilbert space, discussed in Chap. 3. According to this method, we should find a minimum of the misfit functional, $\varphi(\alpha_1, \alpha_2, ...\alpha_K)$:

$$\varphi(\alpha_1, \alpha_2, ...\alpha_K) = \left\| m^{(i)} - \sum_k \alpha_k e_k \right\|_{L_2}^2$$

$$= \left(\left[m^{(i)} - \sum_{k=1}^{K} \alpha_k e_k \right], \left[m^{(i)} - \sum_{k=1}^{K} \alpha_k e_k \right] \right)_{L_2} = \min. \tag{8.29}$$

The minimum can be determined by taking the first variation of the misfit functional, φ, with respect to coefficient α_l and equal it to zero as follows:

$$\delta_{\alpha_l}\varphi = -2 \left(\left[m^{(i)} - \sum_{k=1}^{K} \alpha_k e_k \right], e_l \right)_{L_2} =$$

$$-2\left(m^{(i)}, e_l\right)_{L_2} + 2\left(\sum_{k=1}^{K} \alpha_k e_k, e_l\right)_{L_2} = 0.$$

From the last equation, we have

$$\left(m^{(i)}, e_l\right)_{L_2} = \sum_k \alpha_k \left(e_k, e_l\right)_{L_2}, \ l = 1, 2, ..K. \tag{8.30}$$

We can write the last system of equations more compactly as follows:

$$\sum_{k}^{K} \Gamma_{lk}\alpha_k = \left(m^{(i)}, e_l\right)_{L_2}, \quad l = 1, 2, ..K, \tag{8.31}$$

where Γ_{lk} are the elements of the Gram matrix, $\mathbf{\Gamma}$, formed by inner products of the basis functions

$$\Gamma_{lk} = (e_k, e_l)_{L_2}. \tag{8.32}$$

Thus, Eq. (8.30) can be written in matrix notations as follows:

$$\mathbf{\Gamma}\boldsymbol{\alpha} = \boldsymbol{M}^{(i)}, \tag{8.33}$$

where α and $\mathbf{M}^{(i)}$ are the vector-columns formed by the unknown coefficients, α_k, $k = 1, 2, ...K$; and the inner products, $\left(m^{(i)}, e_l\right)_{L_2}$, $l = 1, 2, ..K$; respectively.

Assuming that the basis functions e_k, $k = 1, 2, ...K$; are linearly independent, we conclude that the Gram matrix is nonsingular. Therefore, there is also exist the inverse matrix $\mathbf{\Gamma}^{-1}$; and the solution of Eq. (8.33) can be written in the following explicit form:

$$\boldsymbol{\alpha} = \mathbf{\Gamma}^{-1}\mathbf{M}^{(i)}. \tag{8.34}$$

We can also write formula (8.34) using scalar notations:

$$\alpha_k = \sum_{l=1}^{K} \Gamma_{kl}^{-1} \left(m^{(i)}, e_l\right)_{L_2}, \quad k = 1, 3, ...K, \tag{8.35}$$

where Γ_{kl}^{-1} are the elements of the inverse Gram matrix, $\mathbf{\Gamma}^{-1}$. Approximation (8.28) with the coefficients determined by formula (8.35) can be used for solving the interpolation problem.

Formula (8.35) provides an analytical solution for the approximation problem. Indeed, by substituting expressions (8.35) for coefficients α_k into Eq. (8.28), we have

$$m^{(i)}(x, y, z) \approx \sum_{k=1}^{K}\sum_{l=1}^{K} \Gamma_{kl}^{-1} \left(m^{(i)}, e_l\right)_{L_2} e_k(x, y, z). \tag{8.36}$$

8.3.3 *Interpolation and Subspace Representation*

We assume that the vector of model parameter $\mathbf{m}^{(i)}$ is given by the values of function $m^{(i)}(x, y, z)$ in the nodes of some rectangular discretization grid, $\Omega^{(i)}$:

$$m^{(i)}_{qps} = m^{(i)}\left(x_q, y_p, z_s\right), \; q = 1, ..Q; \; p = 1, ..P, \; s = 1, ..S; \tag{8.37}$$

where x_q, y_p, z_s are the centers of the corresponding cells of the grid, $\Omega^{(i)}$.

The interpolated data, $\widehat{m}^{(i)}(x, y, z)$, can be described by the following formula:

$$\widehat{m}^{(i)}(x, y, z) = \sum_{qps} P_{qps}(x, y, z)\, m^{(i)}_{qps}, \tag{8.38}$$

where $P_{qps}(x, y, z)$ are some interpolation functions.

These interpolation functions can be determined using the approximation formula (8.36). To this end, let us calculate the inner product $\left(m^{(i)}, e_l\right)_{L_2}$ approximately on the grid $\Omega^{(i)}$:

$$\left(m^{(i)}, e_l\right)_{L_2} = \iiint_D m^{(i)}(x, y, z)\, e_l(x, y, z)\, dv$$

$$\approx \sum_{qps} m^{(i)}_{qps} e_{l,qps} \Delta v, \tag{8.39}$$

where Δv is the volume of the rectangular cell, and $e_{l,qps}$ are the values of the corresponding basis functions in the center of the cells

$$e_{l,qps} = e_l\left(x_q, y_p, z_s\right).$$

Substituting Eqs. (8.39) into (8.36), we have

$$m^{(i)}(x, y, z) \approx \sum_{k=1}^{K} \sum_{l=1}^{K} \Gamma^{-1}_{kl} \sum_{qps} m^{(i)}_{qps} e_{l,qps} \Delta v e_k(x, y, z). \tag{8.40}$$

Changing the order of summations, we can write

$$m^{(i)}(x, y, z) \approx \sum_{qps} \left\{ \sum_{k=1}^{K} \sum_{l=1}^{K} \Gamma^{-1}_{kl} e_{l,qps} \Delta v e_k(x, y, z) \right\} m^{(i)}_{qps}. \tag{8.41}$$

A comparison between Eqs. (8.41) and (8.38) shows that the interpolation functions can be expressed via the basis functions using the following formula:

$$P_{qps}(x, y, z) = \sum_{k=1}^{K} \sum_{l=1}^{K} \Gamma^{-1}_{kl} e_{l,qps} e_k(x, y, z)\, \Delta v. \tag{8.42}$$

Interpolation formula (8.38) allows us to resample the model parameters from discretization grid $\Omega^{(i)}$ to another grid with different cell's size, $\widehat{\Omega}$, which is used for all model parameters considered in joint inversion. Indeed, we can find the values of the model parameters $m^{(i)}(x, y, z)$ in the nodes (x_n, y_k, z_l) from Eq. (8.38):

$$\widehat{m}^{(i)}(x_n, y_k, z_l) = \sum_{qps} P_{qps}(x_n, y_k, z_l)\, m_{qps}^{(i)}. \tag{8.43}$$

Equation (8.43) shows that the vector of model parameters $\widehat{\mathbf{m}}^{(i)}$ interpolated on the new grid, $\widehat{\Omega}$, is related to the vector of model parameters $\mathbf{m}^{(i)}$ on the original grid, $\Omega^{(i)}$, by linear transformation:

$$\widehat{\mathbf{m}}^{(i)} = \mathbf{P}^{(i)}\mathbf{m}^{(i)}, \tag{8.44}$$

where $\mathbf{P}^{(i)}$ is the matrix formed by interpolation functions $P_{qps}(x_n, y_k, z_l)$. The upper index (i) at matrix $\mathbf{P}^{(i)}$ indicates that it depends on the parameters of the discretization grid $\Omega^{(i)}$.

The action of $\mathbf{P}^{(i)}$ on vector $\mathbf{m}^{(i)}$ resamples the model to the corresponding grid. Equation (8.44) can be used, for example, for resampling the model parameters from a fine grid to the coarse grid, which may be required in a joint inversion if we use the same size of the grids for the high-resolution and low-resolution parameters. In this case, we can call transformation described by formula (8.44) a *subspace representation* because it reduces the space of model parameter $\mathbf{m}^{(i)}$ representation. A review of the sampling theory can be found in Garcia (2000).

There are many types of basis functions that can be used for subspace representation, for example, pseudo-linear and spline functions, trigonometric and exponential functions, Legendre and Chebyshev polynomials. Cubic B-spline basis functions are particularly useful in geophysical applications, since it offers differentiable continuity, and can be used for an irregular distribution of nodes (Rawlinson 2008; Tu and Zhdanov 2021).

8.3.4 Subspace Inversion with Resampling

We can now rewrite the parametric functional with the constraint stabilizer (8.7), used for the solution of the multimodal inverse problem, as follows:

$$P_C^{\alpha}(\widetilde{\mathbf{m}}^{(1)}, \widetilde{\mathbf{m}}^{(2)}, \ldots, \widetilde{\mathbf{m}}^{(n)}) = \sum_{i=1}^{N} \left\| \widetilde{\mathbf{A}}^{(i)}(\widetilde{\mathbf{m}}^{(i)}) - \widetilde{\mathbf{d}}^{(i)} \right\|_D^2 +$$

$$+\alpha c_1 \sum_{i=1}^{n} S_{MN,\,MS,\,MGS}^{(i)} + \alpha c_2 S_C(\widehat{\mathbf{m}}^{(1)}, \widehat{\mathbf{m}}^{(2)}, \ldots, \widehat{\mathbf{m}}^{(n)}), \tag{8.45}$$

where $\widehat{\mathbf{m}}^{(1)}, \widehat{\mathbf{m}}^{(2)}, \ldots, \widehat{\mathbf{m}}^{(n)}$ are the model parameters resampled according to formula (8.44) at the same grid,

The idea of resampling can be explained as follows. For forward modeling, we can use different discretization grids for various physical models, subject to the resolution of the data with regard to the corresponding model parameters. For example, in geophysical applications, we discretize the density model on a coarse grid, while the velocity model on a fine grid. At the same time, we can use another relatively coarse grid for coupling stabilizer S_C calculation to resample both the density and velocity models so that they are coupled at the same scale.

In this chapter, we have discussed the application of resampling for joint inversion based on functional relationships between different model parameters. In the following chapters, we will consider the various joint inversion methods with structural, Gramian, focusing, and other constraints. All these methods will require resampling the multimodal parameters on the same discretization grid for computing the corresponding joint stabilizing functionals.

References and Recommended Reading to This Chapter

Colombo D, De Stefano M (2007) Geophysical modeling via simultaneous joint inversion of seismic, gravity, and electromagnetic data. Application prestack depth imaging. Lead Edge 26:326–331

Dell'Aversana P (2013) Cognition in Geosciences–The feeding loop between geo-disciplines, cognitive sciences and epistemology. EAGE Publications, 204 pp

De Stefano M, Andreasi FG, Re S, Virgilio M, Snyder FF (2011) Multiple-domain, simultaneous joint inversion of geophysical data with application to subsalt imaging. Geophys 76:R69–R80

Fregoso E, Gallardo LA (2009) Cross-gradients joint 3D inversion with applications to gravity and magnetic data. Geophys 74:L31–L42

Gallardo LA (2007) Multiple cross-gradient joint inversion for geospectral imaging. Geophys Res Lett 34:L19301

Gallardo LA, Meju MA (2003) Characterization of heterogeneous near-surface materials by joint 2D inversion of DC resistivity and seismic data. Geophys Res Let 30:1658–1661

Gallardo LA, Meju MA (2004) Joint two-dimensional DC resistivity and seismic travel time inversion with cross-gradients constraints. J Geophys Res: Solid Earth 109(B3)

Gallardo LA, Meju MA (2007) Joint two-dimensional cross-gradient imaging of magnetotelluric and seismic traveltime data for structural and lithological classification. Geophys J Int 169(3):1261–1272

Gallardo LA, Meju MA (2011) Structure-coupled multi-physics imaging in geophysical sciences. Rev Geophys 49:RG1003

Garcia AG (2000) Orthogonal sampling formulas: a unified approach. SIAM Rev 42(3):499–512

Haber E, Oldenburg D (1997) Joint inversion: a structural approach. Inverse Probl 13:63–67

Haber E, Modersitzki J (2007) Intensity gradient based registration and fusion of multimodal images. Methods Inf Med 46:292–299

Hoversten GM, Gritto R, Washbournez J, Daley T (2003) Pressure and fluid saturation prediction in a multicomponent reservoir using combined seismic and electromagnetic imaging. Geophysics 68:1580–1591

Hoversten GM, Cassassuce F, Gasperikova E, Newman GA, Chen J, Rubin Y, Hou Z, Vasco D (2006) Direct reservoir parameter estimation using joint inversion of marine seismic AVA and CSEM data. Geophysics 71:C1–C13

Hu WY, Abubakar A, Habashy TM (2009) Joint electromagnetic and seismic inversion using structural constraints. Geophysics 74:R99–R109

Jegen MD, Hobbs RW, Tarits P, Chave A (2009) Joint inversion of marine magnetotelluric and gravity data incorporating seismic constraints: preliminary results of sub-basalt imaging off the Faroe Shelf. Earth Planet Sci Lett 282:47–55

Jupp DLB, Vozoff K (1975) Joint inversion of geophysical data. Geophys J Roy Astron Soc 42:977–991

Meju MA (2011) Joint multi-geophysical inversion: effective model integration, challenges and directions for future research. Presented at international workshop on gravity. Electrical and Magnetic Methods and their Applications, Bejing, China

Moorkamp M, Heincke B, Jegen M, Robert AW, Hobbs RW (2011) A framework for 3-D joint inversion of MT, gravity and seismic refraction data. Geophys J Int 184:477–493

Rawlinson N, Hauser J, Sambridge M (2008) Seismic ray tracing and wavefront tracking in laterally heterogeneous media. Adv Geophys 49:203–273

Tu X, Zhdanov MS (2021) Joint Gramian inversion of geophysical data with different resolution capabilities: case study in Yellowstone. Geophys J Int 226(2):1058–1085

Zhdanov MS (2015) Inverse theory and applications in geophysics, 2nd edn. Elsevier

Zhdanov MS, Gribenko AV, Wilson G (2012) Generalized joint inversion of multimodal geophysical data using Gramian constraints. Geophys Res Lett 39(9)

Chapter 9
Joint Inversion Based on Structural Similarities

Abstract An important approach to joint inversion is based on enforcing the structural similarity between different images of the target. This chapter considers two types of mathematical criteria for structural similarity. One is based on the structural similarity index (SSI). Another one requires the parallelism of the model parameter gradients. The concepts of the structural similarity index and structural similarity conditions are discussed in detail. The cross-gradient and dot-gradient similarity conditions are also introduced. This chapter demonstrates how the structural similarity index and the method of parallel gradients can be used in the joint inversion of multiphysics data

Keywords Structural similarity index · Similarity function · Cross-gradient condition · Dot-gradient condition

9.1 Concept of Structural Similarity and Its Mathematical Formulation

The common practice in many physical, geophysical, and other natural science experiments or in medical imaging is collecting multiphysics data to study the target. Still, these data reflect different physical properties of the target, e.g., its density (if we measure the gravity field), its magnetization (if we measure the magnetic field), its conductivity (if we measure the electric or electromagnetic fields), etc. At the same time, the geometrical structure of the target is the same for all kinds of physical measurements. For example, in geophysical applications, we have geophysical signatures of the same geologic structures underground; in optical applications, we analyze different optical images of the same object having unique structural characteristics analyzed by the human visual system. The medical images produced by various imaging instruments also represent the examined organ having a specific geometrical structure. Thus, one important approach to joint inversion and image analyses can be based on preserving this structural similarity between different images of the

M. S. Zhdanov, *Advanced Methods of Joint Inversion and Fusion of Multiphysics Data*, Advances in Geological Science,
https://doi.org/10.1007/978-981-99-6722-3_9

target in joint inversion or image reconstruction. This approach requires developing a proper mathematical formulation of the structural similarity concept.

This chapter considers two types of mathematical criteria for structural similarity. One was introduced in image processing theory for image quality assessment (Wang et al. 2004). It is based on the structural similarity index (SSI). Another one was proposed for joint inversion of geophysical data. This approach requires the parallelism of the model parameter gradients (Droske and Rumpf 2003; Gallardo and Meju 2003).

In the following Chaps. 10 and 11, I will discuss additional techniques to evaluate the structural similarity of different inverse images based on joint focusing and minimum entropy functionals.

9.1.1 Structural Similarity Index (SSI)

The structural similarity index (SSI) was introduced as the quality measure of optical images. Let us assume that we have two images described by two non-negative continuous functions, $m^{(1)}$ and $m^{(2)}$, which are distributed on the surface (two-dimensional images $m^{(1)}(x, y)$ and $m^{(2)}(x, y)$) or within a volume (three-dimensional images $m^{(1)}(x, y, z)$ and $m^{(2)}(x, y, z)$). Let us assume that both images represent the same object but have been taken under different conditions. There are three factors that can be evaluated for these images. The first factor is related to the luminance (or brightness) of the images. The second factor is related to the contrast of the images. Finally, the third factor characterizes the similarity of the images.

These factors can be numerically evaluated using the statistical estimates if we consider the functions $m^{(1)}(x, y, z)$ and $m^{(2)}(x, y, z)$ being realizations of some random variables (see Chap. 2). Indeed, the *luminances or brightness* of the images are obviously proportional to the mean values of the functions $m^{(1)}(x, y, z)$ and $m^{(2)}(x, y, z)$, defined according to formula (2.7) of Chap. 2. The corresponding standard deviations can evaluate the contrasts between the images.

The difference in luminances between two images can be measured as the square of the difference between the mean values of each function,

$$\Delta_l^2\left(m^{(1)}, m^{(2)}\right) = \left[\langle m^{(1)}\rangle - \langle m^{(2)}\rangle\right]^2, \tag{9.1}$$

where $\langle m^{(1)}\rangle$ and $\langle m^{(2)}\rangle$ stand for mean values of the corresponding functions, defined according to formula (2.7) of Chap. 2.

It is convenient to introduce a normalized square luminance difference,

$$\widetilde{\Delta}_l^2\left(m^{(1)}, m^{(2)}\right) = \frac{\left[\langle m^{(1)}\rangle - \langle m^{(2)}\rangle\right]^2}{\langle m^{(1)}\rangle^2 + \langle m^{(2)}\rangle^2} = 1 - \frac{2\langle m^{(1)}\rangle \cdot \langle m^{(2)}\rangle}{\langle m^{(1)}\rangle^2 + \langle m^{(2)}\rangle^2}. \tag{9.2}$$

The second term in the right side of Eq. (9.2) is called the *luminance comparison function*, $l\left(m^{(1)},\ m^{(2)}\right)$:

$$l\left(m^{(1)},\ m^{(2)}\right) = \frac{2\langle m^{(1)}\rangle \cdot \langle m^{(2)}\rangle}{\langle m^{(1)}\rangle^2 + \langle m^{(2)}\rangle^2}. \tag{9.3}$$

Substituting equation (9.3) back into (9.2), we have

$$\widetilde{\Delta}_l^2\left(m^{(1)},\ m^{(2)}\right) = 1 - l\left(m^{(1)},\ m^{(2)}\right). \tag{9.4}$$

From the last two equations and conditions that $m^{(1)}$ and $m^{(2)}$ are non-negative functions, it follows at once that

$$0 \le l\left(m^{(1)},\ m^{(2)}\right) \le 1. \tag{9.5}$$

Thus, we conclude that the luminance comparison function is equal to 1 if there is no luminance difference between the two images. It becomes equal to zero when we observe the maximum difference in brightness between two images.

Similarly, the difference in contrasts between two images can be measured as the square of the difference between the standard deviations of the corresponding functions:

$$\Delta_c^2\left(m^{(1)},\ m^{(2)}\right) = [\sigma_1 - \sigma_2]^2, \tag{9.6}$$

where σ_1 and σ_2 are the standard deviations of the corresponding functions, defined according to formula (2.8) of Chap. 2.

Introducing a normalized square contrast difference, $\Delta_c^2\left(m^{(1)},\ m^{(2)}\right)$, we write

$$\widetilde{\Delta}_c^2\left(m^{(1)},\ m^{(2)}\right) = \frac{[\sigma_1 - \sigma_2]^2}{\sigma_1^2 + \sigma_2^2} = 1 - \frac{2\sigma_1\sigma_2}{\sigma_1^2 + \sigma_2^2} = 1 - c\left(m^{(1)},\ m^{(2)}\right), \tag{9.7}$$

where $c\left(m^{(1)},\ m^{(2)}\right)$ is the *contrast comparison function* :

$$c\left(m^{(1)},\ m^{(2)}\right) = \frac{2\sigma_1\sigma_2}{\sigma_1^2 + \sigma_2^2}. \tag{9.8}$$

This function has the same property as the *luminance comparison function*,

$$0 \le c\left(m^{(1)},\ m^{(2)}\right) \le 1. \tag{9.9}$$

In other words, the contrast comparison function is equal to one if there is no contrast difference between two images. It goes to zero when this difference reaches the maximum.

Note that, in practical calculations, it is convenient to write these functions in the following form (Wang et al. 2004):

(1) the *luminance comparison function*:

$$l\left(m^{(1)},\ m^{(2)}\right)=\frac{2\langle m^{(1)}\rangle\cdot\langle m^{(2)}\rangle+\varepsilon_l}{\langle m^{(1)}\rangle^2+\ \langle m^{(2)}\rangle^2+\varepsilon_l};\tag{9.10}$$

(2) the *contrast comparison function*:

$$c\left(m^{(1)},\ m^{(2)}\right)=\frac{2\sigma_1\,\sigma_2+\varepsilon_c}{\sigma_1^2+\ \sigma_2^2+\varepsilon_c}.\tag{9.11}$$

The parameters ε_l and ε_c in Eqs. (9.10) and (9.11) are small positive numbers introduced to avoid numerical instability of division by a small value when both mean values and standard deviations are close to zero.

Finally, the similarity of the images can be analyzed by calculating the correlation coefficient between the two functions, $m^{(1)}\,(x, y, z)$ and $m^{(2)}\,(x, y, z)$.
(3) The *similarity function*, $s\left(m^{(1)},\ m^{(2)}\right)$, is equal to:

$$s\left(m^{(1)},\ m^{(2)}\right)=\frac{\left|cov\left(m^{(1)},\ m^{(2)}\right)\right|}{\sigma_1\,\sigma_2}=\left|\eta\left(m^{(1)},\ m^{(2)}\right)\right|,\tag{9.12}$$

where $cov\left(m^{(1)},\ m^{(2)}\right)$ and $\eta\left(m^{(1)},\ m^{(2)}\right)$ are the covariance and correlation coefficient between the corresponding functions defined according to formulas (2.17) and (2.21) of Chap. 2. Note that the similarity function satisfies the same inequality as the luminance and contrast comparison functions due to the property of the correlation coefficient:

$$0\le s\left(m^{(1)},\ m^{(2)}\right)\le 1.\tag{9.13}$$

The strongest similarity manifests itself by the linear correlation between two images with the correlation coefficient and the similarity function equal to 1.

The *structural similarity index* (SSI) is introduced as a product of all these three functions:

$$I_{SS}\left(m^{(1)},\ m^{(2)}\right)=l^{\alpha}\left(m^{(1)},\ m^{(2)}\right)\,c^{\beta}\left(m^{(1)},\ m^{(2)}\right)\,s^{\gamma}\left(m^{(1)},\ m^{(2)}\right),\tag{9.14}$$

where α, β, and γ are the positive coefficients used to tune up the contributions of each of these functions.

Based on inequalities (9.5), (9.9), and (9.13), we conclude that

$$0\le I_{SS}\left(m^{(1)},\ m^{(2)}\right)\le 1.\tag{9.15}$$

Thus, the structural similarity index (SSI) equals 1 when all three image comparison functions, luminance l, contrast c, and similarity s, are equal to 1 as well. This means that functions $m^{(1)}(x, y, z)$ and $m^{(2)}(x, y, z)$ have the same mean values and standard deviations and they are related linearly to each other:

$$\langle m^{(1)} \rangle = \langle m^{(2)} \rangle; \tag{9.16}$$

$$\sigma_1 = \sigma_2; \tag{9.17}$$

$$m^{(1)}(x, y, z) = am^{(2)}(x, y, z) + b. \tag{9.18}$$

Substituting equation (9.18) into (9.16), we find immediately that

$$a = 1, \quad b = 0. \tag{9.19}$$

Thus, we have arrived at a very important conclusion that the structural similarity index equals one, $I_{SS} = 1$, if and only if two images are identical:

$$m^{(1)}(x, y, z) = m^{(2)}(x, y, z). \tag{9.20}$$

Oppositely, the smaller the structural similarity index is, the more significant the difference between the two images becomes.

9.1.2 *Structural Similarity Conditions*

A similarity function (9.12) of two images describing various model parameters introduced above manifests the degree of closeness of the corresponding geometrical boundaries shown in the two images, $m^{(1)}(x, y, z)$ and $m^{(2)}(x, y, z)$. In an ideal case, these boundaries represent the surfaces (in 3D cases) or lines (in 2D cases) separating the volumes or areas with different values of the corresponding parameters. However, the functions $m^{(1)}(x, y, z)$ and $m^{(2)}(x, y, z)$ describing the inverse model parameters are usually continuous and differentiable. Therefore, we should treat these boundaries as thin layers with rapid changes in the model parameters. The gradients characterize the directions of the model parameter changes' maximum rate; they are directed perpendicular to interfaces. Therefore, the closeness of the interfaces expressed in different model parameters can be measured by the degree of gradients being parallel.

Figure 9.1 illustrates this idea. We assume that the geophysical study aims to image the salt dome structure. The salt forming this structure is characterized by specific density and seismic field velocity values different from those of the surrounding medium. The left panels in Fig. 9.1 show the vertical sections of the seismic velocity (panel a) and density (panel b) distributions in the subsurface. One can clearly see the geometric structure of the salt dome in these images, which is, of course, the same for both the density and velocity models.

Let us denote by $m^{(1)}(x, y, z)$ the velocity distribution of this model, and by $m^{(2)}(x, y, z)$ the corresponding density distribution. It is also obvious that the

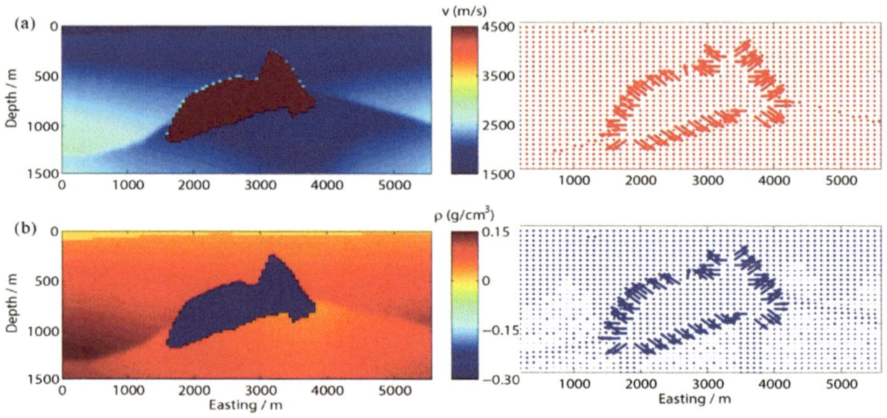

Fig. 9.1 Illustration of the concept of structural similarity. The left panels **a** and **b** show the vertical sections of the seismic velocity and density, respectively. The *red* and *blue* arrows in the right panels indicate the velocity and density gradients, respectively. The gradient vectors are parallel for the models with the same geometrical structure

directions of the gradients of the velocity distribution, $\nabla m^{(1)}(x, y, z)$, and those of the density distribution, $\nabla m^{(2)}(x, y, z)$, are perpendicular to the surface of the salt dome as shown in the right panels of Fig. 9.1. The red arrows in this figure indicate the velocity gradients, $\nabla m^{(1)}$, while the blue arrows show the density gradients, $\nabla m^{(2)}$, respectively. Therefore the gradients $\nabla m^{(1)}$ and $\nabla m^{(2)}$ have to be parallel.

In more general mathematical terms, the structural similarity of the models described by two continuously differentiable functions, $m^{(1)}(x, y, z)$ and $m^{(2)}(x, y, z)$, can be analyzed by comparing the isosurfaces of these functions.

The isosurfaces, $\Sigma^{(1)}$ and $\Sigma^{(2)}$ of functions $m^{(1)}(x, y, z)$ and $m^{(2)}(x, y, z)$ represent the points of constant values of these functions within a volume. The following equations define them:

$$\Sigma^{(1)} : m^{(1)}(x, y, z) = c^{(1)} = const;$$

$$\Sigma^{(2)} : m^{(2)}(x, y, z) = c^{(2)} = const.$$

Obviously, these isosurfaces describe the geometrical structure of the corresponding models. Therefore, we consider that two models have a similar structure if their isosurfaces are alike. We can now give a formal definition of the *structural similarity conditions*:

Definition 9.1 We say that two models $m^{(1)}$ and $m^{(2)}$ described by continuously differentiable functions, $m^{(1)}(x, y, z)$ and $m^{(2)}(x, y, z)$ have the same geometrical structure if isosurfaces of functions $m^{(1)}(x, y, z)$ and $m^{(2)}(x, y, z)$ passing through the same point in the volume coincide.

Thus, to compare the structural similarity of the two models, we should develop a measure of the difference between two sets of isosurfaces, $\Sigma^{(1)}$ and $\Sigma^{(2)}$. Since the gradients, $\nabla m^{(1)}$ and $\nabla m^{(2)}$, are always directed along the normal to isosurfaces, the gradient vectors should be parallel for the models with the same geometrical structure. Therefore, we can use the difference between gradients, $\nabla m^{(1)}$ and $\nabla m^{(2)}$, as a structural similarity measure.

Droske and Rumpf (2003) proposed to analyze the behavior of the normalized gradients:

$$\mathbf{u}^{(1)}(\mathbf{r}) = \frac{\nabla m^{(1)}(\mathbf{r})}{\left|\nabla m^{(1)}(\mathbf{r})\right|}, \quad \mathbf{u}^{(2)}(\mathbf{r}) = \frac{\nabla m^{(2)}(\mathbf{r})}{\left|\nabla m^{(1)}(\mathbf{r})\right|}, \tag{9.21}$$

where $\mathbf{u}^{(1)}(\mathbf{r})$ and $\mathbf{u}^{(2)}(\mathbf{r})$ are the unit vectors of the corresponding gradient directions at point $\mathbf{r} = (x, y, z)$.

The structural similarity between two models, $m^{(1)}$ and $m^{(2)}$, can be estimated by the L_2 norm of the difference between $\mathbf{u}^{(1)}(\mathbf{r})$ and $\mathbf{u}^{(2)}(\mathbf{r})$:

$$S_{ng}\left(m^{(1)}, m^{(2)}\right) = \left\|\mathbf{u}^{(1)} - \mathbf{u}^{(2)}\right\|_{L_2}^2 = \int_V \left|\mathbf{u}^{(1)}(\mathbf{r}) - \mathbf{u}^{(2)}(\mathbf{r})\right|^2 dv, \tag{9.22}$$

where $\|....\|_{L_2}$ denotes L_2 norm calculated over inversion domain V.

We call $S_{ng}\left(m^{(1)}, m^{(2)}\right)$ *a normalized gradient similarity functional*. The smaller the S_{ng} similarity functional is, the closer the geometrical structure of two images are to each other. Functional S_{ng} provides a good quantitative estimate of the degree of structural similarity between two images. However, the calculation of this functional is complicated by singularities which may occur in the points or areas where gradients are close to zero.

Gallardo and Meju (2003) proposed an idea of measuring the structural similarity of two models by calculating the cross products of the gradients of the corresponding functions, $m^{(1)}(x, y, z)$ and $m^{(2)}(x, y, z)$, describing the spatial distributions of their properties:

$$S_{cg}\left(m^{(1)}, m^{(2)}\right) = \left\|\nabla m^{(1)} \times \nabla m^{(2)}\right\|_{L_2}^2$$

$$= \int_V \left|\nabla m^{(1)}(\mathbf{r}) \times \nabla m^{(2)}(\mathbf{r})\right|^2 dv = \min \tag{9.23}$$

Equation (9.23) is called a *cross-gradient condition*. It enforces the structural similarity between the isosurfaces (or isolines in 2D case) of the different model parameters, $m^{(1)}(x, y, z)$ and $m^{(2)}(x, y, z)$. This condition is widely used in geophysical applications (e.g., Gallardo and Meju 2004, 2007, 2011, Gallardo et al. 2005, 2007, 2009); however, it can also be applied to enforce the structural similarities of the optical images, providing an alternative approach to the method based on the structural similarity index.

We should note that in numerical implementation, it is convenient to present the cross-gradient condition (9.23) using dot-product operation.

Indeed, the absolute value of the cross product of two vectors, **a** and **b**, is defined as follows:

$$|\mathbf{a} \times \mathbf{b}| = |\mathbf{a}|\ |\mathbf{b}|\ \sin\varphi, \tag{9.24}$$

where φ is an angle between two vectors.

The dot product of two vectors, **a** and **b**, is equal to

$$\mathbf{a} \cdot \mathbf{b} = |\mathbf{a}|\ |\mathbf{b}|\ \cos\varphi. \tag{9.25}$$

From Eqs. (9.24) and (9.25), we can see at once that

$$|\mathbf{a} \times \mathbf{b}|^2 = |\mathbf{a}|^2\,|\mathbf{b}|^2\,\sin^2\varphi = |\mathbf{a}|^{\ 2}\,|\mathbf{b}|^2\left(1 - \cos^2\varphi\right) = |\mathbf{a}|^2\,|\mathbf{b}|^2 - |\mathbf{a}\cdot\mathbf{b}|^2. \tag{9.26}$$

Therefore, based on vector identity (9.26), the cross-gradient condition (9.23) can be written in an equivalent form using the dot product of gradients as follows (Haber and Gazit 2013):

$$S_{cg}\left(m^{(1)}, m^{(2)}\right) = S_{dg}\left(m^{(1)}, m^{(2)}\right) =$$

$$= \int_V \left[\left|\nabla m^{(1)}(\mathbf{r})\right|^2 \left|\nabla m^{(2)}(\mathbf{r})\right|^2 - \left|\nabla m^{(1)}(\mathbf{r}) \cdot \nabla m^{(2)}(\mathbf{r})\right|^2\right] dv = \min. \tag{9.27}$$

Equation (9.27) is called a *dot-gradient condition*. We will find below that the dot-gradient condition can be efficiently used to enforce the structural similarity in the regularized inversion.

There is another equivalent form of this condition which arises from the following algebraic identity:

$$|\mathbf{a}|^2\,|\mathbf{b}|^2 - |\mathbf{a}\cdot\mathbf{b}|^2 = \begin{vmatrix} (\mathbf{a}\cdot\mathbf{a}) & (\mathbf{a}\cdot\mathbf{b}) \\ (\mathbf{b}\cdot\mathbf{a}) & (\mathbf{b}\cdot\mathbf{b}) \end{vmatrix}. \tag{9.28}$$

Therefore, we can write

$$S_{cg}\left(m^{(1)}, m^{(2)}\right) = S_{dg}\left(m^{(1)}, m^{(2)}\right) = S_{Gram}\left(m^{(1)}, m^{(2)}\right) =$$

$$= \int_V \begin{vmatrix} \left(\nabla m^{(1)}(\mathbf{r}) \cdot \nabla m^{(1)}(\mathbf{r})\right) & \left(\nabla m^{(1)}(\mathbf{r}) \cdot \nabla m^{(2)}(\mathbf{r})\right) \\ \left(\nabla m^{(2)}(\mathbf{r}) \cdot \nabla m^{(1)}(\mathbf{r})\right) & \left(\nabla m^{(2)}(\mathbf{r}) \cdot \nabla m^{(2)}(\mathbf{r})\right) \end{vmatrix} dv = \min. \tag{9.29}$$

We will show in Chap. 12 that the integrand in Eq. (9.29) is *Gramian*, $G_\nabla\left(\nabla m^{(1)}, \nabla m^{(2)}\right)$, formed by the gradients of the model parameters, $\nabla m^{(1)}(\mathbf{r})$ and $\nabla m^{(2)}(\mathbf{r})$:

$$G_\nabla\left(\nabla m^{(1)}, \nabla m^{(2)}\right) = \begin{vmatrix} \left(\nabla m^{(1)} \cdot \nabla m^{(1)}\right) & \left(\nabla m^{(1)} \cdot \nabla m^{(2)}\right) \\ \left(\nabla m^{(2)} \cdot \nabla m^{(1)}\right) & \left(\nabla m^{(2)} \cdot \nabla m^{(2)}\right) \end{vmatrix}. \tag{9.30}$$

The Gramian is equal to zero if vectors $\nabla m^{(1)}$ and $\nabla m^{(2)}$ are linearly dependent, which means they are parallel. Thus, substituting equation (9.30) into (9.29), we arrive at the following equivalent condition enforcing the structural similarity of the two models:

$$S_G\left(m^{(1)}, m^{(2)}\right) = \int_V G_\nabla\left(\nabla m^{(1)}, \nabla m^{(2)}\right) dv = \min . \tag{9.31}$$

This representation of the cross-gradient condition provides a link to the Gramian constraint for joint structural inversion introduced by Zhdanov et al. (2012). We discuss the general Gramian method of joint structural inversion in Chap. 12.

Note that all three forms of the structural similarity conditions, cross-gradient, dot-gradient, and Gramian, are mathematically equivalent; however, there is a significant difference in the numerical implementation of these conditions, which we will discuss below.

9.2 The Maximum Structural Similarity Index Inversion

This section demonstrates how the structural similarity index can be used in the joint inversion of multiphysics data. For example, let us consider an inverse problem for two data sets. The following system of operator equations describes this problem:

$$d^{(1)} = A^{(1)}(m^{(1)}), \text{ and } d^{(2)} = A^{(2)}(m^{(2)}), \tag{9.32}$$

where functions $m^{(1)}(x, y, z)$ and $m^{(2)}(x, y, z)$ are the volume distributions of two different model parameters.

The regularized joint solution of equations (9.32) can be obtained by minimization of the following parametric functional:

$$P^{\alpha}_{I_{SS}}(m^{(1)}, m^{(2)}) = \sum_{i=1}^{2} \left\| A^{(i)}(m^{(i)}) - d^{(i)} \right\|_D^2 + \alpha S_{I_{SS}}\left(m^{(1)}, m^{(2)}\right) = \min . \tag{9.33}$$

Note that in expression (9.33) and everywhere below in this chapter, we consider $m^{(1)}$ and $m^{(2)}$ being the dimensionless weighted model parameters, as defined in Chap. 8.

In the last formula, we have introduced stabilizing functional $S_{I_{SS}}$ as follows:

$$S_{I_{SS}}\left(m^{(1)}, m^{(2)}\right) = 1 - I_{SS}\left(m^{(1)}, m^{(2)}\right). \tag{9.34}$$

According to inequality (9.15), functional $S_{I_{SS}}$ is always positive and it reaches zero value if and only if two solutions, $m^{(1)}(x, y, z)$ and $m^{(2)}(x, y, z)$, coincide:

$$m^{(1)}(x, y, z) = m^{(2)}(x, y, z). \tag{9.35}$$

Thus, the inversion based on the maximum structural similarity index imposes a very strong requirement on the joint inversion to produce identical results for both parameters. This requirement is helpful in image reconstruction (see Chap. 14); however, it can rarely be used in multiphysics inversion when considering the different physical properties of the target. These properties should have a structural similarity but do not need to have the same values.

I should also note that functional $S_{I_{SS}}\left(m^{(1)}, m^{(2)}\right)$ is not quadratic, which makes minimization of parametric functional $P^{\alpha}_{I_{SS}}(m^{(1)}, m^{(2)})$ a computationally challenging problem. I will discuss in Chap. 14 how this difficulty can be effectively resolved.

9.3 The Method of Parallel Gradients

We consider again the joint inverse problem for two data sets described by equation (9.32). The structural similarities between two functions, $m^{(1)}(x, y, z)$ and $m^{(2)}(x, y, z)$, can be measured by the cross-gradient condition (9.23), the dot-gradient condition (9.27), or by the Gramian condition (9.31). All these conditions are based on the concept of parallel gradients of functions representing geometrically similar inverse models. We call the inversion algorithms based on these conditions *the parallel gradient methods.*

The corresponding regularized inversion can be represented as a minimization of the following parametric functional:

$$P^{\alpha}(m^{(1)}, m^{(2)}) = \sum_{i=1}^{2} \left\| A^{(i)}(m^{(i)}) - d^{(i)} \right\|_D^2 + \alpha S_{cg,dg,G}\left(m^{(1)}, m^{(2)}\right) = \min, \quad (9.36)$$

where $S_{cg,dg,Gram}$ represents one of the structural similarity stabilizing functionals, S_{cg}, S_{dg}, or S_G, introduced above in Eqs. (9.23), (9.27), or (9.31), respectively.

9.3.1 *Steepest Ascent Directions for the Structural Similarity Functionals*

One can find the minimum of parametric functional (9.36) using any of the minimization methods discussed in Chap. 7. The main problem with the practical implementation of these methods is related to computing the steepest ascent direction (gradient) of the corresponding stabilizing functionals, S_{cg}, S_{dg}, or S_G. In the case of the cross-gradient functional, S_{cg}, one has to apply some approximation of this functional to find the steepest descent direction (Meju and Gallardo 2016). This results in lower accuracy and slower convergence of the inversion procedure.

I will demonstrate below that this problem can be solved rigorously without an approximation by using the Gramian form of structural similarity conditions (9.31):

$$S_G = \int_V \left[\left(\nabla m^{(1)}(\mathbf{r}) \cdot \nabla m^{(1)}(\mathbf{r}) \right) \left(\nabla m^{(2)}(\mathbf{r}) \cdot \nabla m^{(2)}(\mathbf{r}) \right) - \right.$$

$$\left. - \left(\nabla m^{(1)}(\mathbf{r}) \cdot \nabla m^{(2)}(\mathbf{r}) \right)^2 \right] dv. \tag{9.37}$$

Let us calculate the first variation of the functional S_{Gram}:

$$\delta S_G(m^{(1)}, m^{(2)}) = \delta_{m^{(1)}} S_G(m^{(1)}, m^{(2)}) + \delta_{m^{(2)}} S_G(m^{(1)}, m^{(2)}). \tag{9.38}$$

The partial variations of the Gramian stabilizer, $\delta_{m^{(1)}} S_G$ and $\delta_{m^{(2)}} S_G$ can be calculated as follows:

$$\delta_{m^{(1)}} S_G = 2 \int_V \left[\left(\delta \nabla m^{(1)}(\mathbf{r}) \cdot \nabla m^{(1)}(\mathbf{r}) \right) \left(\nabla m^{(2)}(\mathbf{r}) \cdot \nabla m^{(2)}(\mathbf{r}) \right) - \right.$$

$$\left. - \left(\delta \nabla m^{(1)}(\mathbf{r}) \cdot \nabla m^{(2)}(\mathbf{r}) \right) \left(\nabla m^{(1)}(\mathbf{r}) \cdot \nabla m^{(2)}(\mathbf{r}) \right) \right] dv, \tag{9.39}$$

$$\delta_{m^{(2)}} S_G = 2 \int_V \left[\left(\delta \nabla m^{(2)}(\mathbf{r}) \cdot \nabla m^{(2)}(\mathbf{r}) \right) \left(\nabla m^{(1)}(\mathbf{r}) \cdot \nabla m^{(1)}(\mathbf{r}) \right) - \right.$$

$$\left. - \left(\delta \nabla m^{(2)}(\mathbf{r}) \cdot \nabla m^{(1)}(\mathbf{r}) \right) \left(\nabla m^{(1)}(\mathbf{r}) \cdot \nabla m^{(2)}(\mathbf{r}) \right) \right] dv. \tag{9.40}$$

Taking $\delta \nabla m^{(1)}$ out of the brackets in Eq. (9.39), we arrive at the following formula:

$$\delta_{m^{(1)}} S_G = 2 \int_V \left(\delta \nabla m^{(1)}(\mathbf{r}) \cdot \left[\nabla m^{(1)}(\mathbf{r}) \left(\nabla m^{(2)}(\mathbf{r}) \cdot \nabla m^{(2)}(\mathbf{r}) \right) - \right. \right.$$

$$\left. \left. - \nabla m^{(2)}(\mathbf{r}) \left(\nabla m^{(1)}(\mathbf{r}) \cdot \nabla m^{(2)}(\mathbf{r}) \right) \right] dv \right.$$

$$= 2 \int_V \left(\nabla \delta m^{(1)}(\mathbf{r}) \cdot \mathbf{b}_{m^{(1)}}(\mathbf{r}) \right) dv, \tag{9.41}$$

where

$$\mathbf{b}_{m^{(1)}}(\mathbf{r}) = \nabla m^{(1)}(\mathbf{r}) \left(\nabla m^{(2)}(\mathbf{r}) \cdot \nabla m^{(2)}(\mathbf{r}) \right) - \nabla m^{(2)}(\mathbf{r}) \left(\nabla m^{(1)}(\mathbf{r}) \cdot \nabla m^{(2)}(\mathbf{r}) \right), \tag{9.42}$$

and we take into account the following equality:

$$\delta \nabla m^{(1)}(\mathbf{r}) = \nabla \delta m^{(1)}(\mathbf{r}).$$

In order to obtain the steepest ascent direction of the Gramian functional, S_{Gram}, we have to find the variation of the model, $\delta m^{(1)}(\mathbf{r})$, which would result in positive variation of this functional:

$$\delta_{m^{(1)}} S_G = 2\int_V \left(\nabla\delta m^{(1)}(\mathbf{r}) \cdot \mathbf{b}_{m^{(1)}}(\mathbf{r})\right) dv > 0. \tag{9.43}$$

Inequality (9.43), however, contains the gradient of the variation, $\nabla\delta m^{(1)}$. We can transform the integrand in the equation into the form which would explicitly include $\delta m^{(1)}(\mathbf{r})$ by moving the del operator from the first to the second multiplier in the dot product. This can be done by integrating by parts using the Gauss theorem.

Indeed, let us write the following differential identity for two continuously differentiable in domain V vector fields, $\delta m^{(1)}(\mathbf{r})$ and $\mathbf{b}_{m^{(1)}}(\mathbf{r})$:

$$\nabla \cdot \left[\delta m^{(1)}(\mathbf{r})\, \mathbf{b}(\mathbf{r})\right] = \nabla\delta m^{(1)}(\mathbf{r}) \cdot \mathbf{b}_{m^{(1)}}(\mathbf{r}) + \delta m^{(1)}(\mathbf{r}) \left[\nabla \cdot \mathbf{b}(\mathbf{r})\right] \tag{9.44}$$

Integrating both sides of identity (9.44) over domain V and applying the Gauss theorem to the integral over divergence, $\nabla \cdot \left[\delta m^{(1)}(\mathbf{r})\, \mathbf{b}(\mathbf{r})\right]$, we arrive at the following integral equality:

$$\int_{\partial V} \left[\delta m^{(1)}(\mathbf{r})\, \mathbf{b}_{m^{(1)}}(\mathbf{r})\right] \cdot \mathbf{n} ds$$

$$= \int_V \nabla\delta m^{(1)}(\mathbf{r}) \cdot \mathbf{b}_{m^{(1)}}(\mathbf{r})\, dv + \int_V \delta m^{(1)}(\mathbf{r}) \left[\nabla \cdot \mathbf{b}_{m^{(1)}}(\mathbf{r})\right] dv, \tag{9.45}$$

where ∂V is a boundary of domain V, and $\mathbf{n}$ is a unit vector of the normal to ∂V directed outward from domain V. We assume that at the boundary, ∂V, both model parameters do not change, with their gradients equal to zero:

$$\nabla m^{(1)}\big|_{\partial V} = 0, \;\; \nabla m^{(2)}\big|_{\partial V} = 0.$$

Therefore, function $\mathbf{b}_{m^{(1)}}$ is equal to zero at the boundary as well, and the surface integral in the left side of Eq. (9.45) vanishes. Thus, we obtain the following integral equality:

$$\int_V \nabla\delta m^{(1)}(\mathbf{r}) \cdot \mathbf{b}_{m^{(1)}}(\mathbf{r})\, dv = -\int_V \delta m^{(1)}(\mathbf{r}) \left[\nabla \cdot \mathbf{b}_{m^{(1)}}(\mathbf{r})\right] dv. \tag{9.46}$$

Substituting equation (9.46) into Eq. (9.41) for the partial variation, $\delta_{m^{(1)}} S_{Gram}$, we obtain the following important result:

$$\delta_{m^{(1)}} S_G = -2\int_V \delta m^{(1)}(\mathbf{r}) \left[\nabla \cdot \mathbf{b}_{m^{(1)}}(\mathbf{r})\right] dv = 2\left(\delta m^{(1)}, l^{(1)}_{Gram}\right)_{L_2(V)}, \tag{9.47}$$

where

$$l_{Gram}^{(1)}(\mathbf{r}) = -\nabla \cdot \mathbf{b}_{m^{(1)}}(\mathbf{r})$$

$$= \nabla \cdot \left[\nabla m^{(2)}(\mathbf{r}) \left(\nabla m^{(1)}(\mathbf{r}) \cdot \nabla m^{(2)}(\mathbf{r})\right)\right] - \nabla \cdot \left[\nabla m^{(1)}(\mathbf{r}) \left|\nabla m^{(2)}(\mathbf{r})\right|^2\right] \quad (9.48)$$

Function $l_{Gram}^{(1)}(\mathbf{r})$ represents the direction of the steepest ascent with respect to model $m^{(1)}$. Indeed, we can select the variation $\delta m^{(1)}(\mathbf{r})$ to be proportional to function $l_G^{(1)}(\mathbf{r})$:

$$\delta m^{(1)}(\mathbf{r}) = k^{(1)} l_{Gram}^{(1)}(\mathbf{r}). \quad (9.49)$$

Substituting equation (9.49) into (9.47), we have

$$\delta_{m^{(1)}} S_G = 2\left(\delta m^{(1)}, l_{Gram}^{(1)}\right)_{L_2(V)}$$

$$= 2k^{(1)}\left(l_{Gram}^{(1)}, l_{Gram}^{(1)}\right)_{L_2(V)} = 2k^{(1)}\left\|l_G^{(1)}\right\|_{L_2(V)}^2 > 0, \quad (9.50)$$

if $k^{(1)} > 0$.

The steepest descent direction is opposite to vector $l_{Gram}^{(1)}(\mathbf{r})$. Therefore, the following equation provides the expression for the variation of the model $m^{(1)}$ corresponding to the steepest descent direction for the Gramian stabilizing functional:

$$\delta m^{(1)}(\mathbf{r}) = -k^{(1)} l_{Gram}^{(1)}(\mathbf{r}). \quad (9.51)$$

We can apply similar derivations to Eq. (9.40) for the partial variation, $\delta_{m^{(2)}} S_G$, of the Gramian stabilizer with respect to parameter $m^{(2)}$.

$$\delta_{m^{(2)}} S_G = 2\int_V \left(\delta \nabla m^{(2)}(\mathbf{r}) \cdot \left[\nabla m^{(2)}(\mathbf{r}) \left(\nabla m^{(1)}(\mathbf{r}) \cdot \nabla m^{(1)}(\mathbf{r})\right) - \right.\right.$$

$$\left.\left. - \nabla m^{(1)}(\mathbf{r}) \left(\nabla m^{(1)}(\mathbf{r}) \cdot \nabla m^{(2)}(\mathbf{r})\right)\right] dv\right.$$

$$= 2\int_V \left(\nabla \delta m^{(2)}(\mathbf{r}) \cdot \mathbf{b}_{m^{(2)}}(\mathbf{r})\right) dv, \quad (9.52)$$

where

$$\mathbf{b}_{m^{(1)}}(\mathbf{r}) = \nabla m^{(2)}(\mathbf{r}) \left(\nabla m^{(1)}(\mathbf{r}) \cdot \nabla m^{(1)}(\mathbf{r})\right) - \nabla m^{(1)}(\mathbf{r}) \left(\nabla m^{(1)}(\mathbf{r}) \cdot \nabla m^{(2)}(\mathbf{r})\right) \quad (9.53)$$

Applying integration by parts, we move the del operator from the first to the second multiplier in Eq. (9.52):

$$\delta_{m^{(2)}} S_G = -2 \int_V \delta m^{(2)} (\mathbf{r}) \left[\nabla \cdot \mathbf{b}_{m^{(2)}} (\mathbf{r})\right] dv. \tag{9.54}$$

From the last formula, we obtain the following expression for the steepest ascent direction, $l_G^{(2)}$, with respect to parameter $m^{(2)}$:

$$l_G^{(2)} = -\nabla \cdot \mathbf{b}_{m^{(2)}} (\mathbf{r})$$

$$= \nabla \cdot \left[\nabla m^{(1)} (\mathbf{r}) \left(\nabla m^{(1)} (\mathbf{r}) \cdot \nabla m^{(2)} (\mathbf{r})\right)\right] - \nabla \cdot \left[\nabla m^{(2)} (\mathbf{r}) \left|\nabla m^{(1)} (\mathbf{r})\right|^2\right]. \tag{9.55}$$

Substituting equation (9.55) into (9.54), we can write the expression for the variation of the Gramian structural stabilizer with respect to $m^{(2)}$ as follows:

$$\delta_{m^{(2)}} S_G = 2 \left(\delta m^{(2)}, l_G^{(2)}\right)_{L_2(V)}. \tag{9.56}$$

Finally, taking into account expressions (9.47) into (9.56), we arrive at the compact form of the first variation of the functional S_{Gram}:

$$\delta S_G = 2 \sum_{i=1}^{2} \left(\delta m^{(i)},\ l_{Gram}^{(i)}\right)_{L_2(V)}, \tag{9.57}$$

9.3.2 Joint Structural Inversion of Two Data Sets

We can now revisit the minimization problem (9.36):

$$P^{\alpha}(m^{(1)}, m^{(2)}) = \sum_{i=1}^{2} \| A^{(i)}(m^{(i)}) - d^{(i)} \|_D^2 + \alpha S_G \left(m^{(1)}, m^{(2)}\right) = \min, \tag{9.58}$$

where we keep Gramian functional, S_{Gram}, only, considering that all three structural similarity stabilizing functionals, S_{cg}, S_{dg}, or S_G, are equal.

In order to solve this minimization problem, we calculate the first variation of the parametric functional $P^{\alpha}(m^{(1)}, m^{(2)})$:

$$\delta P^{\alpha}(m^{(1)}, m^{(2)})$$

$$= 2 \sum_{i=1}^{2} \left(\delta A^{(i)}(m^{(i)}),\ A^{(i)}(m^{(i)}) - d^{(i)}\right)_D +$$

$$+ \alpha \delta S_G(m^{(1)}, m^{(2)}). \tag{9.59}$$

Taking into consideration that operators $A^{(i)}$ are differentiable, we can write

$$\delta A^{(i)}(m^{(i)}) = F_m^{(i)} \delta m^{(i)}, \; i = 1, 2, \tag{9.60}$$

where $F_m^{(i)}$ are linear operators of the Fréchet derivative of $A^{(i)}$.

We have obtained the above Eq. (9.57) for the variation of the Gramian stabilizer. Substituting expressions (9.60) and (9.57) into formula (9.59), we find the equation for the variation of the parametric functional:

$$\delta P^{\alpha}(m^{(1)}, m^{(2)}) =$$

$$= 2 \sum_{i=1}^{2} \left(\delta m^{(i)}, \left[F_m^{(i)\star} \left(A^{(i)}(m^{(i)}) - d^{(i)} \right) + \alpha l_G^{(i)} \right] \right)_{L_2(V)}, \tag{9.61}$$

where $F_m^{(i)\star}$ are the adjoint Fréchet derivative operators.

From Eq. (9.61), we obtain at once the directions of the steepest ascent of the functional P^{α}:

$$l_{\alpha}^{(i)} = F_m^{(i)\star} \left(A^{(i)}(m^{(i)}) - d^{(i)} \right) + \alpha l_G^{(i)}. \tag{9.62}$$

We can now apply one of the gradient-type algorithms discussed in Chap. 7 to solve the minimization problem (9.36).

9.3.3 *Joint Structural Inversion of Multiple Data Sets*

Thus, we have developed a unified approach to joint inversion based on the concept of parallel gradients. This approach can be naturally expanded to the case of joint inversion of multiple data sets considered above in Chap. 8:

$$d^{(i)} = A^{(i)}(m^{(i)}), \; i = 1, 2, 3, \ldots, N. \tag{9.63}$$

In this case, we can expand the parametric functional (9.58) as follows:

$$P^{\alpha}(m^{(1)}, m^{(2)}, \ldots, m^{(N)}) = \sum_{i=1}^{N} \left\| A^{(i)}(m^{(i)}) - d^{(i)} \right\|_D^2 + \alpha \sum_{j=2}^{N} S_G\left(m^{(1)}, m^{(j)}\right) = \min, \tag{9.64}$$

where the stabilizing term in parametric functional, $\sum_{j=2}^{N} S_G\left(m^{(1)}, m^{(j)}\right)$, is a superposition of the structural Gramians between the first and all other physical model parameters:

$$\sum_{j=2}^{N} S_G\left(m^{(1)}, m^{(j)}\right) = S_G\left(m^{(1)}, m^{(2)}\right) + S_G\left(m^{(1)}, m^{(3)}\right) + \cdots + S_G\left(m^{(1)}, m^{(N)}\right). \tag{9.65}$$

Minimization of this stabilizing functional keeps all gradient vectors, $\nabla m^{(1)}, \nabla m^{(2)}, \ldots, \nabla m^{(N)}$, parallel to each other. Considering that the gradient directions are orthogonal to the interfaces between the structures with contrasting physical properties, this condition results in structural similarities between the inverse models describing different physical properties (Zhdanov et al. 2021).

References and Recommended Reading to This Chapter

Colombo D, De Stefano M (2007) Geophysical modeling via simultaneous joint inversion of seismic, gravity, and electromagnetic data. Appl prestack Depth Imaging: Lead Edge 26:326–331

Droske M, Rumpf M (2003) A variational approach to nonrigid morphological image registration. SIAM J Appl Math 64(2):668–687

Gallardo LA (2007) Multiple cross-gradient joint inversion for geospectral imaging. Geophys Res Lett 34:L19301

Gallardo LA, Meju MA (2003) Characterization of heterogeneous near-surface materials by joint 2D inversion of DC resistivity and seismic data. Geophys Res Let 30:1658–1661

Gallardo LA, Meju, MA (2004) Joint two-dimensional DC resistivity and seismic travel-time inversion with cross-gradients constraints: Journal of Geophysical Research 109:B03311

Gallardo LA, Meju MA (2007) Joint two-dimensional cross-gradient imaging of magnetotelluric and seismic traveltime data for structural and lithological classification. Geophys J Int 169(3):1261–1272

Gallardo LA, Meju MA (2011) Structure-coupled multi-physics imaging in geophysical sciences. Rev Geophys 49:RG1003

Gallardo LA, Meju MA, Pérez-Flores M (2005) A quadratic programming approach for joint image reconstruction: mathematical and geophysical examples. Inverse Probl 21:435–452

Haber E, Gazit MH (2013) Model fusion and joint inversion. Surv Geophys 34:675–695

Hu WY, Abubakar A, Habashy TM (2009) Joint electromagnetic and seismic inversion using structural constraints. Geophysics 74:R99–R109

Meju MA, Gallardo LA (2016) Structural coupling approaches in integrated geophysical imaging. In: Integrated imaging of the earth, Chap 4. American Geophysical Union (AGU), pp 49–67

Wang Z, Bovik A, Sheikh H, Simoncelli E (2004) Image quality assessment: from error visibility to structural similarity. Proc IEEE Trans Image Proc 13(4):600–612

Zhdanov MS, Jorgensen M, Cox L (2021) Advanced methods of joint inversion of multiphysics data for mineral exploration. Geosciences 11:262

Chapter 10
Joint Focusing Inversion of Multiphysics Data

Abstract This chapter discusses the method of joint structural inversion based on joint focusing stabilizers. A family of joint focusing stabilizers is introduced, including joint minimum support, Lp-norm joint minimum support, joint total variation, and joint minimum gradient support stabilizing functionals. The properties of these stabilizers and their first variations are analyzed in detail. These stabilizers force the domains corresponding to different physical models to coincide. The chapter concludes with a description of the inversion methods based on joint focusing stabilizers.

Keywords Joint focusing stabilizer · Joint minimum support · Joint minimum gradient support · Joint total variation

In Chap. 9, we introduced two approaches to the joint inversion that provided structural similarity between different inverse models. One maximized the structural similarity index (SSI). Another enforced the parallelism of the model parameter gradients. The current chapter discusses another method of joint structural inversion based on joint focusing stabilizers. We have shown in Chap. 4 that focusing stabilizers help produce the inverse images with the sharp model parameter contrast boundaries. Molodtsov and Troyan (2017) and Zhdanov and Čuma (2018) introduced joint focusing stabilizers designed to ensure that these sharp boundaries for different model parameters coincide. I will present below some examples of joint focusing stabilizers and discuss the methods of joint focusing inversion using these stabilizers.

10.1 Joint Focusing Stabilizers and Their Properties

We consider a multimodal inverse problem for multiple physical data sets, which is described by the following operator equations:

$$\mathbf{d}^{(i)} = \mathbf{A}^{(i)}(\mathbf{m}^{(i)}), \quad i = 1, 2, 3, \ldots, N, \tag{10.1}$$

M. S. Zhdanov, *Advanced Methods of Joint Inversion and Fusion of Multiphysics Data*, Advances in Geological Science,
https://doi.org/10.1007/978-981-99-6722-3_10

where $\mathbf{A}^{(i)}$ are nonlinear operators, $\mathbf{d}^{(i)}$ $(i = 1, 2, 3, \ldots, N)$ are different observed data sets; $\mathbf{m}^{(i)}$ $(i = 1, 2, 3, \ldots, N)$ are the sets of model parameters, represented by functions $m^{(i)} = m^{(i)}(x, y, z) = m^{(i)}(\mathbf{r})$, defined in some domain, $\mathbf{r} \in V$.

We should note that, for convenience, we denote by the bold fonts, $\mathbf{m}^{(i)}$, the discrete representation on some 3D grid of the corresponding continuously differentiable function, $m^{(i)}(\mathbf{r})$, describing the volume distribution of the model parameter. In Eq. (10.1), the observed data shown by the bold fonts, $\mathbf{d}^{(i)}$, also denotes the vector of discrete observed data. Note also that in expression (10.1) and everywhere below in this Chapter, we consider $d^{(1)}, d^{(2)}, \ldots, d^{(n)}$ and $m^{(1)}, m^{(2)}, \ldots, m^{(n)}$ being the dimensionless weighted data and model parameters, as defined in Chap. 8.

We now introduce a family of joint focusing stabilizing functionals, which enforce the structural similarity of the multimodal inverse images.

10.1.1 Joint Minimum Support (JMS) Stabilizer

The joint minimum support (JMS) stabilizer is introduced as follows (Zhdanov and Čuma 2018; Tu and Zhdanov 2022):

$$S_{JMS} = \int_V \frac{\sum_{i=1}^N \left(m^{(i)} - m_{apr}^{(i)}\right)^2}{\sum_{i=1}^N \left(m^{(i)} - m_{apr}^{(i)}\right)^2 + e^2} dv, \tag{10.2}$$

where e is a small number (focusing parameter) introduced to avoid singularity when $m^{(i)} \approx m_{apr}^{(i)}$.

For every model parameter $m^{(i)}$ $(i = 1, 2, \ldots, N)$, we can introduce support of this specific parameter (denoted spt $m^{(i)}$) as the combined closed subdomains of V where $m^{(i)} \neq m_{apr}^{(i)}$. The joint support of all model parameters, spt m, is a union set of all model parameter supports:

$$\text{spt } m = \bigcup_{i=1}^{N} \text{spt } m^{(i)}. \tag{10.3}$$

In other words, the joint support, spt m, is the volume of the subdomain where at least one of the model parameters deviates from the a priori model. It can be demonstrated that the joint minimum support stabilizer is proportional (for a small e) to the joint model parameter support, spt m. Indeed, by adding and subtracting parameter e^2 in the nominator of the integrand in Eq. (10.2), we can write S_{JMS} in the following form:

$$S_{JMS} = \int_V \frac{\sum_{i=1}^{N} \left(m^{(i)}-m_{apr}^{(i)}\right)^2 + e^2 - e^2}{\sum_{i=1}^{N} \left(m^{(i)}-m_{apr}^{(i)}\right)^2 + e^2} dv$$

$$= \int_{\text{spt } m} \left[1 - \frac{e^2}{\sum_{i=1}^{N} \left(m^{(i)}-m_{apr}^{(i)}\right)^2 + e^2} \right] dv, \tag{10.4}$$

where the volume of integration was reduced to the joint model parameter support because outside spt m, we have

$$m^{(i)}=m_{apr}^{(i)} \text{ for all } i = 1, 2, \ldots, N,$$

and integrand in Eq. (10.2) for JMS stabilizer is zero.

From Eq. (10.4), we obtain at once that

$$S_{JMS} = \text{spt } m - e^2 \int_{\text{spt } m} \frac{1}{\sum_{i=1}^{N} \left(m^{(i)}-m_{apr}^{(i)}\right)^2 + e^2} dv. \tag{10.5}$$

Thus, we can see that

$$S_{JMS} \to \text{spt } m, \text{ if } e \to 0, \tag{10.6}$$

which was to be proved.

Therefore, the JMS functional can be used to minimize the total volume of the domain with the nonzero departure of any of the multimodal parameters from the given *a priori* model. In other words, the JMS functional, similar to the minimum support stabilizing functional introduced in Chap. 4, provides the models with the smallest possible domains of anomalous parameters distribution. In addition, all these domains corresponding to different physical model parameters should coincide in order to minimize the joint model parameter support, spt m.

Figure 10.1 illustrates the action of the JMS stabilizer. This stabilizer is proportional to the combined subdomain volume where at least one of the model parameters deviates from the a priori model, m_{apr}. In this example, we have three subdomains, A, B, and C, where $m^{(1)} \neq m_{apr}^{(1)}$, $m^{(2)} \neq m_{apr}^{(2)}$, and $m^{(3)} \neq m_{apr}^{(3)}$, respectively. These subdomains may overlap (e.g., subdomains A and B) or be separate (e.g., subdomain C). The joint model parameter support, spt m, is a combined volume of subdomains A, B, and C: spt m = spt $m^{(1)} \cup$ spt $m^{(2)} \cup$ spt $m^{(3)}$. Minimizing the JMS functional decreases the joint support and, therefore, shrinks the combined volume. This forces all three subdomains to merge into one domain with the smallest possible volume, as shown on the right side of Fig. 10.1. Thus, the JMS functional enforces the structural similarities of the models representing different physical properties in the inversion.

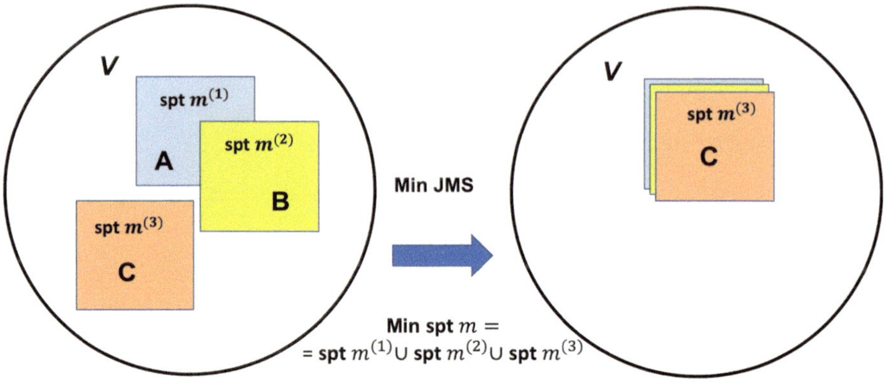

Fig. 10.1 Schematic illustration of the property of the joint minimum support (JMS) stabilizer. This stabilizer is proportional to the combined volume of subdomains A, B, and C with anomalous model parameters. Minimizing the JMS functional decreases the joint support and, therefore, shrinks the combined volume. This forces all three subdomains to merge into one anomalous domain with the smallest possible volume, as shown on the right side of the Figure

10.1.2 L_p-Norm Joint Minimum Support Stabilizers ($JMSL_p$)

We can also introduce L_p-norm joint minimum support functionals (JMSL_p) as follows:

$$S_{JMSL_p} = \int_V \frac{\sum_{i=1}^N \left|m^{(i)} - m_{apr}^{(i)}\right|^p}{\sum_{i=1}^N \left|m^{(i)} - m_{apr}^{(i)}\right|^p + e^p} dv, \; 0 \le p < \infty. \tag{10.7}$$

Repeating the derivation above for the JMS functional, we can show that JMSL_p functional is also equal (for a small e) to the joint model parameter support, spt m:

$$S_{JMSL_p} = \int_{\text{spt } m} \left[1 - \frac{e^p}{\sum_{i=1}^N \left|m^{(i)} - m_{apr}^{(i)}\right|^p + e^p} \right] dv$$

$$= \text{spt } m - e^p \int_{\text{spt } m} \frac{1}{\sum_{i=1}^N \left|m^{(i)} - m_{apr}^{(i)}\right|^p + e^p} dv \to \text{spt } m, \; \text{if } e \to 0. \tag{10.8}$$

By changing factor p and focusing parameter e, one can control the inverse images' degree of focusing. Increasing factor p or decreasing parameter e results in sharper inverse images. At the same time, by minimizing the JMSL_p functional, we merge the domains occupied by different anomalous model parameters. Therefore, all inverse models produced by the joint focusing inversion will have a similar structure.

10.1.3 Joint Total Variation (JTV) Stabilizer

The joint total variation (JTV) stabilizer can be introduced as a generalization of the total variation (TV) method (Haber and Gazit 2013; Molodtsov and Troyan 2017). Indeed, we can apply expression (4.25) to the superposition of the gradients of different model parameters as follows:

$$S_{JTV} = \int_V \sqrt{\sum_{i=1}^{N} \left(\nabla m^{(i)} \cdot \nabla m^{(i)}\right) + e^2 dv}, \tag{10.9}$$

where e is a small number.

This stabilizer, similar to the conventional TV stabilizer, tends to generate a sparse model (Aster et al. 2013). In places where all models experience discontinuities or have sharp boundaries, the JTV stabilizer forces these discontinuities to coincide.

Similarly, we can introduce a joint L_p-norm of the gradient stabilizing functional (JTVL_p) by extending formula (4.23) to the case of multi-modal parameters:

$$S_{JTVL_p} = \int_V \left[\sum_{i=1}^{N} \left(\nabla m^{(i)} \cdot \nabla m^{(i)}\right) + e^2\right]^{p/2} dv. \tag{10.10}$$

The application of the JTVL_p stabilizer to the regularized solution of the inverse problem also results in models with sharp physical property contrasts and consistent positions of the contrast boundaries for all model parameters.

10.1.4 Joint Minimum Gradient Support (JMGS) Stabilizer

The focusing effect of the stabilizing functionals can be increased by using a joint minimum gradient support functional (JMGS):

$$S_{JMGS} = \int_V \frac{\sum_{i=1}^{N} \left(\nabla m^{(i)} \cdot \nabla m^{(i)}\right)}{\sum_{i=1}^{N} \left(\nabla m^{(i)} \cdot \nabla m^{(i)}\right) + e^2} dv, \tag{10.11}$$

where we have assumed for simplicity that $\nabla m_{apr}^{(i)} = 0,\ i = 1, 2, \ldots, N$.

We call *a joint model gradient support*, spt ∇m, the combined closed subdomains of V where $\nabla m^{(i)} \neq 0,\ i = 1, 2, \ldots, N$:

$$\text{spt}\, \nabla m = \bigcup_{i=1}^{N} \text{spt}\, \nabla m^{(i)}. \tag{10.12}$$

Equation (10.11) for the JMGS functional can be modified as follows:

$$S_{JMGS} = \int_{\text{spt}\,\nabla m} \left[1 - \frac{e^2}{\sum_{i=1}^{N} \left(\nabla m^{(i)} \cdot \nabla m^{(i)} \right) + e^2} \right] dv$$

$$= \text{spt}\,\nabla m - e^2 \int_{\text{spt}\,\nabla m} \frac{1}{\sum_{i=1}^{N} \left(\nabla m^{(i)} \cdot \nabla m^{(i)} \right) + e^2} dv. \qquad (10.13)$$

Therefore, the JMGS functional is approximately equal (for a small e) to the joint model gradient support:

$$S_{JMGS} \rightarrow \text{spt}\,\nabla m, \ \text{if}\ e \rightarrow 0. \qquad (10.14)$$

We have learned in Chap. 4 that the minimization of the minimum gradient support functional results in narrowing the areas of the rapid changes of the model parameters, thus producing the sharp boundaries in the inverse images. The minimization of the joint minimum gradient support functional not only sharpens the boundaries of the domains with the different values of the specific model parameter but also forces these boundaries to merge for multimodal parameters. The latter property is obvious because, by merging the boundaries, the JMGS functional minimizes the joint model gradient support of the inverse images.

10.2 Steepest Ascent Directions of the Joint Focusing Functionals

We have already demonstrated in the previous chapters that the numerical implementation of the regularized inversion requires the calculation of the steepest ascent directions (gradients) of the corresponding stabilizing functionals. Considering that the joint focusing stabilizers introduced above are non-quadratic, finding their gradients is a non-trivial problem. In this section, we will find the expressions for the corresponding steepest ascent directions for JMS, JMSL_p, and JMGS stabilizers in the analytical form as examples. Similar expressions can be found for all other joint focusing functionals.

10.2.1 The First Variation of the Joint Minimum Support (JMS) Stabilizer

Let us calculate the first variation of the joint minimum support (JMS) stabilizer (10.2):

$$\delta S_{JMS} = \sum_{n=1}^{N} \delta_{m^{(n)}} S_{JMS} = \sum_{n=1}^{N} \int_V \delta_{m^{(n)}} \left[\frac{\sum_{i=1}^{N} \left(m^{(i)} - m_{apr}^{(i)}\right)^2}{\sum_{j=1}^{N} \left(m^{(j)} - m_{apr}^{(j)}\right)^2 + e^2} \right] dv. \quad (10.15)$$

The variations of the integrands in Eq. (10.15) can be calculated as follows:

$$\delta_{m^{(n)}} \left[\frac{\sum_{i=1}^{N} \left(m^{(i)} - m_{apr}^{(i)}\right)^2}{\sum_{j=1}^{N} \left(m^{(j)} - m_{apr}^{(j)}\right)^2 + e^2} \right] =$$

$$= 2 \frac{\left(m^{(n)} - m_{apr}^{(n)}\right) \delta m^{(n)}}{\sum_{j=1}^{N} \left(m^{(j)} - m_{apr}^{(j)}\right)^2 + e^2} - 2 \frac{\sum_{i=1}^{N} \left(m^{(i)} - m_{apr}^{(i)}\right)^2 \left(m^{(n)} - m_{apr}^{(n)}\right) \delta m^{(n)}}{\left[\sum_{j=1}^{N} \left(m^{(j)} - m_{apr}^{(j)}\right)^2 + e^2\right]^2}$$

$$= 2 \frac{e^2 \left(m^{(n)} - m_{apr}^{(n)}\right) \delta m^{(n)}}{\left[\sum_{j=1}^{N} \left(m^{(j)} - m_{apr}^{(j)}\right)^2 + e^2\right]^2}. \quad (10.16)$$

Substituting equation (10.16) back in the integral formula (10.15) for the variation of the JMS functional, we find

$$\delta S_{JMS} = 2 \sum_{n=1}^{N} \left[\int_V \frac{e^2 \left(m^{(n)} - m_{apr}^{(n)}\right)}{\left[\sum_{j=1}^{N} \left(m^{(j)} - m_{apr}^{(j)}\right)^2 + e^2\right]^2} \delta m^{(n)} \right] dv$$

$$= 2 \sum_{n=1}^{N} \left(\delta m^{(n)}, l_{JMS}^{(n)}\right)_{L_2}, \quad (10.17)$$

where functions $l_{JMS}^{(n)}(\mathbf{r})$ represent the directions of the steepest ascent of the JMS functional with respect to model $m^{(n)}$:

$$l_{JMS}^{(n)} = \frac{e^2 \left(m^{(n)} - m_{apr}^{(n)}\right)}{\left[\sum_{j=1}^N \left(m^{(j)} - m_{apr}^{(j)}\right)^2 + e^2\right]^2}. \tag{10.18}$$

10.2.2 The First Variation of the L_p-Norm Joint Minimum Support ($JMSL_p$) Stabilizer

We now turn to the derivation of the first variation of the L_p-norm joint minimum support (JMS) stabilizer (10.2):

$$\delta S_{JMSL_p} = \sum_{n=1}^N \delta_{m^{(n)}} S_{JMSL_p} = \sum_{n=1}^N \int_V \delta_{m^{(n)}} \left[\frac{\sum_{i=1}^N \left|m^{(i)} - m_{apr}^{(i)}\right|^p}{\sum_{j=1}^N \left|m^{(j)} - m_{apr}^{(j)}\right|^p + e^p}\right] dv. \tag{10.19}$$

The variations of the integrands in Eq. (10.19) can be calculated as follows:

$$\delta_{m^{(n)}} \left[\frac{\sum_{i=1}^N \left|m^{(i)} - m_{apr}^{(i)}\right|^p}{\sum_{j=1}^N \left|m^{(j)} - m_{apr}^{(j)}\right|^p + e^p}\right] =$$

$$= p\frac{\left|m^{(n)} - m_{apr}^{(n)}\right|^{p-1} \operatorname{signum}\left(m^{(n)} - m_{apr}^{(n)}\right) \delta m^{(n)}}{\sum_{j=1}^N \left|m^{(j)} - m_{apr}^{(j)}\right|^p + e^2} -$$

$$- p\frac{\sum_{i=1}^N \left|m^{(i)} - m_{apr}^{(i)}\right|^p \left|m^{(n)} - m_{apr}^{(n)}\right|^{p-1} \operatorname{signum}\left(m^{(n)} - m_{apr}^{(n)}\right) \delta m^{(n)}}{\left[\sum_{j=1}^N \left|m^{(j)} - m_{apr}^{(j)}\right|^p + e^2\right]^2}$$

$$= p\frac{e^2 \left|m^{(n)} - m_{apr}^{(n)}\right|^{p-1} \operatorname{signum}\left(m^{(n)} - m_{apr}^{(n)}\right) \delta m^{(n)}}{\left[\sum_{j=1}^N \left|m^{(j)} - m_{apr}^{(j)}\right|^p + e^2\right]^2}, \tag{10.20}$$

where symbol $\operatorname{signum}\left(m^{(n)} - m_{apr}^{(n)}\right)$ denotes the sign of the difference $\left(m^{(n)} - m_{apr}^{(n)}\right)$:

$$\operatorname{signum}\left(m^{(n)} - m_{apr}^{(n)}\right) = \begin{cases} 1, & m^{(n)} > m_{apr}^{(n)} \\ 0, & m^{(n)} = m_{apr}^{(n)} \\ -1, & m^{(n)} < m_{apr}^{(n)} \end{cases}.$$

Substituting equation (10.20) back in the integral formula (10.19) for the variation of the $JMSL_p$ functional, we find:

$$\delta S_{JMSL_p} = p \sum_{n=1}^{N} \left[\int_V \frac{e^2 \left| m^{(n)} - m_{apr}^{(n)} \right|^{p-1} \text{signum} \left(m^{(n)} - m_{apr}^{(n)} \right)}{\left[\sum_{j=1}^{N} \left| m^{(j)} - m_{apr}^{(j)} \right|^p + e^2 \right]^2} \delta m^{(n)} \right] dv$$

$$= 2 \sum_{n=1}^{N} \left(\delta m^{(n)}, l_{JMSL_p}^{(n)} \right)_{L_2}, \tag{10.21}$$

where functions $l_{JMSL_p}^{(n)}(\mathbf{r})$ represent the directions of the steepest ascent of the JMS functional with respect to model $m^{(n)}$:

$$l_{JMSL_p}^{(n)} = \frac{p}{2} \frac{e^2 \left| m^{(n)} - m_{apr}^{(n)} \right|^{p-1} \text{signum} \left(m^{(n)} \right)}{\left[\sum_{j=1}^{N} \left| m^{(j)} - m_{apr}^{(j)} \right|^p + e^2 \right]^2}. \tag{10.22}$$

10.2.3 The First Variation of the Joint Minimum Gradient Support (JMGS) Stabilizer

Finally, we derive the first variation of the joint minimum gradient support (JMGS) stabilizer (10.2):

$$\delta S_{JMGS} = \sum_{n=1}^{N} \delta_{m^{(n)}} S_{JMGS} = \sum_{n=1}^{N} \int_V \delta_{m^{(n)}} \left[\frac{\sum_{i=1}^{N} \left(\nabla m^{(i)} \cdot \nabla m^{(i)} \right)}{\sum_{j=1}^{N} \left(\nabla m^{(j)} \cdot \nabla m^{(j)} \right) + e^2} \right] dv. \tag{10.23}$$

Similar to the case of the JMS functional, we calculate the variations of the integrands in Eq. (10.23) as follows:

$$\delta_{m^{(n)}} \left[\frac{\sum_{i=1}^{N} \left(\nabla m^{(i)} \cdot \nabla m^{(i)} \right)}{\sum_{j=1}^{N} \left(\nabla m^{(j)} \cdot \nabla m^{(j)} \right) + e^2} \right] =$$

$$= 2 \frac{\nabla m^{(n)} \cdot \nabla \delta m^{(n)}}{\sum_{j=1}^{N} \left(\nabla m^{(j)} \cdot \nabla m^{(j)} \right) + e^2} - 2 \frac{\sum_{i=1}^{N} \left(\nabla m^{(i)} \cdot \nabla m^{(i)} \right) \nabla m^{(n)} \cdot \nabla \delta m^{(n)}}{\left[\sum_{j=1}^{N} \left(\nabla m^{(j)} \cdot \nabla m^{(j)} \right) + e^2 \right]^2}$$

$$= 2 \frac{e^2 \nabla m^{(n)} \cdot \nabla \delta m^{(n)}}{\left[\sum_{j=1}^{N} \left(\nabla m^{(j)} \cdot \nabla m^{(j)} \right) + e^2 \right]^2}. \tag{10.24}$$

Substituting equation (10.24) back in the integral formula (10.23) for the variation of the JMGS functional, we find

$$\delta S_{JMGS} = 2\sum_{n=1}^{N}\int_V \frac{e^2 \nabla m^{(n)} \cdot \nabla \delta m^{(n)}}{\left[\sum_{j=1}^{N}\left(\nabla m^{(j)} \cdot \nabla m^{(j)}\right) + e^2\right]^2} dv \tag{10.25}$$

Let us examine the integrals in Eq. (10.25):

$$\int_V \frac{e^2 \nabla m^{(n)} \cdot \nabla \delta m^{(n)}}{\left[\sum_{j=1}^{N}\left(\nabla m^{(j)} \cdot \nabla m^{(j)}\right) + e^2\right]^2} dv = \int_V \mathbf{C}^{(n)} \cdot \nabla \delta m^{(n)} dv, \tag{10.26}$$

where

$$\mathbf{C}^{(n)} = b^2 \nabla m^{(n)}, \tag{10.27}$$

and

$$b = \frac{e}{\left[\sum_{j=1}^{N}\left(\nabla m^{(j)} \cdot \nabla m^{(j)}\right) + e^2\right]}.$$

Using the identity

$$\nabla \cdot \left(\mathbf{C}^{(n)} \delta m^{(n)}\right) = \left(\nabla \cdot \mathbf{C}^{(n)}\right) \delta m^{(n)} + \mathbf{C}^{(n)} \cdot \nabla \delta m^{(n)},$$

we can take integral (10.26) by parts:

$$\int_V \mathbf{C}^{(n)} \cdot \nabla \delta m^{(n)} dv = -\int_V \left(\nabla \cdot \mathbf{C}^{(n)}\right) \delta m^{(n)} dv + \int_V \nabla \cdot \left(\mathbf{C}^{(n)} \delta m^{(n)}\right) dv$$

$$= -\int_V \left(\nabla \cdot \mathbf{C}^{(n)}\right) \delta m^{(n)} dv + \int_{\partial V} \mathbf{C}^{(n)} \delta m^{(n)} \cdot \mathbf{n} ds, \tag{10.28}$$

where we have applied the Gauss theorem, and $\mathbf{n}$ is a unit vector of the normal directed outward from the domain V.

We assume homogeneous Neumann (i.e., no flux) boundary conditions:

$$\int_{\partial V} \mathbf{C}^{(n)} \delta m^{(n)} \cdot \mathbf{n} ds = \int_{\partial V} \delta m^{(n)} b^2 \left(\nabla m^{(n)} \cdot \mathbf{n}\right) ds = 0.$$

The last condition can be reformulated as follows:

$$\nabla m^{(n)} \cdot \mathbf{n} = \frac{\partial m^{(n)}}{\partial \mathbf{n}} = 0, \quad r \in \partial V,$$

where $\partial m^{(n)} / \partial \mathbf{n}$ is a directional derivative in the direction of vector $\mathbf{n}$. In other words, we assume that at the boundary the model parameters do not change.

Therefore, we have from Eq. (10.28)

$$\int_V \mathbf{C}^{(n)} \cdot \nabla \delta m^{(n)} dv = -\int_V \left(\nabla \cdot \mathbf{C}^{(n)}\right) \delta m^{(n)} dv = -\left(\delta m^{(n)}, \left(\nabla \cdot \mathbf{C}^{(n)}\right)\right)_{L_2}. \tag{10.29}$$

Taking into account Eqs. (10.26) and (10.29), we can write expression (10.25) for the first variation of the JMGS functional as follows:

$$\delta S_{JMGS} = 2\sum_{n=1}^{N} \left(\delta m^{(n)}, l_{JMGS}^{(n)}\right)_{L_2}, \tag{10.30}$$

where functions $l_{JMGS}^{(n)}(\mathbf{r})$ represent the directions of the steepest ascent of the JMGS functional with respect to model $m^{(n)}$:

$$l_{JMGS}^{(n)} = \nabla \cdot \mathbf{C}^{(n)} = \nabla \cdot b^2 \nabla m^{(n)}. \tag{10.31}$$

This concludes our analysis of the steepest ascent directions of the joint focusing stabilizers.

10.3 Inversion Based on the Joint Focusing Stabilizers

We now revisit again the multimodal inverse problem (10.1). According to the basic principles of the regularization method, we have to find the models $\mathbf{m}_\alpha^{(1)}$,.... $\mathbf{m}_\alpha^{(2)}$, ...$\mathbf{m}_\alpha^{(N)}$, a quasi-solution of this inverse problem, which minimize the parametric functional:

$$P^\alpha(\mathbf{m}^{(1)}, \mathbf{m}^{(2)}, ...\mathbf{m}^{(N)}) = \sum_{i=1}^{N} \left\| \mathbf{A}^{(i)}(\mathbf{m}^{(i)}) - \mathbf{d}^{(i)} \right\|_D^2 + \alpha S_{JMS,\ JMSL_p,\ JMGS} = \min, \tag{10.32}$$

where α is the regularization parameter. The terms $S_{JMS,\ JMSL_p,\ JMGS}$ are the joint stabilizing functionals based on minimum support, S_{JMS}, L_p-norm minimum support, S_{JMSL_p}, or minimum gradient support, S_{JMGS}, constraints, respectively.

In order to solve this minimization problem, we calculate the first variation of the parametric functional with joint focusing stabilizers:

$$\begin{aligned}\delta P^\alpha(\mathbf{m}^{(1)}, \mathbf{m}^{(2)}, ...\mathbf{m}^{(N)}) = & 2\sum_{i=1}^{N} \left(\delta \mathbf{A}^{(i)}(\mathbf{m}^{(i)}),\ \mathbf{A}^{(i)}(\mathbf{m}^{(i)}) - \mathbf{d}^{(i)}\right)_D + \\ & + 2\alpha \delta S_{JMS,\ JMSL_p,\ JMGS}.\end{aligned} \tag{10.33}$$

Taking into consideration that operators $\mathbf{A}^{(i)}$ are differentiable, we can write

$$\delta \mathbf{A}^{(i)}(\mathbf{m}^{(i)}) = \mathbf{F}_m^{(i)} \delta \mathbf{m}^{(i)}, \tag{10.34}$$

where $\mathbf{F}_m^{(i)}$ is a linear Fréchet derivative operator of $\mathbf{A}^{(i)}$.

It was shown above that,

$$\delta S_{JMS,\ JMSL_p,\ JMGS} = 2 \sum_{i=1}^{N} \left(\delta \mathbf{m}^{(i)},\ \mathbf{l}_{JMS,\ JMSL_p,\ JMGS}^{(i)} \right), \tag{10.35}$$

where vectors $\delta \mathbf{m}^{(i)}$ and $\mathbf{l}_{JMS,JMSL_p,\ JMGS}^{(i)}$ are the discrete analogs of the functions, $\delta m^{(i)}$ and $l_{JMS,JMSL_p,\ JMGS}^{(i)}$ representing the variation of the model parameter $m^{(i)}$ and the corresponding directions of the steepest ascent for the stabilizing functionals, based on joint minimum support, L_p-norm joint minimum support, and joint minimum gradient support constraints, described by formulas (10.18), (10.22), and (10.31), respectively.

Substituting expressions (10.34) and (10.35) into formula (10.33), we obtain:

$$\delta P^{\alpha}(\mathbf{m}^{(1)}, \mathbf{m}^{(2)}, \ldots\mathbf{m}^{(N)})$$

$$= 2 \sum_{i=1}^{N} \left(\delta \mathbf{m}^{(i)},\ \left[\mathbf{F}_m^{(i)*} \left(\mathbf{A}^{(i)} \left(\mathbf{m}^{(i)} \right) - \mathbf{d}^{(i)} \right) + \mathbf{l}_{JMS,\ JMSL_p,\ JMGS}^{(i)} \right] \right),$$

where $\mathbf{F}_m^{(i)\star}$are the adjoint Fréchet derivative operators.

Let us select

$$\delta \mathbf{m}^{(i)} = -k^{\alpha} \mathbf{l}^{\alpha(i)}, \tag{10.36}$$

where k^{α} is some positive real number, and $\mathbf{l}^{\alpha(i)}$ are the directions of the steepest ascent of the functional P^{α}:

$$\mathbf{l}^{\alpha(i)} = \mathbf{F}_m^{(i)\star} \left(\mathbf{A}^{(i)}(\mathbf{m}^{(i)}) - \mathbf{d}^{(i)} \right) + \alpha \mathbf{l}_{JMS,\ JMGS}^{(i)}. \tag{10.37}$$

Then

$$\delta P^{\alpha}(\mathbf{m}^{(1)}, \mathbf{m}^{(2)}, \ldots\mathbf{m}^{(N)}) = -2k^{\alpha} \sum_{i=1}^{N} \left\| \mathbf{l}^{\alpha(i)} \right\|^2. \tag{10.38}$$

The last expression confirms that the selection of the perturbations of the model parameters based on formula (10.36) ensures a decrease in the parametric functional.

We can construct an iterative process for the regularized conjugate gradient (RCG) algorithm for solving minimization problem (10.32), which can be summarized as follows:

$$
\begin{aligned}
&\mathbf{r}_k = \mathbf{A}(\mathbf{m}_k) - \mathbf{d}, \quad \mathbf{l}_k^{\alpha} = \mathbf{l}^{\alpha}(m_{\mathbf{k}}) && (a) \\
&\beta_k^{\alpha} = \left\|\mathbf{l}_k^{\alpha}\right\|^2 / \left\|\mathbf{l}_{k-1}^{\alpha}\right\|^2, \quad \tilde{\mathbf{l}}_k^{\alpha} = \mathbf{l}_k^{\alpha} + \beta_k^{\alpha}\tilde{\mathbf{l}}_{k-1}^{\alpha}, \quad \tilde{\mathbf{l}}_0^{\alpha} = \mathbf{l}_0^{\alpha}, && (b) \\
&s_k^{\alpha} = \left(\tilde{\mathbf{l}}_k^{\alpha}, \mathbf{l}_k^{\alpha}\right) / \left\{\left\| F_{m_k}\tilde{\mathbf{l}}_k^{\alpha}\right\|^2 + \alpha \left\| W\tilde{\mathbf{l}}_k^{\alpha}\right\|^2\right\}, && (c) \\
&\mathbf{m}_{k+1} = \mathbf{m}_k - s_k^{\alpha}\, \tilde{\mathbf{l}}_k^{\alpha}. && (d)
\end{aligned}
\tag{10.39}
$$

In the last formula, we used the following notations:
$\mathbf{d}$ is a combined vector of the observed data

$$
\mathbf{d} = \left(\mathbf{d}^{(1)}, \mathbf{d}^{(2)}, ...\mathbf{d}^{(N)}\right)^T; \tag{10.40}
$$

$\mathbf{m}_k$ is a combined vector of different model parameters computed at iteration number k,

$$
\mathbf{m}_k = \left(\mathbf{m}_k^{(1)}, \mathbf{m}_k^{(2)}, ...\mathbf{m}_k^{(N)}\right)^T; \tag{10.41}
$$

$\mathbf{A}(\mathbf{m}_k)$ is a vector of the predicted data computed at iteration number k; and $\mathbf{l}_k^{\alpha}$ is a combined vector of the directions of the steepest ascent calculated at iteration number k,

$$
\mathbf{l}_k^{\alpha} = \left(\mathbf{l}_{Ck}^{\alpha(1)}, \mathbf{l}_{Ck}^{\alpha(2)}, ...\mathbf{l}_{Ck}^{\alpha(N)}\right)^T. \tag{10.42}
$$

The expressions for the steepest ascent directions are shown above in formula (10.37).

The iterative process (10.39) is terminated when the misfit reaches the required level:

$$
\varphi\left(\widetilde{\mathbf{m}}_{k+1}\right) = \|\widetilde{\mathbf{r}}_{k+1}\|_D^2 = \delta_d. \tag{10.43}
$$

10.4 Re-weighted Gradient-Type Methods of Joint Focusing Inversion

In the previous sections, we provided the exact analytical expressions for the steepest ascent directions of the joint focusing stabilizing functionals. In numerical calculations, however, it is convenient to use the approximate approach based on representations of these functionals in the pseudo-quadratic form. This representation makes it possible to apply the standard minimization algorithms developed for quadratic functional (see Chap. 4), simplifying the calculations.

10.4.1 Representation of Joint Stabilizing Functionals in the Form of Pseudo-quadratic Functionals

The joint focusing stabilizing functionals introduced above can be expressed as pseudo-quadratic functionals of the model parameters in a similar way as it was done in Chap. 4 for individual focusing stabilizers:

$$s(\mathbf{m}) = \left(\mathbf{W}_e\left(\mathbf{m}-\mathbf{m}_{apr}\right), \mathbf{W}_e\left(\mathbf{m}-\mathbf{m}_{apr}\right)\right)_{L_2} = \int_V \left|w_e(\mathbf{r})\left(m(\mathbf{r})-m_{apr}(\mathbf{r})\right)\right|^2 dv, \tag{10.44}$$

where $\mathbf{W}_e$ is a linear operator of the multiplication of the model parameters function $m(\mathbf{r})$ by function $w_e(\mathbf{r})$. We call this representation a *pseudo-quadratic functional* of $\mathbf{m}$ because function $w_e(\mathbf{r})$ may depend on $\mathbf{m}$. The stabilizers' presentation in a pseudo-quadratic form simplifies the calculation of the regularized steepest ascent directions in the solution of the minimization problem of the corresponding parametric functionals.

For example, the joint minimum support functional, $S_{JMS}(\mathbf{m})$, can be written as follows:

$$\begin{aligned} S_{JMS}(\mathbf{m}) &= \sum_{i=1}^{N}\int_V \frac{\left(m^{(i)}-m_{apr}^{(i)}\right)^2}{\sum_{j=1}^{n}\left(m^{(j)}-m_{apr}^{(j)}\right)^2+e^2}dv \\ &= \sum_{i=1}^{N}\int_V \left|w_e^{(i)JMS}(\mathbf{r})\left(m^{(i)}-m_{apr}^{(i)}\right)\right|^2 dv \\ &= \sum_{i=1}^{N}\left(\mathbf{W}_e^{JMS}\left(\mathbf{m}^{(i)}-\mathbf{m}_{apr}^{(i)}\right), \mathbf{W}_e^{JMS}\left(\mathbf{m}^{(i)}-\mathbf{m}_{apr}^{(i)}\right)\right)_{L_2} \end{aligned} \tag{10.45}$$

where $\mathbf{W}_e^{(i)JMS}$ $(i = 1, 2, \ldots, N)$ are the linear operators of multiplication of the model parameter functions $\left(m^{(i)}-m_{apr}^{(i)}\right)$ by the following functions, $w_e^{(i)JMS}(\mathbf{r})$:

$$w_e^{(i)JMS}(\mathbf{r}) = \frac{1}{\left[\sum_{j=1}^{N}\left(m^{(j)}(\mathbf{r})-m_{apr}^{(j)}(\mathbf{r})\right)^2+e^2\right]^{1/2}}. \tag{10.46}$$

We can derive a similar quadratic representation for L_p-norm joint minimum support functionals, $S_{JMSL_p}(\mathbf{m})$, as follows:

$$
\begin{aligned}
S_{JMSL_p}(\mathbf{m}) &= \sum_{i=1}^{N} \int_V \frac{\left|m^{(i)} - m_{apr}^{(i)}\right|^p}{\sum_{j=1}^{N}\left|m^{(j)} - m_{apr}^{(j)}\right|^p + e^p} dv \\
&= \sum_{i=1}^{N} \int_V \frac{\left|m^{(i)} - m_{apr}^{(i)}\right|^p \left[\left(m^{(i)} - m_{apr}^{(i)}\right)^2 + \beta^2\right]}{\left[\sum_{j=1}^{N}\left|m^{(j)} - m_{apr}^{(j)}\right|^p + e^p\right]\left[\left(m^{(i)} - m_{apr}^{(i)}\right)^2 + \beta^2\right]} dv \\
&\approx \sum_{i=1}^{N} \int_V \left|w_e^{(i)JMSL_p}(\mathbf{r})\left(m^{(i)} - m_{apr}^{(i)}\right)\right|^2 dv \\
&= \sum_{i=1}^{N} \left(\mathbf{W}_e^{(i)JMSL_p}\left(\mathbf{m}^{(i)} - \mathbf{m}_{apr}^{(i)}\right), \mathbf{W}_e^{(i)JMSL_p}\left(\mathbf{m}^{(i)} - \mathbf{m}_{apr}^{(i)}\right)\right)_{L_2}
\end{aligned}
\tag{10.47}
$$

where $\mathbf{W}_e^{(i)JMSL_p}$ $(i = 1, 2, \ldots, N)$ are the linear operators of multiplication of the model parameter functions, $\left(m^{(i)} - m_{apr}^{(i)}\right)$, by the following functions:

$$
w_e^{(i)JMSL_p}(\mathbf{r}) = \frac{\left|m^{(i)} - m_{apr}^{(i)}\right|^{p/2}}{\left[\sum_{j=1}^{N}\left|m^{(j)} - m_{apr}^{(j)}\right|^p + e^p\right]^{1/2}\left[\left(m^{(i)} - m_{apr}^{(i)}\right)^2 + \beta^2\right]^{1/2}},
\tag{10.48}
$$

and β is a small number introduced to avoid division by zero.

For the joint minimum gradient support functional, $S_{JMGS}(\mathbf{m})$, we find:

$$
\begin{aligned}
S_{JMGS}(\mathbf{m}) &= \sum_{i=1}^{N} \int_V \frac{\left(\nabla m^{(i)} \cdot \nabla m^{(i)}\right)}{\sum_{j=1}^{N}\left(\nabla m^{(j)} \cdot \nabla m^{(j)}\right) + e^2} dv \\
&= \sum_{i=1}^{N} \int_V \frac{\left(\nabla m^{(i)} \cdot \nabla m^{(i)}\right)\left[\left(m^{(i)}\right)^2 + \beta^2\right]}{\left[\sum_{j=1}^{N}\left(\nabla m^{(j)} \cdot \nabla m^{(j)}\right) + e^2\right]\left[\left(m^{(i)}\right)^2 + \beta^2\right]} dv \\
&\approx \sum_{i=1}^{N} \int_V \left|w_e^{(i)JMGS}(\mathbf{r})\, m^{(i)}(\mathbf{r})\right|^2 dv = \sum_{i=1}^{N}\left(\mathbf{W}_e^{(i)JMGS}\mathbf{m}^{(i)}, \mathbf{W}_e^{(i)JMGS}\mathbf{m}^{(i)}\right)_{L_2},
\end{aligned}
\tag{10.49}
$$

where $\mathbf{W}_e^{(i)JMGS}$ $(i = 1, 2, \ldots, N)$ are the linear operators of multiplication of the model parameters functions, $m^{(i)}$, by the following functions:

$$
w_e^{(i)JMGS}(\mathbf{r}) = \frac{\nabla m^{(i)}(\mathbf{r})}{\left[\sum_{j=1}^{N}\left(\nabla m^{(j)} \cdot \nabla m^{(j)}\right) + e^2\right]^{1/2}\left[\left(m^{(i)}\right)^2 + \beta^2\right]^{1/2}},
\tag{10.50}
$$

and β is a small number introduced to avoid division by zero.

Using the pseudo-quadratic form (10.44) of stabilizing functionals, we can present the corresponding parametric functional (10.32) as follows:

$$P^{\alpha}(\mathbf{m}^{(1)}, \mathbf{m}^{(2)}, \ldots \mathbf{m}^{(N)}) = (A(\mathbf{m}) - \mathbf{d}, A(\mathbf{m}) - \mathbf{d})_D +$$

$$+\sum_{i=1}^{N} \alpha^{(i)} \left(\mathbf{W}_e^{(i)} \left(\mathbf{m}^{(i)} - \mathbf{m}_{apr}^{(i)}\right), \mathbf{W}_{ei} \left(\mathbf{m}^{(i)} - \mathbf{m}_{apr}^{(i)}\right)\right)_{L_2}, \tag{10.51}$$

where $\mathbf{W}_e^{(i)}$ are linear operators of multiplication of the model parameters function $m^{(i)}(\mathbf{r})$ by the function $w_e^{(i)}(\mathbf{r})$:

$$\begin{cases} w_e^{(i)JMS}(\mathbf{r}), & \text{for joint minimum support functional,} \\ w_e^{(i)JMSL_p}(\mathbf{r}), & \text{for } L_p\text{-norm joint minimum support functional,} \\ w_e^{(i)JMGS}(\mathbf{r}), & \text{for joint minimum gradient support functional.} \end{cases} \tag{10.52}$$

Therefore, the problem of minimization of the parametric functional introduced by Eq. (10.32) can be treated in a similar way to the minimization of the conventional Tikhonov parametric functional. The only difference is that now we introduce some variable weighting operators, $\mathbf{W}_e^{(i)}$, which depend on the model parameters. We will discuss in the next section a practical technique of minimizing the parametric functional (10.32).

10.4.2 Re-weighted Steepest Descent Method of Joint Focusing Inversion

We have discussed in Chap. 7 (Sect. 7.5.2) the importance of using data weighting and model weighting matrices in regularized inversion. In this case, parametric functional (10.32) can be written using matrix notations as follows:

$$P^{\alpha}(\mathbf{m}^{(1)}, \mathbf{m}^{(2)}, \ldots \mathbf{m}^{(N)}) =$$

$$\sum_{i=1}^{N}(\mathbf{W}_d^{(i)}\mathbf{A}^{(i)}(\mathbf{m}^{(i)}) - \mathbf{W}_d^{(i)}\mathbf{d}^{(i)})^T(\mathbf{W}_d^{(i)}\mathbf{A}^{(i)}(\mathbf{m}^{(i)}) - \mathbf{W}_d^{(i)}\mathbf{d}^{(i)}) +$$

$$+\sum_{i=1}^{N}\alpha^{(i)}(\mathbf{W}_e^{(i)}\mathbf{W}_m^{(i)}\mathbf{m}^{(i)} - \mathbf{W}_e^{(i)}\mathbf{W}_m^{(i)}\mathbf{m}_{apr}^{(i)})^T(\mathbf{W}_e^{(i)}\mathbf{W}_m^{(i)}\mathbf{m}^{(i)} - \mathbf{W}_e^{(i)}\mathbf{W}_m^{(i)}\mathbf{m}_{apr}^{(i)}), \tag{10.53}$$

where matrix $\mathbf{W}_e^{(i)}$ is a variable matrix of the focusing stabilizer, which depends on $\mathbf{m}^{(1)}, \mathbf{m}^{(2)}, \ldots \mathbf{m}^{(N)}$, and $\mathbf{W}_d^{(i)}$ and $\mathbf{W}_m^{(i)}$ are the conventional fixed diagonal matrices for weighting the data, $\mathbf{d}^{(i)}$, and model parameters, $\mathbf{m}^{(i)}$, respectively.

Therefore, the problem of minimizing the parametric functional, given by Eq. (10.53), can be treated in a similar way to the minimization of the conventional Tikhonov functional. The only difference is that now we introduce some variable weighting matrices $\mathbf{W}_e^{(i)}$ for the model parameters. The minimization problem for the parametric functional introduced by Eq. (10.53) can be solved using the traditional gradient-type methods discussed in Chap. 7.

For example, following Zhdanov (2002, 2015) we can describe the re-weighted steepest decent method. In the framework of this approach, the variable weighting matrices $\mathbf{W}_e^{(i)}$, $i = 1, 2, \ldots, N$, are precomputed on each iteration, $\mathbf{W}_e^{(i)} = \mathbf{W}_{en}^{(i)} = \mathbf{W}_e^{(i)}\left(\mathbf{m}_n^{(1)}, \mathbf{m}_n^{(2)}, \ldots \mathbf{m}_n^{(N)}\right)$ based on the values $\mathbf{m}_n^{(1)}, \mathbf{m}_n^{(2)}, \ldots \mathbf{m}_n^{(N)}$, obtained on the previous iteration. As a result, they are treated as fixed matrices on each iteration. Under this assumption, we calculate the first variation of the parametric functional (10.53) as follows

$$\delta P^{\alpha}(\mathbf{m}^{(1)}, \mathbf{m}^{(2)}, \ldots \mathbf{m}^{(N)}) =$$

$$= 2\sum_{i=1}^{N} \delta\mathbf{m}^{(i)T}\mathbf{F}_m^{(i)T}\mathbf{W}_d^{(i)2}\left(\mathbf{A}^{(i)}(\mathbf{m}^{(i)}) - \mathbf{d}^{(i)}\right) +$$

$$+ 2\sum_{i=1}^{N} \alpha^{(i)}\delta\mathbf{m}^{(i)T}\mathbf{W}_e^{(i)2}\mathbf{W}_m^{(i)2}\left(\mathbf{m}^{(i)} - \mathbf{m}_{apr}^{(i)}\right).$$

After some algebra, we obtain:

$$\delta P^{\alpha} =$$

$$= 2\sum_{i=1}^{N} \delta\mathbf{m}^{(i)T}\left[\mathbf{F}_m^{(i)T}\mathbf{W}_d^{(i)2}\left(\mathbf{A}^{(i)}(\mathbf{m}^{(i)}) - \mathbf{d}^{(i)}\right) + \alpha^{(i)}\mathbf{W}_e^{(i)2}\mathbf{W}_m^{(i)2}\left(\mathbf{m}^{(i)} - \mathbf{m}_{apr}^{(i)}\right)\right]. \tag{10.54}$$

Following the general scheme of the steepest descent method, we can again select

$$\delta\mathbf{m}^{(i)} = -k^{\alpha(i)}\mathbf{l}^{\alpha(i)}(\mathbf{m}^{(i)}),$$

where $k^{\alpha(i)}$ are some positive real numbers (lengths of steps) and $\mathbf{l}^{\alpha(i)}(\mathbf{m}^{(i)})$ are column matrices defining the direction of the steepest ascent :

$$\mathbf{l}^{\alpha(i)}(\mathbf{m}^{(i)}) =$$

$$= \mathbf{F}_m^{(i)T}\mathbf{W}_d^{(i)2}\left(\mathbf{A}^{(i)}(\mathbf{m}^{(i)}) - \mathbf{d}^{(i)}\right) + \alpha^{(i)}\mathbf{W}_e^{(i)2}\mathbf{W}_m^{(i)2}\left(\mathbf{m}^{(i)} - \mathbf{m}_{apr}^{(i)}\right). \tag{10.55}$$

Thus, the regularized re-weighted steepest descent method is based on the successive line search in the gradient direction $\mathbf{l}^{\alpha(i)}(\mathbf{m}_{(n)}^{(i)})$:

$$\mathbf{m}_{n+1}^{(i)} = \mathbf{m}_n^{(i)} + \mathbf{m}^{(i)} = \mathbf{m}_n^{(i)} - k^{\alpha(i)}\mathbf{l}^{\alpha(i)}(\mathbf{m}_n^{(i)}), \tag{10.56}$$

where

$$\mathbf{l}^{\alpha(i)}(\mathbf{m}_n^{(i)}) = \mathbf{F}_{m_n}^{(i)T}\mathbf{W}_d^{(i)2}\left(\mathbf{A}^{(i)}(\mathbf{m}_n^{(i)}) - \mathbf{d}^{(i)}\right) + \alpha^{(i)}\mathbf{W}_{en}^{(i)2}\mathbf{W}_m^{(i)2}\left(\mathbf{m}_n^{(i)} - \mathbf{m}_{apr}^{(i)}\right). \tag{10.57}$$

10.4.3 Re-weighted Conjugate Gradient Method of Joint Focusing Inversion

The regularized re-weighted conjugate gradient (RRCG) method can be developed in the same way as the steepest descent method; however, the model parameters are updated based on the successive line search in the conjugate gradient direction $\tilde{\mathbf{l}}^{\alpha(i)}(\mathbf{m}_n^{(i)})$:

$$\mathbf{m}_{n+1}^{(i)} = \mathbf{m}_n^{(i)} + \mathbf{m}^{(i)} = \mathbf{m}_n^{(i)} - k^{\alpha(i)}\tilde{\mathbf{l}}^{\alpha(i)}(\mathbf{m}_n^{(i)}).$$

The conjugate gradient directions $\tilde{\mathbf{l}}^{\alpha(i)}(\mathbf{m}_n^{(i)})$ are selected as follows. In the initial step, we use the "direction" of regularized steepest ascent for the initial model $\mathbf{m}_0$:

$$\tilde{\mathbf{l}}_0^{\alpha(i)} = \tilde{\mathbf{l}}^{\alpha(i)}(\mathbf{m}_0^{(i)}) = \mathbf{l}^{\alpha(i)}(\mathbf{m}_0^{(i)}) =$$

$$= \mathbf{F}_{m_0}^{(i)T}\mathbf{W}_d^{(i)2}\left(\mathbf{A}^{(i)}(\mathbf{m}_0^{(i)}) - \mathbf{d}^{(i)}\right) + \alpha^{(i)}\mathbf{W}_{e0}^{(i)2}\mathbf{W}_m^{(i)2}\left(\mathbf{m}_0^{(i)} - \mathbf{m}_{apr}^{(i)}\right), \tag{10.58}$$

where $\mathbf{F}_{m_0}^{(i)}$ is the Fréchet derivative matrix for the initial model and $\mathbf{W}_{e0}^{(i)} = \mathbf{W}_{e0}^{(i)}\left(\mathbf{m}_0^{(i)}\right)$.

In the next step, the "direction" of ascent is a linear combination of the regularized steepest ascent on this step and the "direction" of ascent $\tilde{\mathbf{l}}_0^{\alpha(i)}$ on the previous step:

$$\tilde{\mathbf{l}}_1^{\alpha(i)} = \mathbf{l}_1^{\alpha(i)} + \beta_1^{\alpha(i)}\tilde{\mathbf{l}}_0^{\alpha(i)}.$$

In the $(n+1)$th step, we have

$$\tilde{\mathbf{l}}_{n+1}^{\alpha(i)} = \mathbf{l}_{n+1}^{\alpha(i)} + \beta_{n+1}^{\alpha(i)}\tilde{\mathbf{l}}_n^{\alpha(i)}, \tag{10.59}$$

where the regularized steepest ascent directions are determined now according to formula (10.58), and

$$\tilde{\mathbf{l}}_n^{\alpha(i)} = \tilde{\mathbf{l}}^{\alpha(i)}(\mathbf{m}_n^{(i)});\ \mathbf{l}_n^{\alpha(i)} = \mathbf{l}^{\alpha(i)}(\mathbf{m}_n^{(i)}).$$

The length of each iteration step, the coefficients $k_n^{\alpha(i)}$, can be determined with a linear or parabolic line search:

$$P^{\alpha}(\mathbf{m}_{n+1}^{(1)}, \mathbf{m}_{n+1}^{(2)}, \ldots \mathbf{m}_{n+1}^{(N)}) =$$

$$P^{\alpha}(\mathbf{m}_n^{(1)} - k_n^{\alpha(1)}\widetilde{\mathbf{l}}_n^{\alpha(1)}, \ldots \mathbf{m}_n^{(N)} - k_n^{\alpha(N)}\widetilde{\mathbf{l}}_n^{\alpha(N)}) = \min .$$

Solution of this minimization problem gives the following best estimate for the lengths of the step using a linear line search:

$$k_n^{\alpha(i)} = \frac{\widetilde{\mathbf{l}}_n^{\alpha(i)T}\mathbf{l}_n^{\alpha(i)}}{\widetilde{\mathbf{l}}_n^{\alpha(i)T}\left(\mathbf{F}_{m_n}^{(i)T}\mathbf{W}_d^2\mathbf{F}_{m_n}^{(i)} + \alpha\mathbf{W}_{en}^{(i)2}\mathbf{W}_m^{(i)2}\right)\widetilde{\mathbf{l}}_n^{\alpha(i)}}. \tag{10.60}$$

One can use a parabolic line search also to improve the convergence rate of the RRCG method (Zhdanov 2002).

The CG method requires that the vectors $\widetilde{\mathbf{l}}_n^{\alpha(i)}$ introduced above will be mutually conjugate. This requirement is fulfilled if the coefficients $\beta_n^{\alpha(i)}$ are determined by the formula

$$\beta_{n+1}^{\alpha(i)} = \frac{\|\mathbf{l}_{n+1}^{\alpha(i)}\|^2}{\|\mathbf{l}_n^{\alpha(i)}\|^2}.$$

Using Eqs. (10.56), (10.58) and (10.60), we can obtain $\mathbf{m}^{(i)}$ iteratively.

We call this algorithm *conjugate gradient re-weighted optimization* because the weighting matrix $\mathbf{W}_{en}^{(i)}$ is updated on every iteration (Portniaguine and Zhdanov 1999).

Note that due to re-weighting, the stabilizing functional can change and even increase from iteration to iteration,

$$s\left(\mathbf{m}_{n+1}^{(i)}\right) =$$

$$= (\mathbf{m}_{n+1}^{(i)} - \mathbf{m}_{apr}^{(i)})^T\mathbf{W}_{e(n+1)}^{(i)2}\mathbf{W}_m^{(i)2}(\mathbf{m}_{n+1}^{(i)} - \mathbf{m}_{apr}^{(i)}) = \gamma_n^{(i)}s\left(\mathbf{m}_n^{(i)}\right), \tag{10.61}$$

where

$$\gamma_n^{(i)} = \frac{s\left(\mathbf{m}_{n+1}^{(i)}\right)}{s\left(\mathbf{m}_n^{(i)}\right)} =$$

$$= \frac{(\mathbf{m}_{n+1}^{(i)} - \mathbf{m}_{apr}^{(i)})^T\mathbf{W}_{e(n+1)}^{(i)2}\mathbf{W}_m^{(i)2}(\mathbf{m}_{n+1}^{(i)} - \mathbf{m}_{apr}^{(i)})}{(\mathbf{m}_n^{(i)} - \mathbf{m}_{apr}^{(i)})^T\mathbf{W}_{en}^{(i)2}\mathbf{W}_m^{(i)2}(\mathbf{m}_n^{(i)} - \mathbf{m}_{apr}^{(i)})}. \tag{10.62}$$

In order to ensure the convergence of the parametric functional to the global minimum, we use adaptive regularization and decrease parameters α_{n+1}, if $\gamma_n^{(i)} > 1$:

$$\alpha_{n+1}^{(i)} = \begin{cases} \alpha_n^{(i)}, \text{ if } \gamma_n^{(i)} \leq 1, \\ \alpha_n^{(i)}/\gamma_n^{(i)}, \text{ if } \gamma_n^{(i)} > 1. \end{cases} \tag{10.63}$$

So, the product of the regularization parameter $\alpha_{n+1}^{(i)}$ and the stabilizer $s\left(\mathbf{m}_{n+1}\right)$ decreases or does not change:

$$\alpha_{n+1}^{(i)} s\left(\mathbf{m}_{n+1}^{(i)}\right) = \begin{cases} \alpha_n^{(i)} s\left(\mathbf{m}_{n+1}^{(i)}\right) = \alpha_n^{(i)} \gamma_n^{(i)} s\left(\mathbf{m}_n^{(i)}\right), & \text{if } \gamma_n^{(i)} \leq 1, \\ \alpha_n^{(i)} s\left(\mathbf{m}_{n+1}^{(i)}\right) / \gamma_n^{(i)} = \alpha_n^{(i)} s\left(\mathbf{m}_n^{(i)}\right), & \text{if } \gamma_n^{(i)} > 1. \end{cases} \tag{10.64}$$

We also decrease the regularization parameter α_{n+1},

$$\alpha_{n+1}^{(i)\prime} = q\alpha_{n+1}^{(i)}, \; q < 1, \tag{10.65}$$

if the total misfit for all data does not decrease fast enough:

$$\sum_{i=1}^{N} \left\| \mathbf{W}_d^{(i)} \mathbf{A}^{(i)}(\mathbf{m}_n^{(i)}) - \mathbf{W}_d^{(i)} \mathbf{d}^{(i)} \right\|^2 - \sum_{i=1}^{N} \left\| \mathbf{W}_d^{(i)} \mathbf{A}^{(i)}(\mathbf{m}_{n+1}^{(i)}) - \mathbf{W}_d^{(i)} \mathbf{d}^{(i)} \right\|^2$$

$$< 0.01 \sum_{i=1}^{N} \left\| \mathbf{W}_d^{(i)} \mathbf{A}^{(i)}(\mathbf{m}_n^{(i)}) - \mathbf{W}_d^{(i)} \mathbf{d}^{(i)} \right\|^2. \tag{10.66}$$

Numerical experiments demonstrate that the recommended choice of the empirical coefficient q is within an interval $(0.5; 0.9)$.

The algorithm of the RRCG method can be summarized as follows:

$$\begin{aligned}
&\mathbf{r}_n^{(i)} = \mathbf{A}^{(i)}(\mathbf{m}_n^{(i)}) - \mathbf{d}^{(i)}, \qquad \mathbf{g}_n^{(i)} = \mathbf{W}_{en}^{(i)} \mathbf{W}_m^{(i)} (\mathbf{m}_n^{(i)} - \mathbf{m}_{apr}^{(i)}), && (a)\\
&\mathbf{l}_n^{\alpha_n(i)} = \mathbf{F}_{m_n}^{(i)T} \mathbf{W}_d^{(i)2} \mathbf{r}_n^{(i)} + \alpha_n \mathbf{W}_{en}^{(i)} \mathbf{W}_m^{(i)} \mathbf{g}_n^{(i)}, && (b)\\
&\beta_n^{\alpha_n(i)} = \left\| \mathbf{l}_n^{\alpha_n(i)} \right\|^2 / \left\| \mathbf{l}_{n-1}^{\alpha_{n-1}(i)} \right\|^2, \qquad \tilde{\mathbf{l}}_n^{\alpha_n(i)} = \mathbf{l}_n^{\alpha_n(i)} + \beta_n^{\alpha_n(i)} \tilde{\mathbf{l}}_{n-1}^{\alpha_{n-1}(i)}, \; \tilde{\mathbf{l}}_0^{\alpha_0(i)} = \mathbf{l}_0^{\alpha_0(i)}, && (c)\\
&\widetilde{k}_n^{\alpha_n(i)} = \left(\tilde{\mathbf{l}}_n^{\alpha_n(i)T} \mathbf{l}_n^{\alpha_n(i)} \right) / \left[\tilde{\mathbf{l}}_n^{\alpha_n(i)T} \left(\mathbf{F}_{m_n}^{(i)T} \mathbf{W}_d^{(i)2} \mathbf{F}_{m_n}^{(i)} + \alpha_n \mathbf{W}_{en}^{(i)2} \mathbf{W}_m^{(i)2} \right) \tilde{\mathbf{l}}_n^{\alpha_n(i)} \right], && (d)\\
&\mathbf{m}_{n+1}^{(i)} = \mathbf{m}_n^{(i)} - \widetilde{k}_n^{\alpha_n(i)} \tilde{\mathbf{l}}_n^{\alpha_n(i)}, \qquad \gamma_n^{(i)} = \left\| \mathbf{g}_{n+1} \right\|^2 / \left\| \mathbf{g}_n \right\|^2, && (e)\\
&\alpha_{n+1}^{(i)} = \alpha_n^{(i)}, \text{ if } \gamma_n^{(i)} \leq 1, \text{ and } \alpha_{n+1}^{(i)} = \alpha_n^{(i)} / \gamma_n^{(i)}, \text{ if } \gamma_n^{(i)} > 1, && (f)\\
&\alpha_{n+1}^{(i)\prime} = q\alpha_{n+1}^{(i)}, \; q < 1, \text{ if } \left\| \mathbf{W}_d^{(i)} \mathbf{r}_n^{(i)} \right\|^2 - \left\| \mathbf{W}_d^{(i)} \mathbf{r}_{n+1}^{(i)} \right\|^2 < 0.01 \left\| \mathbf{W}_d^{(i)} \mathbf{r}_n^{(i)} \right\|^2, && (g)
\end{aligned} \tag{10.67}$$

where $\alpha_n^{(i)}$ are the subsequent values of the regularization parameter. The iterative process (10.67) is terminated when the misfit reaches the given level ε_0:

$$\phi(\mathbf{m}_N^{(i)}) = \left\| \mathbf{r}_N^{(i)} \right\|^2 \leq \varepsilon_0.$$

This method is called the RRCG method with adaptive regularization (Zhdanov 2002, 2015)

References and Recommended Reading to This Chapter

Aster RC, Borchers B, Thurber CH (2013) Parameter estimation and inverse problems. Academic

Haber E, Gazit MH (2013) Model fusion and joint inversion. Surv Geophys 34:675–695

Molodtsov DM, Troyan V (2017) Multiphysics joint inversion through joint sparsity regularization. In: Expanded Abstracts: Proceedings of 87th annual international meeting. Society of Exploration Geophysicists: Tulsa, OK, USA, pp 1262–1267

Portniaguine O, Zhdanov MS (1999) Focusing geophysical inversion images. Geophysics 64(3):874–887

Tu X, Zhdanov MS (2022) Joint focusing inversion of marine controlled-source electromagnetic and full tensor gravity gradiometry data. Geophysics 87(5):K35–K47

Zhdanov MS (2002) Geophysical inverse theory and regularization problems. Elsevier, 628 pp

Zhdanov MS (2015) Inverse theory and applications in geophysics. Elsevier, 704 pp

Zhdanov MS, Čuma M (2018) Joint inversion of multimodal data using focusing stabilizers and Gramian constraints: In: Expanded Abstracts, proceedings of the 88th SEG international exposition and annual meeting. Society of Exploration Geophysicists, Tulsa, OK, USA, pp 1430–1434

Chapter 11
Joint Minimum Entropy Inversion

Abstract In this chapter, we consider an approach to joint inversion, which requires the minimum of joint entropy in the distribution of different model parameters. The entropy functional is introduced as a measure of the disorder in the distribution of the model parameters. By minimizing the entropy functionals, one decreases the uncertainty in the inverse problem solution, producing a more stable and robust solution of the inverse problem. The joint minimum entropy stabilizing functional characterizes the degree of joint disorder or uncertainty in the distribution of the different model parameters. By minimizing this functional in the framework of the regularized inversion, we produce a consistent image of the same geological structure expressed in various geophysical data.

Keywords Entropy · Joint minimum entropy · Uncertainty · Disorder

11.1 Concept of Entropy in the Inverse Problem Solution

In this chapter, we consider an approach to joint inversion, which requires the minimum of joint entropy in the distribution of different model parameters. This requirement forces the simplest multiphysics solution that fits the multimodal data.

Shannon (1948) introduced the concept of entropy as the average level of "uncertainty" in a system of random variables (see Chap. 2). In the framework of a probabilistic approach to inverse problem solution (Franklin 1970; Tarantola 1987, 2005; Zhdanov 2002), one can treat the unknown physical model parameter, $\mathbf{m}$, as some random variable, with the volume distribution described by function $m = m(\mathbf{r})$, where $\mathbf{r} \in V$ is a position of the observation point. In this case, the entropy of the model parameter distribution can be evaluated using the following formula for a *differential entropy* (2.53):

$$h(\mathbf{m}) = -\int_V p(\mathbf{r}) \ln p(\mathbf{r}) dv. \tag{11.1}$$

M. S. Zhdanov, *Advanced Methods of Joint Inversion and Fusion of Multiphysics Data*, Advances in Geological Science,
https://doi.org/10.1007/978-981-99-6722-3_11

In the last formula, $p(\mathbf{r})$ can be introduced as an analog of the probability density function of the random variable, $\mathbf{m}$,

$$p(\mathbf{r}) = |m(\mathbf{r})| / Q, \tag{11.2}$$

where Q is a volume integral of the absolute value of the distribution, $m(\mathbf{r})$:

$$Q = \int_V |m(\mathbf{r})| \, dv. \tag{11.3}$$

It is important to note that, in a general case, we can use any transformation of the model parameters, $T[m(\mathbf{r})]$, in formula (11.2) as an analog of the probability density function:

$$p_T(\mathbf{r}) = |T[m(\mathbf{r})]| / Q_T, \quad \text{where } Q_T = \int_V |T[m(\mathbf{r})]| \, dv. \tag{11.4}$$

For example, we may consider different powers of m or the results of the application of varying differential operators to m, e. g., gradient, ∇m, or Laplacian, Δm, operators. By selecting different transformations of the model parameters, we impose particular conditions on the behavior of the corresponding inverse models in order to satisfy the a priori known requirements. Some examples of these transformations will be discussed below.

It was originally shown by Amato and Hughes (1991) and Kopec (1991) that the concept of entropy could be used in the regularized solution of Fredholm integral equations. Ramos (1999) and Zhdanov (2002) suggested using the minimum entropy stabilizer based on formula (11.1) to produce a focused solution of geophysical inverse problems:

$$s_{ME}(\mathbf{m}) = -\int_V \frac{|m - m_{apr}| + \beta}{Q} \ln \frac{|m - m_{apr}| + \beta}{Q} dv, \tag{11.5}$$

where

$$Q = \int_V \left(|m - m_{apr}| + \beta\right) dv,$$

and $\beta > 0$ is a small positive number introduced to avoid a singularity in the above formulas.

Note that the application of this stabilizer has some similarities with the maximum entropy principle in spectral analysis, considered, for example, in Ables (1974), Burg (1975), and Wernecke and D'Addario (1977). However, in the framework of Tikhonov regularization, the goal is to minimize a stabilizing functional, which justifies the "minimum entropy" name for this stabilizer.

The entropy functionals defined by formulas (11.1) or (11.5) can be treated as a measure of the disorder in the distribution of the model parameters. By minimizing the entropy functionals, we decrease the uncertainty in the inverse problem solution, producing a more stable and robust solution of the inverse problem.

11.2 Joint Minimum Entropy Stabilizers

The concept of reducing uncertainty in the distribution of the inverse model parameters could also be extended to the case of joint inversion (Zhdanov 2022; Zhdanov et al. 2022).

11.2.1 Joint Minimum Entropy Stabilizer of Order q

Indeed, we introduce a joint minimum entropy stabilizing functional of order q of multimodal parameters, S_{qJME}, as follows (Zhdanov et al. 2022):

$$S_{qJME} = -\int_V \frac{\sum_{i=1}^N \left|m^{(i)} - m_{apr}^{(i)}\right|^q + \beta}{Q_J} \ln \frac{\sum_{i=1}^N \left|m^{(i)} - m_{apr}^{(i)}\right|^q + \beta}{Q_J} dv, \quad (11.6)$$

where q is an integer number ($q = 1, 2, ...N$), and

$$Q_J = \int_V \left(\sum_{i=1}^N \left|m^{(i)} - m_{apr}^{(i)}\right|^q + \beta \right) dv. \quad (11.7)$$

The joint minimum entropy stabilizing functional characterizes the degree of joint disorder or uncertainty in the distribution of the different model parameters. By minimizing this functional in the framework of the regularized inversion, we produce a consistent image of the same geological structure expressed in various geophysical data.

11.2.2 Joint Minimum Entropy of Gradient Stabilizer of Order q

The above approach can be extended to impose a structural similarity between inverse models corresponding to different physical parameters. We can achieve this by considering a spatial gradient, $\nabla m = \nabla m(\mathbf{r})$, of the model parameter distribution function. Following Zhdanov (2002, 2015), we can introduce the qth order *minimum entropy of gradient stabilizer*, S_{qMEG}:

$$S_{qMEG}(\mathbf{m}) = -\int_V \frac{|\nabla m|^q + \beta}{Q'} \ln \frac{|\nabla m|^q + \beta}{Q'} dv, \tag{11.8}$$

where

$$Q' = \int_V \left(|\nabla m|^q + \beta\right) dv.$$

Note that the spatial gradient, ∇m, increases at the interfaces between the different blocks, where the corresponding physical property changes rapidly and is directed perpendicular to these interfaces (see Chap. 9). Therefore, minimizing the entropy of the gradient stabilizer, S_{qMEG}, should produce a simpler geometrical structure of the inverse model.

Similarly, we can introduce a joint minimum entropy of gradient functional S_{qJMEG} (JMEG), as follows:

$$S_{qJMEG} =$$

$$= -\int_V \frac{\sum_{i=1}^N \left(\nabla m^{(i)} \cdot \nabla m^{(i)}\right)^{q/2} + \beta}{Q'_J} \ln \frac{\sum_{i=1}^N \left(\nabla m^{(i)} \cdot \nabla m^{(i)}\right)^{q/2} + \beta}{Q'_J} dv, \tag{11.9}$$

where

$$Q'_J = \int_V \left(\sum_{i=1}^N \left(\nabla m^{(i)} \cdot \nabla m^{(i)}\right)^{q/2} + \beta\right) dv.$$

This joint entropy of gradient functional characterizes the degree of joint disorder of the spatial gradients of the different model parameters, $\nabla m^{(1)}$, $\nabla m^{(2)}$, ...$\nabla m^{(N)}$. Applying this functional enforces the structural (geometric) similarity between the different images by minimizing the joint entropy of their structure.

11.3 Steepest Ascent Directions of the Joint Minimum Entropy Functional

Following the same logic we discussed in Chap. 10 for joint focusing inversion, we now derive the expressions for the steepest ascent directions of the joint minimum entropy functionals. Let us rewrite Eq. (11.6) in the following form:

$$S_{qJME} = -\int_V \frac{p_J}{Q_J} \ln \frac{p_J}{Q_J} dv, \tag{11.10}$$

where

$$p_J = \sum_{i=1}^N \left|m^{(i)} - m^{(i)}_{apr}\right|^q + \beta, \tag{11.11}$$

and Q_J is defined above in Eq. (11.7).

We can take the first variation of functional S_{qJME} given by Eq. (11.10):

$$\delta S_{qJME} = \sum_{n=1}^{N} \delta_{m^{(n)}} S_{qJME} = -\sum_{n=1}^{N} \int_V \delta_{m^{(n)}} \left[\frac{p_J}{Q_J} \ln \frac{p_J}{Q_J} \right] dv. \tag{11.12}$$

One has to find the variations of the integrands in Eq. (11.12). Applying the standard rules of the variational calculus, we have

$$\delta_{m^{(n)}} \left[\frac{p_J}{Q_J} \ln \frac{p_J}{Q_J} \right] =$$

$$= \delta_{m^{(n)}} \left[\frac{p_J}{Q_J} \right] \ln \frac{p_J}{Q_J} + \frac{p_J}{Q_J} \delta_{m^{(n)}} \left[\ln \frac{p_J}{Q_J} \right]$$

$$= \delta_{m^{(n)}} \left[\frac{p_J}{Q_J} \right] \ln \frac{p_J}{Q_J} + \frac{p_J}{Q_J} \frac{Q_J}{p_J} \delta_{m^{(n)}} \left[\frac{p_J}{Q_J} \right]$$

$$= \left(\ln \frac{p_J}{Q_J} + 1 \right) \delta_{m^{(n)}} \left[\frac{p_J}{Q_J} \right]. \tag{11.13}$$

We can now derive the variations of the ration $\frac{p_J}{Q_J}$:

$$\delta_{m^{(n)}} \left[\frac{p_J}{Q_J} \right] = \frac{\delta_{m^{(n)}} p_J}{Q_J} - \frac{p_J \delta_{m^{(n)}} Q_J}{Q_J^2}$$

$$= \frac{Q_J \delta_{m^{(n)}} p_J - p_J \delta_{m^{(n)}} Q_J}{Q_J^2} \tag{11.14}$$

The variations $\delta_{m^{(n)}} p_J$ and $\delta_{m^{(n)}} Q_J$ in the last formula are calculated as follows:

$$\delta_{m^{(n)}} p_J = \delta_{m^{(n)}} \left[\sum_{i=1}^{N} |m^{(i)} - m_{apr}^{(i)}|^q + \beta \right]$$

$$= q \left| m^{(n)} - m_{apr}^{(n)} \right|^{q-1} \text{signum} \left(m^{(n)} - m_{apr}^{(n)} \right) \delta m^{(n)} = q u^{(n)} \delta m^{(n)}, \tag{11.15}$$

and

$$\delta_{m^{(n)}} Q_J = \int_V \delta_{m^{(n)}} \left(\sum_{i=1}^{N} |m^{(i)} - m_{apr}^{(i)}|^q + \beta \right) dv$$

$$= \int_V q \left| m^{(n)} - m_{apr}^{(n)} \right|^{q-1} \text{signum} \left(m^{(n)} - m_{apr}^{(n)} \right) \delta m^{(n)} dv = q \int_V u^{(n)} \delta m^{(n)} dv, \tag{11.16}$$

where we introduced a new function $u^{(n)}$:

$$u^{(n)} = \left|m^{(n)} - m_{apr}^{(n)}\right|^{q-1} \text{signum}\left(m^{(n)} - m_{apr}^{(n)}\right). \tag{11.17}$$

Substituting Eqs. (11.15) and (11.16) into (11.14) and (11.13), we have

$$\delta_{m^{(n)}}\left[\frac{p_J}{Q_J}\ln\frac{p_J}{Q_J}\right] = \left(\ln\frac{p_J}{Q_J}+1\right)\delta_{m^{(n)}}\left[\frac{p_J}{Q_J}\right] =$$

$$= \frac{q}{Q_J^2}\left(\ln\frac{p_J}{Q_J}+1\right)\left[Q_J u^{(n)}\delta m^{(n)} - p_J\int_V u^{(n)}\delta m^{(n)}dv\right]. \tag{11.18}$$

Integrating the last formula over domain V, we find

$$\int_V \delta_{m^{(n)}}\left[\frac{p_J}{Q_J}\ln\frac{p_J}{Q_J}\right]dv =$$

$$= \frac{q}{Q_J^2}\int_V\left(\ln\frac{p_J}{Q_J}+1\right)Q_J u^{(n)}\delta m^{(n)}dv - \frac{q}{Q_J^2}\int_V\left(\ln\frac{p_J}{Q_J}+1\right)p_J dv\int_V u^{(n)}\delta m^{(n)}dv. \tag{11.19}$$

Let us introduce the following notations:

$$a^{(n)} = \left(\ln\frac{p_J}{Q_J}+1\right)Q_J u^{(n)} = Q_J\left|m^{(n)} - m_{apr}^{(n)}\right|^{q-1}\text{signum}\left(m^{(n)} - m_{apr}^{(n)}\right)\left(\ln\frac{p_J}{Q_J}+1\right), \tag{11.20}$$

and

$$b^{(n)} = \int_V\left(\ln\frac{p_J}{Q_J}+1\right)p_J dv. \tag{11.21}$$

Using these notations, we can write

$$\int_V \delta_{m^{(n)}}\left[\frac{p_J}{Q_J}\ln\frac{p_J}{Q_J}\right]dv = \frac{q}{Q_J^2}\int_V\left(a^{(n)} - b^{(n)}\right)\delta m^{(n)}dv = -2\left(\delta m^{(n)}, l_{qJME}^{(n)}\right)_{L_2}, \tag{11.22}$$

where functions $l_{qJME}^{(n)}$ represent the directions of the steepest ascent of the joint minimum entropy functional with respect to model $m^{(n)}$:

$$l_{qJME}^{(n)} = \frac{q}{2Q_J^2}\left(b^{(n)} - a^{(n)}\right). \tag{11.23}$$

Finally, we can write Eq. (11.12) for the first variation of the joint minimum entropy functional, as follows:

$$\delta S_{qJME} = \sum_{n=1}^{N} \delta_{m^{(n)}} S_{qJME} = 2\sum_{n=1}^{N} \left(\delta m^{(n)}, l_{qJME}^{(n)}\right)_{L_2}, \tag{11.24}$$

Similarly, one can calculate the first variation and the directions of the steepest ascent of the joint minimum entropy of gradient functional (11.8).

11.4 Representation of the Minimum Entropy Stabilizing Functionals in Pseudo-Quadratic Form

Expressions (11.6) and (11.9) show that the joint minimum entropy functionals are nonquadratic. In numerical applications, it is convenient to represent them in a pseudo-quadratic form, as it was done in Chap. 10 for focusing stabilizers.

For example, the joint minimum entropy functional, $S_{qJME}(m)$, can be written as follows:

$$S_{qJME} = -\int_V \frac{\sum_{i=1}^{N} \left|m^{(i)} - m_{apr}^{(i)}\right|^q + \beta}{Q_J} \log \frac{\sum_{j=1}^{N} \left|m^{(j)} - m_{apr}^{(j)}\right|^q + \beta}{Q_J} dv$$

$$\approx \sum_{i=1}^{N} \int_V \frac{\left|m^{(i)} - m_{apr}^{(i)}\right|^2 \left|m^{(i)} - m_{apr}^{(i)}\right|^q}{Q_J \left(\left|m^{(i)} - m_{apr}^{(i)}\right|^2 + \beta\right)} \log \frac{Q_J}{\sum_{j=1}^{N} \left|m^{(j)} - m_{apr}^{(j)}\right|^q + \beta} dv$$

$$= \sum_{i=1}^{N} \left(\mathbf{W}_{eq}^{(i)JME} \left(\mathbf{m}^{(i)} - \mathbf{m}_{apr}^{(i)}\right), \mathbf{W}_{eq}^{(i)JME} \left(\mathbf{m}^{(i)} - \mathbf{m}_{apr}^{(i)}\right)\right)_M, \tag{11.25}$$

where β is a small positive number, and $\mathbf{W}_{eq}^{(i)JME}$ is a linear operator of the multiplication of the model parameter function $m^{(i)}(\mathbf{r})$ by the following function, $w_{eq}^{(i)JME}(\mathbf{r})$:

$$w_{eq}^{(i)JME}(\mathbf{r}) =$$

$$= \left[\frac{\left|m^{(i)} - m_{apr}^{(i)}\right|^q}{Q_J \left(\left|m^{(i)} - m_{apr}^{(i)}\right|^2 + \beta\right)} \log \left[Q_J / \left(\sum_{j=1}^{N} \left|m^{(j)} - m_{apr}^{(j)}\right|^q + \beta\right)\right]\right]^{\frac{1}{2}}. \tag{11.26}$$

For the joint minimum entropy of gradient functional, $S_{JMEG}(m)$, we find

$$S_{qJMEG} =$$

$$= \int_V \frac{\sum_{i=1}^N \left(\nabla m^{(i)} \cdot \nabla m^{(i)}\right)^{q/2} + \beta}{Q'_J} \log\left[Q'_J / \left(\sum_{j=1}^N \left(\nabla m^{(j)} \cdot \nabla m^{(j)}\right)^{q/2} + \beta\right)\right] dv$$

$$\approx \sum_{i=1}^N \int_V \frac{\left|m^{(i)} - m_{apr}^{(i)}\right|^2 \left(\nabla m^{(i)} \cdot \nabla m^{(i)}\right)^{q/2}}{Q'_J \left(\left|m^{(i)} - m_{apr}^{(i)}\right|^2 + \beta\right)} \log\left[Q'_J / \left(\sum_{j=1}^N \left(\nabla m^{(j)} \cdot \nabla m^{(j)}\right)^{q/2} + \beta\right)\right] dv$$

$$= \sum_{i=1}^N \left(\mathbf{W}_{eq}^{(i)JMEG}\left(\mathbf{m}^{(i)} - \mathbf{m}_{apr}^{(i)}\right), \mathbf{W}_{eq}^{(i)JMEG}\left(\mathbf{m}^{(i)} - \mathbf{m}_{apr}^{(i)}\right)\right)_M, \tag{11.27}$$

where $\mathbf{W}_{eq}^{(i)JMEG}$ is a linear operator of the multiplication of the model parameter function $m^{(i)}(\mathbf{r})$ by the following function, $w_{eq}^{(i)JMEG}(\mathbf{r})$:

$$w_e^{(i)JMEG}(\mathbf{r})$$

$$= \left[\frac{\left(\nabla m^{(i)} \cdot \nabla m^{(i)}\right)^{q/2}}{Q'_J \left(\left|m^{(i)} - m_{apr}^{(i)}\right|^2 + \beta\right)} \log\left[Q'_J / \left(\sum_{j=1}^N \left(\nabla m^{(j)} \cdot \nabla m^{(j)}\right)^{q/2} + \beta\right)\right]\right]^{\frac{1}{2}}. \tag{11.28}$$

11.5 Joint Minimum Entropy Inversion of Multiphysics Data

We can introduce the multiphysics inverse problem again as follows:

$$\mathbf{d}^{(i)} = \mathbf{A}^{(i)}(\mathbf{m}^{(i)}), \quad i = 1, 2, 3, ..., N. \tag{11.29}$$

The regularized solution of a nonlinear inverse problem (11.29) can be obtained by minimization of the parametric functional with the minimum entropy stabilizers,

$$P^\alpha(\mathbf{m}^{(1)}, \mathbf{m}^{(2)}, ..., \mathbf{m}^{(n)}) = \sum_{i=1}^N \left\|\mathbf{A}^{(i)}(\mathbf{m}^{(i)}) - \mathbf{d}^{(i)}\right\|_D^2 + \alpha S_{qJME,\ qJMEG}, \tag{11.30}$$

where α is the regularization parameter.

The terms S_{qJME} and S_{qJMEG} are the joint stabilizing functionals based on minimum entropy and minimum entropy gradient constraints, respectively, defined above.

Using the pseudo-quadratic forms (11.25) and (11.27) of stabilizing functionals, we can present the corresponding parametric functional (11.30) as follows:

$$P^{\alpha}(\mathbf{m}^{(1)}, \mathbf{m}^{(2)}, ...\mathbf{m}^{(N)}) = \sum_{i=1}^{N} \left(\mathbf{A}^{(i)}(\mathbf{m}^{(i)}) - \mathbf{d}^{(i)}, \mathbf{A}^{(i)}(\mathbf{m}^{(i)}) - \mathbf{d}^{(i)}\right)_D +$$

$$+ \sum_{i=1}^{N} \alpha^{(i)} \left(\mathbf{W}_e^{(i)} \left(\mathbf{m}^{(i)} - \mathbf{m}_{apr}^{(i)}\right), \mathbf{W}_{ei} \left(\mathbf{m}^{(i)} - \mathbf{m}_{apr}^{(i)}\right)\right)_M, \tag{11.31}$$

where $\mathbf{W}_e^{(i)}$ is a linear operator of multiplication of the model parameter function $m^{(i)}(\mathbf{r})$ by the function $w_e^{(i)}(\mathbf{r})$:

$$\begin{cases} w_e^{(i)} = w_{eq}^{JME}(\mathbf{r}), & \text{for joint minimum entropy functional,} \\ w_e^{(i)} = w_{eq}^{(i)JMEG}(\mathbf{r}), & \text{for joint minimum entropy gradient functional.} \end{cases}$$

Therefore, the problem of minimization of the parametric functional introduced by Eq. (11.31) can be treated in a similar way to the minimization of the conventional Tikhonov parametric functional. The only difference is that now we introduce some variable weighting operator $\mathbf{W}_e$, which depends on the model parameters. This problem can be solved using the re-weighted regularized steepest descent or conjugate gradient methods, as discussed in Chap. 10 for joint focusing inversion.

References and Recommended Reading to This Chapter

Ables JG (1974) Maximum entropy spectral analysis. Astron Astrophys Suppl Ser 15:383

Amato U, Hughes W (1991) Maximum entropy regularization of Fredholm integral equations of the first kind. Inverse Probl 7:793–808

Burg JP (1975) Maximum entropy spectral analysis. Stanford University

Cover TM, Thomas JA (2006) Elements of information theory, 2nd edn. Willey, New York

Franklin JN (1970) Well-posed stochastic extensions of ill-posed linear problems. J Math Anal Appl 31:682–716

Kopec S (1991) Properties of maximum entropy approximate solutions to Fredholm integral equations. J Math Phys 32:1269–1272

Ramos FM, Campos Velho HF, Carvalho JC, Ferreira NJ (1999) Novel approaches to entropic regularization. Inverse Probl 15:1139–1148

Shannon CE (1948) A mathematical theory of communication. Bell Sys Tech J 27(3):379–423

Tarantola A (1987) Inverse problem theory. Elsevier, Oxford, New York, Tokyo, Amsterdam, p 613

Tarantola A (2005) Inverse problem theory and methods for model parameter estimation. Society for Industrial and Applied Mathematics

Wernecke SJ, D'Addario LR (1977) Maximum entropy image reconstruction. IEEE Trans Comput 26(04):351–364

Zhdanov MS (2002) Geophysical inverse theory and regularization problems. Elsevier

Zhdanov MS (2009) New advances in 3D regularized inversion of gravity and electromagnetic data. Geophys Prospect 57(4):463–478

Zhdanov MS (2015) Inverse theory and applications in geophysics. Elsevier

Zhdanov MS (2022) Joint minimum entropy method for simultaneous processing and fusion of multi-physics data and images. U.S. Patent Application 17/343,218

Zhdanov MS, Tu X, Čuma M (2022) Cooperative inversion of multiphysics data using joint minimum entropy constraints. Near Surf Geophys 1:1–14

Chapter 12
Gramian Method of Generalized Joint Inversion

Abstract The Gramian space of model parameters is revisited. The norm of the function in the Gramian space is defined as the determinant of the Gram matrix of a system of different model parameters or their attributes. This determinant is called a Gramian. It is demonstrated that minimization of the Gramian results in enforcing the correlation between different transforms (attributes) of the model parameters, producing inverse images with similar patterns. This chapter also presents the methods of joint inversion with Gramian stabilizers. The concept of localized Gramian is introduced to allow for variable relationships between the different physical properties of the models over the area of investigation.

Keywords Gramian space · Gramian functional · Gramian stabilizer · Localized Gramian

In this chapter, we discuss a general approach to joint inversion of multimodal geophysical data using Gramian constraints, which is based on the minimization of the determinant of a Gram matrix of a system of different model parameters or their attributes (Zhdanov et al. 2012a; 2012b; Zhdanov 2019). This approach does not require any a priori knowledge about the types of relationships between the different model parameters. Instead, it determines the form of these relationships in the inversion process. The Gramian constraints make it possible to consider both linear and nonlinear relationships between the different physical parameters of the model. Furthermore, by specifying a type of Gramian constraint, one can enforce polynomial, gradient, or any other complex correlations.

12.1 Gramian Spaces Revisited

We consider inverse problems for multiple data sets. The following operator relationships can describe these problems:

$$d^{(i)} = A^{(i)}(m^{(i)}), \quad i = 1, 2, 3, ..., n, \tag{12.1}$$

M. S. Zhdanov, *Advanced Methods of Joint Inversion and Fusion of Multiphysics Data*, Advances in Geological Science,
https://doi.org/10.1007/978-981-99-6722-3_12

where, in a general case, $A^{(i)}$ is a nonlinear operator, $d^{(i)}$ $(i = 1, 2, 3, ..., n)$ are different observed data sets, and $m^{(i)}$ $(i = 1, 2, 3, ..., n)$ are the unknown sets of model parameters described by integrable functions of a radius-vector $\mathbf{r} = (x, \ y, \ z)$ defined within some volume V of a 3D space. Note that in expression (12.1) and everywhere below in this chapter, we consider $d^{(1)}, d^{(2)}, ..., d^{(n)}$ and $m^{(1)}, m^{(2)}, ..., m^{(n)}$ being the dimensionless weighted data and model parameters, as defined in Chap. 8.

In the framework of regularization theory, the solution of inverse problems (12.1) can be reduced to minimization of the following parametric functional:

$$P^{\alpha}(m^{(1)},m^{(2)}, ...m^{(n)}) = \sum_{i=1}^{n} \| A^{(i)}(m^{(i)}) - d^{(i)} \|_D^2$$

$$+ \alpha S_J(m^{(1)},m^{(2)}, ...m^{(n)}) = \min, \tag{12.2}$$

where α is the regularization parameter and S_J is the joint stabilizing functional, which enforces the mutual relationships between different physical parameters of the joint inversion. This chapter introduces Gramian stabilizing functional, which will serve this purpose.

12.1.1 Gramian Space of Model Parameters

In Chap. 3, we provided a definition of the Gramian space of complex functions of a radius-vector $\mathbf{r} = (x, \ y, \ z)$ defined within some volume V of a 3D space. We considered these functions as the elements of a complex Hilbert space $L_2^C [V]$ with a L_2^C norm, defined by the corresponding inner product:

$$(f, \ g)_{L_2^C} = \int_V f(\mathbf{r}) \ g^*(\mathbf{r}) \, dv, \quad \|f\|^2 = (f, \ f), \tag{12.3}$$

where asterisk "*" denotes the complex conjugate value.

Following the same ideas, we now introduce a space M of model parameters formed by some complex functions of the radius-vector $\mathbf{r} = (x, \ y, \ z)$, $\mathbf{r} \in \mathbf{V}$. Let us consider two arbitrary functions from the model space, $p(\mathbf{r})$ and $q(\mathbf{r}) \in M$. We can define a new inner product operation, $(p, q)_{G^{(n)}}$, between two functions, p and q, as the determinant of the following matrix:

$$(p, q)_{G^{(n)}} =$$

$$= \begin{vmatrix} \left(m^{(1)}, m^{(1)}\right) & \left(m^{(1)}, m^{(2)}\right) & \ldots & \left(m^{(1)}, m^{(n-1)}\right) & \left(m^{(1)}, q\right) \\ \left(m^{(2)}, m^{(1)}\right) & \left(m^{(2)}, m^{(2)}\right) & \ldots & \left(m^{(2)}, m^{(n-1)}\right) & \left(m^{(2)}, q\right) \\ \ldots & \ldots & \ldots & \ldots & \ldots \\ \left(m^{(n-1)}, m^{(1)}\right) & \left(m^{(n-1)}, m^{(2)}\right) & \ldots & \left(m^{(n-1)}, m^{(n-1)}\right) & \left(m^{(n-1)}, q\right) \\ \left(p, m^{(1)}\right) & \left(p, m^{(2)}\right) & \ldots & \left(p, m^{(n-1)}\right) & \left(p, q\right) \end{vmatrix}, \tag{12.4}$$

where matrix elements are formed by the conventional $L_2^C[V]$ inner product between two functions (see Chap. 3, Sect. 3.6.3) as follows:

$$\left(m^{(i)}, m^{(j)}\right) = \left(m^{(i)}, m^{(j)}\right)_{L_2^C} = \int_V m^{(i)}(\mathbf{r})\ m^{(j)*}(\mathbf{r})\,dv,$$

$$\left(m^{(i)}, q\right) = \left(m^{(i)}, q\right)_{L_2^C} = \int_V m^{(i)}(\mathbf{r})\ q^*(\mathbf{r})\,dv,$$

$$\left(p, m^{(j)}\right) = \left(p, m^{(j)}\right)_{L_2^C} = \int_V p(\mathbf{r})\ m^{(j)*}(\mathbf{r})\,dv. \tag{12.5}$$

It is easy to check that all the properties of the inner product hold for Eq. (12.4) (see Chap. 3).

Note that the norm square of a function, $\|p\|^2_{G^{(n)}}$, is equal to the determinant, $G(m^{(1)}, m^{(2)}, \ldots., m^{(n-1)}, p)$, of the Gram matrix, $\mathbf{G}$, of a set of functions, $(m^{(1)}, m^{(2)}, \ldots., m^{(n-1)}, p,)$, which is called a *Gramian*:

$$\|p\|^2_{G^{(n)}} = (p, p)_{G^{(n)}} = G(m^{(1)}, m^{(2)}, \ldots., m^{(n-1)}, p) = |\mathbf{G}| =$$

$$= \begin{vmatrix} \left(m^{(1)}, m^{(1)}\right) & \left(m^{(1)}, m^{(2)}\right) & \ldots & \left(m^{(1)}, m^{(n-1)}\right) & \left(m^{(1)}, p\right) \\ \left(m^{(2)}, m^{(1)}\right) & \left(m^{(2)}, m^{(2)}\right) & \ldots & \left(m^{(2)}, m^{(n-1)}\right) & \left(m^{(2)}, p\right) \\ \ldots & \ldots & \ldots & \ldots & \ldots \\ \left(m^{(n-1)}, m^{(1)}\right) & \left(m^{(n-1)}, m^{(2)}\right) & \ldots & \left(m^{(n-1)}, m^{(n-1)}\right) & \left(m^{(n-1)}, p\right) \\ \left(p, m^{(1)}\right) & \left(p, m^{(2)}\right) & \ldots & \left(p, m^{(n-1)}\right) & (p, p) \end{vmatrix}. \tag{12.6}$$

It was demonstrated in Chap. 3 that Gramian satisfies Gram's inequality (3.105):

$$G(m^{(1)}, m^{(2)}, \ldots., m^{(n-1)}, p) \geq 0. \tag{12.7}$$

Equality holds in (12.7) if and only if the system of functions $\left(m^{(1)}, m^{(2)}, \ldots., m^{(n-1)}, p\right)$ is linearly dependent.

We will call the Hilbert space formed by the integrable functions, defined within some volume V of a 3D space, with the inner product operation determined by formula (12.4), a *Gramian space* of the model parameters, $G^{(n)}$. The main property of the Gramian space is that the norm of the function in the Gramian space provides a measure of correlation between this function and the additional model parameters $m^{(1)}, m^{(2)}, \ldots., m^{(n-1)}$.

It can be shown that one can introduce the Gramian space $G^{(j)}$, where the inner product is defined by an expression similar to (12.6) with the only difference being that functions p and q are located within the row and column with number j, respectively:

$$(p, q)_{G^{(j)}} =$$

$$= \begin{vmatrix} \left(m^{(1)}, m^{(1)}\right) & \left(m^{(1)}, m^{(2)}\right) & \ldots & \left(m^{(1)}, q\right) & \ldots \left(m^{(1)}, m^{(n)}\right) \\ \ldots & \ldots & \ldots & \ldots & \ldots\ldots \\ \left(p, m^{(1)}\right) & \left(p, m^{(2)}\right) & \ldots & (p, q) & \ldots \left(p, m^{(n)}\right) \\ \ldots & \ldots & \ldots & \ldots & \ldots\ldots \\ \left(m^{(n)}, m^{(1)}\right) & \left(m^{(n)}, m^{(2)}\right) & \ldots & \left(m^{(n)}, q\right) \ldots & \left(m^{(n)}, m^{(n)}\right) \end{vmatrix}. \quad (12.8)$$

In the Gramian space $G^{(j)}$, the norm square of a function, $\|p\|^2_{G^{(j)}}$, is equal to the Gramian of a set of functions, $(m^{(1)}, m^{(2)}, \ldots., m^{(j-1)}, p, m^{(j+1)}, \ldots.m^{(n)})$:

$$\|p\|^2_{G^{(j)}} = (p, p)_{G^{(j)}} = G(m^{(1)}, m^{(2)}, \ldots.m^{(j-1)}, p, m^{(j+1)}, \ldots, m^{(n)}). \quad (12.9)$$

Therefore, the norm of the function in the Gramian space $G^{(j)}$ provides a measure of linear dependence between this function and all other model parameters, except for parameter $m^{(j)}$: $m^{(1)}, m^{(2)}, \ldots\ldots. m^{(j-1)}, m^{(j+1)}, \ldots, m^{(n)}$.

One can see from formula (12.8) that the following identities hold:

$$\left\|m^{(i)}\right\|^2_{G^{(i)}} = G(m^{(1)}, m^{(2)}, \ldots, m^{(n)}, \text{ for } i = 1, 2, \ldots, n. \quad (12.10)$$

Therefore, we conclude at once that the Gramian norm has the following important property:

$$\left\|m^{(i)}\right\|^2_{G^{(i)}} = \left\|m^{(j)}\right\|^2_{G^{(j)}}, \text{ for } i, j = 1, 2, \ldots, n. \quad (12.11)$$

The last formula demonstrates that all the functions, $m^{(1)}, m^{(2)}, \ldots\ldots, m^{(n)}$, have the same norm in the corresponding Gramian spaces $G^{(j)}$, $j = 1, 2, \ldots, n$.

12.1.2 Gramian Space of Model Parameter Gradients

In many applications, it is necessary to jointly invert the data, which are produced by unrelated physical phenomena. In this case, one cannot use any correlation between different model parameters but instead should consider the possibility of some structural (geometrical) similarities between the different physical models. The concept of joint structure-coupled inversion was discussed in Chap. 9, where we considered two mathematical criteria for structural similarity. One was introduced in image processing theory for image quality assessment (Wang et al. 2004). It is based on the structural similarity index (SSI). Another approach requires the parallelism of the model parameter gradients (e.g., Fregoso and Gallardo 2009; Gallardo 2007; Gallardo and Meju 2003, 2004, 2007, 2011; Hu et al. 2009; Meju 2011).

In this section, we consider a Gramian-based method to find the inverse models possessing the maximum geometrical similarities between the different physical modalities. This method is based on the Gramian space of model parameter gradients. We introduce this space following Zhdanov (2015).

Let us consider a complex Hilbert space M_∇ formed by gradients of the functions differentiable within volume V with the metric, defined by the following inner product operation:

$$(\nabla p, \nabla q)_{M_\nabla} = \int_V \left(\nabla p\,(\mathbf{r}) \cdot \nabla q^*\,(\mathbf{r})\right) dv, \tag{12.12}$$

where $p(\mathbf{r})$, $q(\mathbf{r}) \in M_\nabla$, and $\nabla p(\mathbf{r})$ and $\nabla q(\mathbf{r})$ are the gradients of functions $p(\mathbf{r})$ and $q(\mathbf{r})$, respectively ($\mathbf{r} \in V$).

Note that the inner product in the space of the gradients, M_∇, can be transformed into the inner product in the space of the original functions, L_2^C.

Indeed, we can integrate integral (12.12) by parts:

$$\int_V \left(\nabla p \cdot \nabla q^*\right) dv = -\int_V p\nabla^2 q^* dv + \int_{\partial V} p\nabla q^* \cdot \mathbf{n} ds,$$

where we have applied the Gauss theorem, and $\mathbf{n}$ is a unit vector of the normal directed outward from domain V.

We assume homogeneous Neumann (i.e., no flux) boundary conditions for the gradients:

$$\int_{\partial V} p\nabla q^* \cdot \mathbf{n} ds = 0.$$

Therefore we have

$$(\nabla p, \nabla q)_{M_\nabla} = \int_V \left(\nabla p \cdot \nabla q^*\right) dv = -\int_V p\nabla^2 q^* dv = -\left(p, \nabla^2 q^*\right)_{L_2^C}. \tag{12.13}$$

We can now introduce a new inner product operation, $(\nabla p, \nabla q)_{G_\nabla^{(n)}}$, between two functions, ∇p and ∇q, from the space, M_∇, of the model parameter gradients, $\nabla p\,(\mathbf{r})$ and $\nabla q\,(\mathbf{r}) \in M_\nabla$, as the determinant of the following matrix:

$$(\nabla p, \nabla q)_{G_\nabla^{(n)}} =$$

$$= \begin{vmatrix} \left(\nabla m^{(1)}, \nabla m^{(1)}\right)_{M_\nabla} & \left(\nabla m^{(1)}, \nabla m^{(2)}\right)_{M_\nabla} & \ldots & \left(\nabla m^{(1)}, \nabla q\right)_{M_\nabla} \\ \left(\nabla m^{(2)}, \nabla m^{(1)}\right)_{M_\nabla} & \left(\nabla m^{(2)}, \nabla m^{(2)}\right)_{M_\nabla} & \ldots & \left(\nabla m^{(2)}, \nabla q\right)_{M_\nabla} \\ \ldots & \ldots & \ldots & \ldots \\ \left(\nabla m^{(n-1)}, \nabla m^{(1)}\right)_{M_\nabla} & \left(\nabla m^{(n-1)}, \nabla m^{(2)}\right)_{M_\nabla} & \ldots & \left(\nabla m^{(n-1)}, \nabla q\right)_{M_\nabla} \\ \left(\nabla p, \nabla m^{(1)}\right)_{M_\nabla} & \left(\nabla p, \nabla m^{(2)}\right)_{M_\nabla} & \ldots & (\nabla p, \nabla q)_{M_\nabla} \end{vmatrix}, \tag{12.14}$$

where

$$\left(\nabla m^{(i)}, \nabla m^{(j)}\right)_{L_2(M_\nabla)} = \int_V \left(\nabla m^{(i)}(\mathbf{r}) \cdot \nabla m^{(j)*}(\mathbf{r})\right) dv,$$

$$\left(\nabla m^{(i)}, \nabla q\right)_{L_2(M_\nabla)} = \int_V \left(\nabla m^{(i)}(\mathbf{r}) \cdot \nabla q^*(\mathbf{r})\right) dv,$$

$$\left(\nabla p, \nabla m^{(j)}\right)_{L_2(M_\nabla)} = \int_V \left(\nabla p(\mathbf{r}) \cdot \nabla m^{(j)*}(\mathbf{r})\right) dv. \tag{12.15}$$

The norm square of a gradient of a function, $\|\nabla p\|^2_{G_\nabla^{(n)}}$, is equal to the Gramian of a set of gradients, $\nabla m^{(1)}, \nabla m^{(2)}, \ldots., \nabla m^{(n-1)}, \nabla p$:

$$\|\nabla p\|^2_{G_\nabla^{(n)}} = G(\nabla m^{(1)}, \nabla m^{(2)}, \ldots., \nabla m^{(n-1)}, \nabla p). \tag{12.16}$$

We will demonstrate in Chap. 13 that the norm of the gradient of function p in the Gramian space provides a measure of the correlation between the gradient of this function and the gradients of the additional model parameters $\nabla m^{(1)}, \nabla m^{(2)}, \ldots.,$ $\nabla m^{(n-1)}$. Minimization of this norm, $\|\nabla p\|_{G_\nabla^{(n)}}$, results in producing multimodal inverse images with correlated directions of the parameter changes, similar to the result of the minimum cross-gradient joint inversion (see Chap. 9).

As it was discussed in the previous section, one could introduce the Gramian space $G_\nabla^{(j)}$, where the inner product is defined by an expression similar to (12.14) with the only difference being that functions ∇p and ∇q are located within the row and column with number j, respectively. The norm square of a gradient of a function, $\|\nabla p\|^2_{G_\nabla^{(j)}}$, in Gramian space $G_\nabla^{(j)}$ is equal to the Gramian of a set of gradients, $\nabla m^{(1)}$, $\nabla m^{(2)}, \ldots., \nabla m^{(j-1)}, \nabla p, \nabla m^{(j+1)}, \ldots.\nabla m^{(n)}$:

$$\|\nabla p\|^2_{G_\nabla^{(j)}} = (\nabla p, \nabla p)_{G^{(j)}}$$

$$= G(\nabla m^{(1)}, \nabla m^{(2)}, \ldots, \nabla m^{(j-1)}, \nabla p, \nabla m^{(j+1)}, \ldots.\nabla m^{(n)}). \tag{12.17}$$

Therefore, the norm of the gradient of the function in the Gramian space $G_\nabla^{(j)}$ provides a measure of the structural correlation between this function, ∇p, and the gradients of all other model parameters, except for parameter $m^{(j)}$: $\nabla m^{(1)}, \nabla m^{(2)}, \ldots, \nabla m^{(j-1)}$, $\nabla m^{(j+1)}, \ldots \nabla m^{(n)}$.

Finally, one can check that the Gramian norm of the gradients has the same property (12.11), as the Gramian norm of the model parameters:

$$\left\|\nabla m^{(i)}\right\|^2_{G_\nabla^{(i)}} = \left\|\nabla m^{(j)}\right\|^2_{G_\nabla^{(j)}}, \text{ for } i, j = 1, 2, \ldots, n. \tag{12.18}$$

12.1.3 Gramian Spaces of Different Transforms of the Model Parameters

The approach to the joint inversion based on Gramian constraints makes it possible to consider different properties (attributes) of the model parameters in the fusion of multimodal inversions. We can use, for example, second derivatives of the model parameters, absolute values of the gradients and/or second derivatives of the model parameters, or any other transforms of the model parameters and their gradients. The idea is that in joint inversion, one could search for inverse images that have similar features expressed by the areas of strong variations of the model parameters or by the boundaries outlying the areas of the strong contrasts in physical properties.

Let us introduce an operator, T, of a transformation of the model parameters from space M into a transformed model space M_T:

$$f_T = Tf, \; g_T = Tg; \quad f, g \in M; \quad f_T, g_T \in M_T.$$

Operator T can be chosen as a differential operator (e.g., gradient or Laplacian of the model parameters) or as an absolute value of the model parameters or their derivatives (e.g., the absolute value of the gradient or Laplacian of the model parameters), or as a Fourier transform or any other transformations which emphasize specific properties of the inverse images. We can treat all these transformations as some "attributes" of the model parameters.

The inner product operation, $(f_T, g_T)_{G_T^{(n)}}$, between two functions, f_T and g_T, is determined as the determinant of the following matrix:

$$(f_T, g_T)_{G_T^{(n)}} =$$

$$= \begin{vmatrix} \left(Tm^{(1)}, Tm^{(1)}\right)_{M_T} & \left(Tm^{(1)}, Tm^{(2)}\right)_{M_T} & \ldots & \left(Tm^{(1)}, g_T\right)_{M_T} \\ \left(Tm^{(2)}, Tm^{(1)}\right)_{M_T} & \left(Tm^{(2)}, Tm^{(2)}\right)_{M_T} & \ldots & \left(Tm^{(2)}, g_T\right)_{M_T} \\ \ldots & \ldots & \ldots & \ldots \\ \left(Tm^{(n-1)}, Tm^{(1)}\right)_{M_T} & \left(Tm^{(n-1)}, Tm^{(2)}\right)_{M_T} & \ldots & \left(Tm^{(n-1)}, g_T\right)_{M_T} \\ \left(f_T, \nabla m^{(1)}\right)_{M_T} & \left(f_T, \nabla m^{(2)}\right)_{M_T} & \ldots & (f_T, g_T)_{M_T} \end{vmatrix}. \quad (12.19)$$

The norm square of a transformed function, $\|Tp\|^2_{G_T^{(n)}}$, is equal to the Gramian of a set of transforms, $Tm^{(1)}, Tm^{(2)}, ..., Tm^{(n-1)}, Tp$:

$$\|Tp\|^2_{G_T^{(n)}} = G(Tm^{(1)}, Tm^{(2)}, ..., Tm^{(n-1)}, \; Tp). \quad (12.20)$$

Therefore, the norm of the transformed function p in the Gramian space provides a measure of the correlation between the transform of this function and similar transforms of the additional model parameters $Tm^{(1)}, Tm^{(2)}, ..., Tm^{(n-1)}$. Minimization of this norm, $\|Tp\|_{G_T^{(n)}}$, results in producing the multimodal inverse images with correlated transformations of the parameters (correlated attributes), which generates the inverse images with similar patterns in the corresponding transformations.

12.2 Gramian Stabilizing Functionals

The *Gramian stabilizing functional*, $S_G = S_G(m^{(1)},m^{(2)}, ...m^{(n)})$, is introduced as the Gramian norm of function $m^{(n)}(\mathbf{r})$ describing the distribution of the model parameters:

$$S_G(m^{(1)},m^{(2)}, ...m^{(n)}) = \left\| m^{(n)} \right\|^2_{G^{(n)}} = G\left(m^{(1)}, m^{(2)}, ..., m^{(n)}\right), \tag{12.21}$$

where according to formula (12.6), Gramian, $G\left(m^{(1)}, m^{(2)}, ..., m^{(n)}\right)$, is defined as follows:

$$G(m^{(1)}, m^{(2)}, ..., m^{(n)})$$

$$= \begin{vmatrix} \left(m^{(1)}, m^{(1)}\right) & \left(m^{(1)}, m^{(2)}\right) & \dots\dots & \left(m^{(1)}, m^{(n)}\right) \\ \left(m^{(2)}, m^{(1)}\right) & \left(m^{(2)}, m^{(2)}\right) & \dots\dots & \left(m^{(2)}, m^{(n)}\right) \\ \dots & \dots & \dots\dots & \dots \\ \left(m^{(n)}, m^{(1)}\right) & \left(m^{(n)}, m^{(2)}\right) & \dots\dots & \left(m^{(n)}, m^{(n)}\right) \end{vmatrix}. \tag{12.22}$$

Note that, according to the properties of the norm, $\|...\|_{G^{(n)}}$, in the Gramian space $G^{(n)}$, minimization of this norm results in enforcing the correlation between the different model parameters $m^{(1)}, m^{(2)}, ...,$ and $m^{(n)}$.

Similarly, one can introduce a Gramian stabilizing functional based on the gradients of the functions describing the corresponding model parameters, as follows:

$$S_{G_\nabla}(m^{(1)},m^{(2)}, ...m^{(n)}) = \left\| \nabla m^{(n)} \right\|^2_{G^{(n)}_\nabla}$$

$$= G(\nabla m^{(1)}, \nabla m^{(2)}, ..., \nabla m^{(n)}). \tag{12.23}$$

Minimization of the norm, $\|...\|_{G^{(n)}_\nabla}$, in the Gramian space $G^{(n)}_\nabla$ results in enforcing the structural correlation between the inverse images obtained for different model parameters.

Finally, we introduce a Gramian stabilizer based on the transformed function describing the corresponding model parameters:

$$S_{G_T}(m^{(1)},m^{(2)}, ...m^{(n)}) = \left\| Tm^{(n)} \right\|^2_{G^{(n)}_T} = G(Tm^{(1)}, Tm^{(2)}, ..., Tm^{(n)}). \tag{12.24}$$

Minimization of the norm, $\|...\|_{G^{(n)}_T}$, in the Gramian space $G^{(n)}_T$ results in enforcing the correlation between different transforms (attributes) of the model parameters, producing the inverse images with similar patterns.

12.3 Steepest Ascent Directions of the Gramian Functionals

As discussed in the previous chapters, the key step in developing the regularized inversion algorithm is the calculation of the steepest ascent directions (gradients) of the corresponding stabilizing functionals. Based on their first variations, I will derive below the expressions for the steepest ascent directions for the Gramian stabilizing functionals.

12.3.1 *The First Variation of the Gramian Stabilizer*

Let us calculate the variation of the Gramian norm:

$$\delta S_G(m^{(1)},m^{(2)}, ...m^{(n)}) = \delta \left\| m^{(n)}-m^{(n)}_{apr} \right\|^2_{G^{(n)}} = \sum_{i=1}^{n} \delta_{\mathbf{m}^{(i)}} \left\| m^{(n)}-m^{(n)}_{apr} \right\|^2_{G^{(n)}} . \tag{12.25}$$

Taking into account property (12.11) of the Gramian norm, Eq. (12.25) can be written as follows:

$$\delta S_G(m^{(1)},m^{(2)}, ...m^{(n)}) = \sum_{i=1}^{n} \delta_{\mathbf{m}^{(i)}} \left\| m^{(i)}-m^{(i)}_{apr} \right\|^2_{G^{(i)}}$$

$$= 2\sum_{i=1}^{n} \left(\delta m^{(i)},\ m^{(i)}-m^{(i)}_{apr}\right)_{G^{(i)}} . \tag{12.26}$$

Assuming first, for simplicity, that $m^{(i)}_{apr} = 0$, one can write expression $\left(\delta m^{(i)},\ m^{(i)}\right)_{G^{(i)}}$ in the following explicit form:

$$\left(\delta m^{(i)},\ m^{(i)}\right)_{G^{(i)}} =$$

$$= \begin{vmatrix} \left(m^{(1)},m^{(1)}\right) & \left(m^{(1)},m^{(2)}\right) & ... & \left(m^{(1)},m^{(i)}\right) & ... \left(m^{(1)},m^{(n)}\right) \\ ... & ... & ... & ... & ...... \\ \left(\delta m^{(i)},m^{(1)}\right) & \left(\delta m^{(i)},m^{(2)}\right) & ... & \left(\delta m^{(i)},m^{(i)}\right) & ... \left(\delta m^{(i)},m^{(n)}\right) \\ ... & ... & ... & ... & .... \\ \left(m^{(n)},m^{(1)}\right) & \left(m^{(n)},m^{(2)}\right) & ... & \left(m^{(n)},m^{(i)}\right) & ... \left(m^{(n)},m^{(n)}\right) \end{vmatrix}$$

$$= \left(\delta m^{(i)}, \sum_{j=1}^{n} (-1)^{i+j}\, G^m_{ij} m^{(j)}\right) = \left(\delta m^{(i)}, l^{(i)}_G\right), \tag{12.27}$$

where G^m_{ij} is the corresponding minor of Gram matrix $G(m^{(1)}, m^{(2)}, ..., m^{(n-1)}, m^{(n)})$ formed by eliminating column i and row j.

In a general case of nonzero $m_{apr}^{(i)}$, we obtain

$$\left(\delta m^{(i)},\ m^{(i)}-m_{apr}^{(i)}\right)_{G^{(i)}} = \left(\delta m^{(i)},\ l_G^{(i)}\right)_M$$

$$= \left(\delta m^{(i)}, \sum_{j=1}^{n} (-1)^{i+j}\, G_{ij}^{m-m_{apr}} \left(m^{(i)}-m_{apr}^{(i)}\right)\right), \tag{12.28}$$

where vectors $l_G^{(i)}$ are the directions of the steepest ascent for the Gramian stabilizing functionals,

$$l_G^{(i)} = \sum_{j=1}^{n} (-1)^{i+j}\, G_{ij}^{m-m_{apr}} \left(m^{(i)}-m_{apr}^{(i)}\right). \tag{12.29}$$

Thus, we obtain the following expression for the first variation of the Gramian stabilizer:

$$\delta S_G(m^{(1)}, m^{(2)}, \dots m^{(n)}) = 2\sum_{i=1}^{n} \left(\delta m^{(i)}, l_G^{(i)}\right). \tag{12.30}$$

12.3.2 The First Variation of the Gramian of a Set of Gradients of the Model Parameters

Let us consider a stabilizing functional formed by the Gramian of a set of gradients of the model parameters:

$$S_{G_\nabla}(m^{(1)}, m^{(2)}, \dots m^{(n)})$$

$$= G(\nabla m^{(1)}, \nabla m^{(2)}, \dots, \nabla m^{n-1)}, \ \nabla p) = \left\| \nabla m^{(n)} \right\|_{G_\nabla^{(n)}}^2. \tag{12.31}$$

Let us calculate the first variation:

$$\delta S_{G_\nabla}(m^{(1)}, m^{(2)}, \dots m^{(n)}) = \delta \left\| \nabla m^{(n)} \right\|_{G_\nabla^{(n)}}^2$$

$$= \sum_{i=1}^{n} \delta_{\mathbf{m}^{(i)}} \left\| \nabla m^{(n)} \right\|_{G_\nabla^{(n)}}^2 = \sum_{i=1}^{n} \delta_{\mathbf{m}^{(i)}} \left\| \nabla m^{(i)} \right\|_{G_\nabla^{(i)}}^2, \tag{12.32}$$

where we take into account the property (12.18) of the Gramian norm.

The first variation of the norm $\left\|\nabla m^{(i)}\right\|^2_{G^{(i)}_\nabla}$ can be calculated as follows:

$$\delta_{\mathbf{m}^{(i)}}\left\|\nabla m^{(i)}\right\|^2_{G^{(i)}_\nabla} = 2\left(\delta\nabla m^{(i)}, \nabla\, m^{(i)}\right)_{G^{(i)}_\nabla}$$

$$= 2\left(\delta\nabla m^{(i)}, \sum_{j=1}^{n}(-1)^{i+j}\, G^{\nabla\tilde{m}}_{ij}\nabla m^{(j)}\right)_{M_\nabla}, \tag{12.33}$$

where $G^{\nabla m}_{ij}$ is the corresponding minor of Gram matrix, $G(\nabla m^{(1)}, \nabla m^{(2)}, \ldots., \nabla m^{(n-1)}, \nabla m^{(n)})$, formed by eliminating column i and row j.

Taking into account property (12.13) of the inner product $(\ldots, \ldots)_{M_\nabla}$, we can write

$$\delta_{\mathbf{m}^{(i)}}\left\|\nabla m^{(i)}\right\|^2_{G^{(i)}_\nabla} = -2\left(\delta m^{(i)}, \sum_{j=1}^{n}(-1)^{i+j}\, G^{\nabla\tilde{m}}_{ij}\nabla^2 m^{(j)}\right) = 2\left(\delta m^{(i)}, l^{(i)}_{G_\nabla}\right), \tag{12.34}$$

where vectors $l^{(i)}_{G_\nabla}$ are the directions of the steepest ascent for the Gramian stabilizing functionals, formed by the Gramian of a set of gradients of the model parameters

$$l^{(i)}_{G_\nabla} = \sum_{j=1}^{n}(-1)^{i+j}\, G^{\nabla m}_{ij}\nabla^2 m^{(j)}. \tag{12.35}$$

12.3.3 The First Variation of the Gramian of the Transformed Model Parameters

Let us consider a stabilizing functional formed by the Gramian of the transformed model parameters:

$$S_{G_T}(m^{(1)}, m^{(2)}, \ldots.m^{(n)})$$

$$= G(Tm^{(1)}, Tm^{(2)}, \ldots., Tm^{(n-1)},\ Tp) = \left\|Tm^{(n)}\right\|^2_{G^{(n)}_\nabla}. \tag{12.36}$$

Let us calculate the first variation:

$$\delta S_{G_T}(m^{(1)}, m^{(2)}, \ldots.m^{(n)}) = \delta\left\|Tm^{(n)}\right\|^2_{G^{(n)}_T}$$

$$= \sum_{i=1}^{n}\delta_{\mathbf{m}^{(i)}}\left\|Tm^{(n)}\right\|^2_{G^{(n)}_T} = \sum_{i=1}^{n}\delta_{\mathbf{m}^{(i)}}\left\|Tm^{(i)}\right\|^2_{G^{(i)}_T}, \tag{12.37}$$

where we take into account the property (12.11) of the Gramian norm.

The first variation of the norm $\left\| Tm^{(i)} \right\|^2_{G_T^{(i)}}$ can be calculated as follows:

$$\delta_{\mathbf{m}^{(i)}} \left\| Tm^{(i)} \right\|^2_{G_\nabla^{(i)}} = 2 \left(\delta Tm^{(i)}, T\, m^{(i)} \right)_{G_T^{(i)}}$$

$$= 2 \left(\delta Tm^{(i)}, \sum_{j=1}^{n} (-1)^{i+j}\, G_{ij}^{Tm} Tm^{(j)} \right)_{L_T}, \tag{12.38}$$

where G_{ij}^{Tm} is the corresponding minor of Gram matrix $G(Tm^{(1)}, Tm^{(2)}, \ldots.., Tm^{(n-1)}, Tm^{(n)})$ formed by eliminating column i and row j.

We assume that operator T is differentiable, and

$$\delta Tm^{(i)} = F_T \delta m^{(i)},$$

where F_T is Fréchet derivative of T.

Taking into account the property of the adjoint operator F_T^*, we can write

$$\delta_{\mathbf{m}^{(i)}} \left\| Tm^{(i)} \right\|^2_{G_T^{(i)}} = 2 \left(\delta m^{(i)}, \sum_{j=1}^{n} (-1)^{i+j}\, G_{ij}^{Tm} F_T^* Tm^{(j)} \right) = 2 \left(\delta m^{(i)},\, l_{G_T}^{(i)} \right), \tag{12.39}$$

where vectors $l_{G_T}^{(i)}$ are the directions of the steepest ascent for the Gramian stabilizing functionals, formed by the Gramian of the transformed model parameters

$$l_{G_T}^{(i)} = \sum_{j=1}^{n} (-1)^{i+j}\, G_{ij}^{Tm} F_T^* Tm^{(j)}. \tag{12.40}$$

12.4 Joint Inversion with Gramian Stabilizers

We consider again the inverse problem (12.1) for multiple geophysical data sets. Following the principles of Tikhonov regularization theory, we introduce a parametric functional with the Gramian stabilizers,

$$P^{\alpha}(m^{(1)}, m^{(2)}, \ldots. m^{(n)}) = \sum_{i=1}^{n} \left\| A^{(i)}(m^{(i)}) - d^{(i)} \right\|^2_D$$

$$+ \alpha c_1 \sum_{i=1}^{n} S^{(i)}_{MN,\ MS,\ MGS} + \alpha c_2 S_{G,\ G_\nabla,\ G_T}(m^{(1)}, m^{(2)}, \ldots. m^{(n)}), \tag{12.41}$$

where α is the regularization parameter, and c_1 and c_2 are the weighting coefficients determining the weights of the different stabilizers in the parametric functional.

The terms $S_{MN}^{(i)}$, $S_{MS}^{(i)}$, and $S_{MGS}^{(i)}$ are the stabilizing functionals based on minimum norm, minimum support, and minimum gradient support constraints, respectively, defined in Eqs. (8.8) and (8.9).

The terms S_G, S_{G_∇}, and S_{G_T} are the Gramian stabilizing functionals,

$$S_G(m^{(1)},m^{(2)},\ldots m^{(n)}) = \left\| m^{(n)} - m_{apr}^{(n)} \right\|_{G^{(n)}}$$

$$= G\left(\left[m^{(1)} - m_{apr}^{(1)}\right], \left[m^{(2)} - m_{apr}^{(2)}\right], \ldots, \left[m^{(n)} - m_{apr}^{(n)}\right]\right), \tag{12.42}$$

$$S_{G_\nabla}(m^{(1)},m^{(2)},\ldots m^{(n)}) = \left\| \nabla m^{(n)} \right\|_{G_\nabla^{(n)}}^2$$

$$= G(\nabla m^{(1)}, \nabla m^{(2)}, \ldots, \nabla m^{(n)}), \tag{12.43}$$

$$S_{G_T}(m^{(1)},m^{(2)},\ldots m^{(n)}) = \left\| T m^{(n)} \right\|_{G_T^{(n)}}^2 = G(Tm^{(1)}, Tm^{(2)}, \ldots, Tm^{(n)}). \tag{12.44}$$

Note that, according to the properties of the norm, $\|\ldots\|_{G^{(n)}}$, in the Gramian space $G^{(n)}$, minimization of this norm results in enforcing the linear dependence between the weighted model parameters. Minimization of the norm, $\|\ldots\|_{G_\nabla^{(n)}}$, in the Gramian space $G_\nabla^{(n)}$ results in enforcing the structural correlation between the inverse images obtained for different model parameters. Finally, minimization of the norm, $\|\ldots\|_{G_T^{(n)}}$, in the Gramian space $G_T^{(n)}$ results in enforcing the linear relationships between different transforms (attributes) of the model parameters, producing the inverse images with similar patterns.

According to the basic principles of the regularization method, we have to find the models $m_\alpha^{(1)}$, $m_\alpha^{(2)}$, $\ldots m_\alpha^{(n)}$, a quasi-solution of the inverse problem, which minimize the parametric functional

$$P^\alpha(m_\alpha^{(1)},m_\alpha^{(2)},\ldots m_\alpha^{(n)}) = \min. \tag{12.45}$$

To solve this minimization problem, we calculate the first variation of the parametric functional with Gramian stabilizers:

$$\delta P^\alpha(m^{(1)},m^{(2)},\ldots m^{(n)})$$

$$= 2\sum_{i=1}^{n} \left(\delta A^{(i)}(m^{(i)}),\ A^{(i)}(m^{(i)}) - d^{(i)}\right)_D +$$

$$+\alpha\left(c_1 \sum_{i=1}^{n} \delta S_{MN,\ MS,\ MGS}^{(i)} + c_2 \delta S_{G,\ G_\nabla,\ G_T}(m^{(1)},m^{(2)},\ldots m^{(n)})\right). \tag{12.46}$$

Taking into consideration that operators $A^{(i)}$ are differentiable, we can write

$$\delta A^{(i)}(m^{(i)}) = F_m^{(i)} \delta m^{(i)}, \tag{12.47}$$

where $F_m^{(i)}$ is a linear operator of the Fréchet derivative of $A^{(i)}$.

It was noted above that,

$$\delta S_{MN,\ MS,\ MGS}^{(i)} = 2\left(\delta m^{(i)},\ l_{MN,\ MS,\ MGS}^{(i)}\right), \tag{12.48}$$

where vectors $l_{MN,\ MS,\ MGS}^{(i)}$ are the directions of the steepest ascent for the stabilizing functionals, based on minimum norm, minimum support, and minimum gradient support constraints, described by formulas (8.16), (8.17), and (8.18), respectively.

We have also shown that

$$\delta S_{G,\ G_\nabla,\ G_T} = 2\sum_{i=1}^{n}\left(\delta m^{(i)},\ l_{G,\ G_\nabla,\ G_T}^{(i)}\right), \tag{12.49}$$

where vectors $l_{G,\ G_\nabla,\ G_T}^{(i)}$ are the directions of the steepest ascent for the Gramian stabilizing functionals, respectively.

Substituting expressions (12.47) through (12.49) into formula (12.46), we obtain

$$\delta P^{\alpha}(m^{(1)}, m^{(2)}, \ldots m^{(n)}) =$$

$$= 2\sum_{i=1}^{n}\left(\delta m^{(i)},\ \left[F_m^{(i)\star}\left(A^{(i)}(m^{(i)}) - d^{(i)}\right) + \right.\right.$$

$$\left.\left. +\alpha\left(c_1 l_{MN,\ MS,\ MGS}^{(i)} + c_2 l_{G,\ G_\nabla,\ G_T}^{(i)}\right)\right]\right), \tag{12.50}$$

where $F_m^{(i)\star}$ are the adjoint Fréchet derivative operators.

Let us select

$$\delta m^{(i)} = -k^{\alpha} l^{\alpha(i)}(m^{(1)}, m^{(2)}, \ldots m^{(n)}), \tag{12.51}$$

where k^{α} is some positive real number, and $l^{\alpha(i)}(m^{(1)}, m^{(2)}, \ldots m^{(n)})$ are the directions of the steepest ascent of the functional P^{α}:

$$l^{\alpha(i)} = F_m^{(i)\star}\left(A^{(i)}(m^{(i)}) - d^{(i)}\right) + \alpha\left(c_1 l_{MN,\ MS,\ MGS}^{(i)} + c_2 l_{G,\ G_\nabla,\ G_T}^{(i)}\right). \tag{12.52}$$

Then

$$\delta P^{\alpha}(m^{(1)}, m^{(2)}, \ldots m^{(n)}) = -2k^{\alpha}\sum_{i=1}^{n}\left\| l^{\alpha(i)}(m^{(1)}, m^{(2)}, \ldots m^{(n)})\right\|^2. \tag{12.53}$$

The last expression confirms that the selection of the perturbations of the model parameters based on formula (12.51) ensures the decrease of the parametric functional.

We can construct an iterative process for the regularized conjugate gradient (RCG) algorithm of solving minimization problem (12.45), similar to one, which was summarized in Chap. 8 in Eq. (8.23). However, in this case, vector $\mathbf{l}_k^\alpha$ of the direction of the steepest ascent computed at iteration number k is calculated based on formula (12.52):

$$\mathbf{l}_k^\alpha = \left(l_k^{\alpha(1)}, l_k^{\alpha(2)}, ...l_k^{\alpha(n)}\right). \tag{12.54}$$

As usual, the iterative process (8.23) is terminated when the misfit reaches the required level:

$$\varphi\left(\mathbf{m}_{k+1}\right) = \|\widetilde{\mathbf{r}}_{+1}\|_D^2 = \delta_d.$$

12.5 Localized Gramian Stabilizer

12.5.1 Definition of the Localized Gramians

The constraints based on the Gramian stabilizer of the model parameters S_G, Eqs. (12.42), or of their gradients S_{G_∇}, Eq. (12.43), can be treated as the global constraints because they enforce similar correlation conditions over the entire inversion domain. In practical applications, however, the specific form of the correlations may vary within the area of investigation. To address this situation, we can subdivide the inversion domain, V, into N subdomains, V_k, with potentially different types of relationships between the model parameters, and define the Gramians, $G_k(m^{(1)}, m^{(2)}, ..., m^{(n)})$, for each of these subdomains separately:

$$G_k(m^{(1)}, m^{(2)}, ..., m^{(n)})$$
$$= \begin{vmatrix} \left(m_k^{(1)}, m_k^{(1)}\right) & \left(m_k^{(1)}, m_k^{(2)}\right) & \ldots\ \ldots & \left(m_k^{(1)}, m_k^{(n)}\right) \\ \left(m_k^{(2)}, m_k^{(1)}\right) & \left(m_k^{(2)}, m_k^{(2)}\right) & \ldots\ \ldots & \left(m_k^{(2)}, m_k^{(n)}\right) \\ \ldots & \ldots & \ldots\ \ldots & \ldots \\ \left(m_k^{(n)}, m_k^{(1)}\right) & \left(m_k^{(n)}, m_k^{(2)}\right) & \ldots\ \ldots & \left(m_k^{(n)}, m_k^{(n)}\right) \end{vmatrix}, \tag{12.55}$$

where $m_k^{(1)}, m_k^{(2)}, ..., m_k^{(n)}$ are the set of model parameters describing the different physical properties of the medium (e. g., density, susceptibility, or conductivity) within subdomain V_k, and matrix elements are formed by the conventional L_2^C inner product between two functions defined within subdomain V_k, respectively:

$$\left(m_k^{(i)}, m_k^{(j)}\right) = \left(m_k^{(i)}, m_k^{(j)}\right)_{L_2^C} = \int_{V_k} m_k^{(i)}(\mathbf{r})\ m_k^{(j)*}(\mathbf{r})\, dv.$$

The *localized Gramian stabilizer*, $S_{LG}(m^{(1)}, m^{(2)}, ..., m^{(n)})$, is introduced as follows:

$$S_{LG}(m^{(1)}, m^{(2)}, ..., m^{(n)}) = \sum_{k=1}^{N} G_k(m^{(1)}, m^{(2)}, ..., m^{(n)}). \tag{12.56}$$

Similarly, we can introduce localized Gramian-based structural constraints using the localized Gramian of model parameter gradients, $G_{\nabla k}$, defined by the following formula:

$$G_{\nabla k}(m^{(1)}, m^{(2)}, ..., m^{(n)}) = \begin{vmatrix} \left(\nabla m_k^{(1)}, \nabla m_k^{(1)}\right) & \left(\nabla m_k^{(1)}, \nabla m_k^{(2)}\right) & \ldots\ldots & \left(\nabla m_k^{(1)}, \nabla m_k^{(n)}\right) \\ \left(\nabla m_k^{(2)}, \nabla m_k^{(1)}\right) & \left(\nabla m_k^{(2)}, \nabla m_k^{(2)}\right) & \ldots\ldots & \left(\nabla m_k^{(2)}, \nabla m_k^{(n)}\right) \\ \ldots & \ldots & \ldots\ldots & \ldots \\ \left(\nabla m_k^{(n)}, \nabla m_k^{(1)}\right) & \left(\nabla m_k^{(n)}, \nabla m_k^{(2)}\right) & \ldots\ldots & \left(\nabla m_k^{(n)}, \nabla m_k^{(n)}\right) \end{vmatrix}, \tag{12.57}$$

where

$$\left(\nabla m_k^{(i)}, \nabla m_k^{(j)}\right) = \left(\nabla m_k^{(i)}, \nabla m_k^{(j)}\right)_{L_2^C} = \int_{V_k} \left(\nabla m_k^{(i)}(\mathbf{r}) \cdot \nabla m_k^{(j)*}(\mathbf{r})\right) dv. \tag{12.58}$$

The corresponding localized Gramian-based structural stabilizing functional, $S_{LG\nabla}(m^{(1)}, m^{(2)}, ..., m^{(n)})$, is written as follows:

$$S_{LG\nabla}(m^{(1)}, m^{(2)}, ..., m^{(n)}) = \sum_{k=1}^{N} G_{\nabla k}(m^{(1)}, m^{(2)}, ..., m^{(n)}). \tag{12.59}$$

For example, in the case of two model parameters, localized Gramian (12.57) takes the form:

$$G_{\nabla k}(m^{(1)}, m^{(2)}) = \begin{vmatrix} \left(\nabla m_k^{(1)}, \nabla m_k^{(1)}\right) & \left(\nabla m_k^{(1)}, \nabla m_k^{(2)}\right) \\ \left(\nabla m_k^{(2)}, \nabla m_k^{(1)}\right) & \left(\nabla m_k^{(2)}, \nabla m_k^{(2)}\right) \end{vmatrix}, \tag{12.60}$$

and the corresponding stabilizing functional is written as follows:

$$S_{LG\nabla}(m^{(1)}, m^{(2)}) = \sum_{k=1}^{N} G_{\nabla k}(m^{(1)}, m^{(2)}). \tag{12.61}$$

The advantage of using the localized Gramian constraints over the global constraints is that the former can be applied in a complex model setting with the variable relationships between different physical properties of the models over the area of investigation.

Note that, in the case of the localized Gramian constraints (12.56) for the model parameters, the size of the subdomain, V_k, should be large enough to provide a meaningful measure of the correlation between these parameters within each subdomain. For example, one cannot reduce subdomains V_k to one point of observation.

At the same time, in the case of Gramian structural constraints (12.59), one can subdivide the inversion domain into an infinite number of infinitesimally small subdomains. Indeed, for the infinitesimally small subdomain, $V_k \to dv$, inner product (12.58) is reduced to the conventional dot product of two vectors:

$$\left(\nabla m_k^{(i)}, \nabla m_k^{(j)}\right) =$$

$$= \int_{V_k} \left(\nabla m_k^{(i)}(\mathbf{r}) \cdot \nabla m_k^{(j)*}(\mathbf{r})\right) dv \to \left(\nabla m_k^{(i)}(\mathbf{r}) \cdot \nabla m_k^{(j)*}(\mathbf{r})\right) dv, \text{ if } V_k \to dv. \tag{12.62}$$

Therefore, the summation over subdomains V_k in the expression for the localized stabilizing functional (12.59) will become an integral over the inversion domain V:

$$S_{LG\nabla}(m^{(1)}, m^{(2)}, ..., m^{(n)}) = \iiint_V G_\nabla(m^{(1)}(\mathbf{r}), m^{(2)}(\mathbf{r}), ..., m^{(n)}(\mathbf{r}))dv, \tag{12.63}$$

where the calculation of the Gramian $G_\nabla(m^{(1)}(\mathbf{r}), m^{(2)}(\mathbf{r}), ..., m^{(n)}(\mathbf{r}))$ is reduced to the following expression:

$$G_{\nabla k}(m^{(1)}(\mathbf{r}), m^{(2)}(\mathbf{r}), ..., m^{(n)}(\mathbf{r}))$$

$$= \begin{vmatrix} \left(\nabla m_k^{(1)} \cdot \nabla m_k^{(1)}\right) & \left(\nabla m_k^{(1)} \cdot \nabla m_k^{(2)}\right) & \dots\dots & \left(\nabla m_k^{(1)} \cdot \nabla m_k^{(n)}\right) \\ \left(\nabla m_k^{(2)} \cdot \nabla m_k^{(1)}\right) & \left(\nabla m_k^{(2)} \cdot \nabla m_k^{(2)}\right) & \dots\dots & \left(\nabla m_k^{(2)} \cdot \nabla m_k^{(n)}\right) \\ \dots & \dots & \dots\dots & \dots \\ \left(\nabla m_k^{(n)} \cdot \nabla m_k^{(1)}\right) & \left(\nabla m_k^{(n)} \cdot \nabla m_k^{(2)}\right) & \dots\dots & \left(\nabla m_k^{(n)} \cdot \nabla m_k^{(n)}\right) \end{vmatrix}. \tag{12.64}$$

For example, for two model parameters, expression (12.64) takes the following form:

$$S_{LG\nabla}(m^{(1)}, m^{(2)}) = \iiint_V G_\nabla(m^{(1)}(\mathbf{r}), m^{(2)}(\mathbf{r}))dv. \tag{12.65}$$

Thus, the localized Gramian stabilizer (12.65) requires the gradients of the different model parameters to be parallel (linearly dependent) vectors at every point while allowing for the coefficients of these linear dependences to vary from point to point. This provides more flexibility to the joint inversion.

12.5.2 Equivalence Between the Localized Gramian-Based Structural Stabilizing Functional for Two Model Parameters and Cross-Gradient Stabilizer

It was shown in Chap. 9 that the localized Gramian-based structural stabilizer (12.65) is equivalent to the cross-gradient functional, S_{cg} (9.23). We reproduce this proof here for convenience.

Indeed, using the integral representation of L_2 norm of the cross-gradient product $\left[\nabla m^{(1)} \times \nabla m^{(2)}\right]$, one can write the cross-gradient functional, S_{cg}, as follows:

$$S_{cg}(m^{(1)}, m^{(2)}) = \iiint_D \left|\nabla m^{(1)}(\mathbf{r}) \times \nabla m^{(2)}(\mathbf{r})\right|^2 dv. \tag{12.66}$$

Let us compare the expressions under the integral sign in formulas (12.65) and (12.66).

Considering that in every point, $\mathbf{r}$, the inner product used in Gramian definition (12.60), is reduced to the dot product of the gradient vectors, we have

$$G_\nabla(m^{(1)}(\mathbf{r}), m^{(2)}(\mathbf{r}))$$
$$= \begin{vmatrix} \left(\nabla m^{(1)}(\mathbf{r}) \cdot \nabla m^{(1)}(\mathbf{r})\right) & \left(\nabla m^{(1)}(\mathbf{r}) \cdot \nabla m^{(2)}(\mathbf{r})\right) \\ \left(\nabla m^{(2)}(\mathbf{r}) \cdot \nabla m^{(1)}(\mathbf{r})\right) & \left(\nabla m^{(2)}(\mathbf{r}) \cdot \nabla m^{(2)}(\mathbf{r})\right) \end{vmatrix}. \tag{12.67}$$

To simplify the derivations, we will use the following notations:

$$\nabla m^{(1)}(\mathbf{r}) = \mathbf{a}; \quad \nabla m^{(2)}(\mathbf{r}) = \mathbf{b}, \tag{12.68}$$

where $\mathbf{a}$ and $\mathbf{b}$ are the corresponding gradient vectors in point $\mathbf{r}$. Using these notations, we express Gramian (12.67) as follows:

$$G_\nabla(m^{(1)}(\mathbf{r}), m^{(2)}(\mathbf{r})) = (\mathbf{a} \cdot \mathbf{a})(\mathbf{b} \cdot \mathbf{b}) - (\mathbf{a} \cdot \mathbf{b})^2. \tag{12.69}$$

According to the definition of the dot product,

$$(\mathbf{a} \cdot \mathbf{b}) = |\mathbf{a}|\,|\mathbf{b}| \cos\varphi,$$

where φ is an angle between two vectors, $\mathbf{a}$ and $\mathbf{b}$. Therefore, we have

$$G_\nabla(m^{(1)}(\mathbf{r}), m^{(2)}(\mathbf{r})) = |\mathbf{a}|^2 |\mathbf{b}|^2 - |\mathbf{a}|^2 |\mathbf{b}|^2 \cos^2\varphi = |\mathbf{a}|^2 |\mathbf{b}|^2 \sin^2\varphi. \tag{12.70}$$

From the last formula it follows at once that

$$G_\nabla(m^{(1)}(\mathbf{r}), m^{(2)}(\mathbf{r})) \geq \mathbf{0}. \tag{12.71}$$

At the same time, the square of absolute value of the gradients' cross product takes the form:

$$|\mathbf{a} \times \mathbf{b}|^2 = |\mathbf{a}|^2 \, |\mathbf{b}|^2 \sin^2 \varphi = G_\nabla(m^{(1)}(\mathbf{r}), m^{(2)}(\mathbf{r})). \tag{12.72}$$

Thus, we have identical expressions under integrals in formulas (12.65) and (12.66). This concludes the proof of equivalence of the localized Gramian stabilizer (12.65) to the cross-gradient functional, S_{cg}, Eq. (9.23), in the case of two model parameters, $m^{(1)}(\mathbf{r})$ and $m^{(2)}(\mathbf{r})$.

References and Recommended Reading to This Chapter

Fregoso E, Gallardo LA (2009) Cross-gradients joint 3D inversion with applications to gravity and magnetic data. Geophysics 74:L31–L42

Gallardo LA (2007) Multiple cross-gradient joint inversion for geospectral imaging. Geophys Res Lett 34:L19301

Gallardo LA, Meju MA (2003) Characterization of heterogeneous near-surface materials by joint 2D inversion of DC resistivity and seismic data. Geophys Res Lett 30:1658–1661

Gallardo LA, Meju MA (2004) Joint two-dimensional DC resistivity and seismic travel-time inversion with cross-gradients constraints. J Geophys Res 109:B03311

Gallardo L A, Meju MA (2007) Joint two-dimensional cross-gradient imaging of magnetotelluric and seismic traveltime data for structural and lithological classification. Geophys J Int 169:1261–1272

Gallardo LA, Meju MA (2011) Structure-coupled multi-physics imaging in geophysical sciences. Rev Geophys 49:RG1003

Hu WY, Abubakar A, and Habashy TM (2009) Joint electromagnetic and seismic inversion using structural constraints. Geophysics 74:R99–R109

Meju MA (2011) Joint multi-geophysical inversion: effective model integration, challenges and directions for future research. Presented at international workshop on gravity, electrical and magnetic methods and their applications, Beijing, China

Wang Z, Bovik A, Sheikh H, Simoncelli E (2004) Image quality assessment: from error visibility to structural similarity. Proce IEEE Trans Image Process 13(4):600–612

Zhdanov MS, Gribenko AV, Wilson G (2012a) Generalized joint inversion of multimodal geophysical data using Gramian constraints. Geophys Res Lett 39:L09301, 1–7

Zhdanov MS, Gribenko AV, Wilson G, Funk C (2012b) 3D joint inversion of geophysical data with Gramian constraints. A case study from the Carrapateena IOCG deposit, South Australia: The Leading Edge, Nov 1382–1388

Zhdanov MS (2015) Inverse theory and applications in geophysics, 2nd ed. Elsevier

Zhdanov MS (2019) Method of simultaneous imaging of different physical properties using joint inversion of multiple datasets. US Patent 10,242,126

Chapter 13
Probabilistic Approach to Gramian Inversion

Abstract The meaning of Gramian and its role in the joint inversion are explained using a probabilistic approach to inverse problem solution. We introduce a Hilbert space of random variables with the metric defined by the covariance matrix between random variables, representing different model parameters. Using this Hilbert space, the probabilistic Gramian is represented as the determinant of the covariance matrix between the different model parameters and their attributes. By minimizing the probabilistic Gramian, we enforce the linear correlation between various inverse models produced by joint inversion. Several gradient-type techniques are considered for solving this minimization problem.

Keywords Random variables · Covariance matrix · Probabilistic Gramian

In Chap. 12, the Gramian method of joint inversion was introduced in the framework of the deterministic approach to the solution of the inverse problem, which considers the data and model parameters characterized by specific functions or vectors with certain (maybe unknown) values. However, as discussed in Chap. 6, there is a probabilistic approach to inverse problems where the observed data and model parameters are treated as realizations of some random variables. This approach was introduced in the pioneering papers of Foster (1961), Franklin (1970), Jackson (1972), Tarantola and Valette (1982), Tarantola (1987), Tarantola (2005).

It can be demonstrated that both these approaches result in similar numerical solutions of the inverse problem (Zhdanov 2002, 2015). At the same time, deterministic or probabilistic interpretation of the observed data and model parameters emphasizes different aspects of the inversion algorithms. This also helps understand better the properties of the inversion parameters.

This chapter introduces an approach to the joint inversion where the Gramian constraints are represented in the probabilistic form as the determinant of the covariance matrix between the different model parameters and their attributes (Zhdanov et al. 2021, 2023). We begin our discussion with the concept of the Gramian space of random variables.

M. S. Zhdanov, *Advanced Methods of Joint Inversion and Fusion of Multiphysics Data*, Advances in Geological Science,
https://doi.org/10.1007/978-981-99-6722-3_13

13.1 Gramian Space of Random Variables

The meaning of Gramian and its role in the joint inversion can be better explained using a probabilistic approach to inverse problem solution. In the framework of this approach, one can treat the observed data and the model parameters as the realizations of some random variables (see Chap. 2).

We can also introduce a Hilbert space of random variables with the metric defined by the covariance between random variables, representing different model parameters. Indeed, let us consider a set, $\Gamma^{(n)}$, of random variables, φ, ψ...., representing different model parameters.

For any two random variables, φ, $\psi \in \Gamma^{(n)}$, we can define an inner product operation, $(\varphi, \psi)_{\Gamma^{(n)}}$, as the determinant of the following covariance matrix:

$$(\varphi, \psi)_{\Gamma^{(n)}} =$$

$$= \begin{vmatrix} cov\left(\gamma^{(1)}, \gamma^{(1)}\right) & cov\left(\gamma^{(1)}, \gamma^{(2)}\right) & \ldots & cov\left(\gamma^{(1)}, \gamma^{(n-1)}\right) & cov\left(\gamma^{(1)}, \psi\right) \\ cov\left(\gamma^{(2)}, \gamma^{(1)}\right) & cov\left(\gamma^{(2)}, \gamma^{(2)}\right) & \ldots & cov\left(\gamma^{(2)}, \gamma^{(n-1)}\right) & cov\left(\gamma^{(2)}, \psi\right) \\ \ldots & \ldots & \ldots & \ldots & \ldots \\ cov\left(\gamma^{(n-1)}, \gamma^{(1)}\right) & cov\left(\gamma^{(n-1)}, \gamma^{(2)}\right) & \ldots & cov\left(\gamma^{(n-1)}, \gamma^{(n-1)}\right) & cov\left(\gamma^{(n-1)}, \psi\right) \\ cov\left(\varphi, \gamma^{(1)}\right) & cov\left(\varphi, \gamma^{(2)}\right) & \ldots & cov\left(\varphi, \gamma^{(n-1)}\right) & cov\left(\varphi, \psi\right) \end{vmatrix}, \tag{13.1}$$

where $\gamma^{(1)}, \gamma^{(2)}, \ldots, \gamma^{(n-1)}$ are some random variables representing a subset of $(n-1)$ known model parameters from $\Gamma^{(n)}$. We will call this set *a core* of the set $\Gamma^{(n)}$.

Let us check that the operation defined by formula (13.1) satisfies all the properties of the inner product in the Hilbert space.

1. The symmetry of operation (13.1):

$$(\varphi, \psi)_{\Gamma^{(n)}} =$$

$$= \begin{vmatrix} cov\left(\gamma^{(1)}, \gamma^{(1)}\right) & \ldots & cov\left(\gamma^{(1)}, \gamma^{(n-1)}\right) & cov\left(\gamma^{(1)}, \psi\right) \\ cov\left(\gamma^{(2)}, \gamma^{(1)}\right) & \ldots & cov\left(\gamma^{(2)}, \gamma^{(n-1)}\right) & cov\left(\gamma^{(2)}, \psi\right) \\ \ldots & \ldots & \ldots & \ldots \\ cov\left(\gamma^{(n-1)}, \gamma^{(1)}\right) & \ldots & cov\left(\gamma^{(n-1)}, \gamma^{(n-1)}\right) & cov\left(\gamma^{(n-1)}, \psi\right) \\ cov\left(\varphi, \gamma^{(1)}\right) & \ldots & cov\left(\varphi, \gamma^{(n-1)}\right) & cov\left(\varphi, \psi\right) \end{vmatrix}$$

$$= \begin{vmatrix} cov\left(\gamma^{(1)}, \gamma^{(1)}\right) & \ldots & cov\left(\gamma^{(n-1)}, \gamma^{(1)}\right) & cov\left(\gamma^{(1)}, \varphi\right) \\ cov\left(\gamma^{(1)}, \gamma^{(2)}\right) & \ldots & cov\left(\gamma^{(n-1)}, \gamma^{(2)}\right) & cov\left(\gamma^{(2)}, \varphi\right) \\ \ldots & \ldots & \ldots & \ldots \\ cov\left(\gamma^{(1)}, \gamma^{(n-1)}\right) & \ldots & cov\left(\gamma^{(n-1)}, \gamma^{(n-1)}\right) & cov\left(\gamma^{(n-1)}, \varphi\right) \\ cov\left(\psi, \gamma^{(1)}\right) & \ldots & cov\left(\psi, \gamma^{(n-1)}\right) & cov\left(\psi, \varphi\right) \end{vmatrix}$$

$$= (\psi, \varphi)_{\Gamma^{(n)}} . \tag{13.2}$$

Equality (13.2) holds because (a) the determinant of the original matrix is equal to the determinant of the transposed matrix and (b) covariance is a symmetric function:

$$cov\,(\varphi, \psi) = cov\,(\psi, \varphi)\,.$$

2. The linearity of operation (13.1):

$$(c_1\varphi_1 + c_2\varphi_2, \psi)_{\Gamma^{(n)}} = c_1\,(\varphi_1, \psi)_{\Gamma^{(n)}} + c_2\,(\varphi_2, \psi)_{\Gamma^{(n)}}\,. \tag{13.3}$$

Equality (13.3) comes immediately from the linearity of the covariance:

$$cov\,(c_1\varphi_1 + c_2\varphi_2, \psi) = c_1 cov\,(\varphi_1, \psi) + c_2 cov\,(\varphi_2, \psi)\,. \tag{13.4}$$

Indeed, according to Eq. (13.1), we have

$$(c_1\varphi_1 + c_2\varphi_2, \psi)_{\Gamma^{(n)}} =$$

$$\begin{vmatrix} cov\left(\gamma^{(1)}, \gamma^{(1)}\right) & \dots & cov\left(\gamma^{(1)}, \gamma^{(n-1)}\right) & cov\left(\gamma^{(1)}, c_1\varphi_1 + c_2\varphi_2\right) \\ cov\left(\gamma^{(2)}, \gamma^{(1)}\right) & \dots & cov\left(\gamma^{(2)}, \gamma^{(n-1)}\right) & cov\left(\gamma^{(2)}, c_1\varphi_1 + c_2\varphi_2\right) \\ \dots & \dots & \dots & \dots \\ cov\left(\gamma^{(n-1)}, \gamma^{(1)}\right) & \dots & cov\left(\gamma^{(n-1)}, \gamma^{(n-1)}\right) & cov\left(\gamma^{(n-1)}, c_1\varphi_1 + c_2\varphi_2\right) \\ cov\left(\psi, \gamma^{(1)}\right) & \dots & cov\left(\psi, \gamma^{(n-1)}\right) & cov\left(\psi, c_1\varphi_1 + c_2\varphi_2\right) \end{vmatrix}$$

$$= D. \tag{13.5}$$

Using the linearity of the covariance, Eq. (13.4), the determinant in the right side of Eq. (13.5) can be written as follows:

$$D =$$

$$= c_1 \begin{vmatrix} cov\left(\gamma^{(1)}, \gamma^{(1)}\right) & \dots & cov\left(\gamma^{(1)}, \gamma^{(n-1)}\right) & cov\left(\gamma^{(1)}, \varphi_1\right) \\ cov\left(\gamma^{(2)}, \gamma^{(1)}\right) & \dots & cov\left(\gamma^{(2)}, \gamma^{(n-1)}\right) & cov\left(\gamma^{(2)}, \varphi_1\right) \\ \dots & \dots & \dots & \dots \\ cov\left(\gamma^{(n-1)}, \gamma^{(1)}\right) & \dots & cov\left(\gamma^{(n-1)}, \gamma^{(n-1)}\right) & cov\left(\gamma^{(n-1)}, \varphi_1\right) \\ cov\left(\psi, \gamma^{(1)}\right) & \dots & cov\left(\psi, \gamma^{(n-1)}\right) & cov\left(\psi, \varphi_1\right) \end{vmatrix} +$$

$$+ c_2 \begin{vmatrix} cov\left(\gamma^{(1)}, \gamma^{(1)}\right) & \dots & cov\left(\gamma^{(1)}, \gamma^{(n-1)}\right) & cov\left(\gamma^{(1)}, \varphi_2\right) \\ cov\left(\gamma^{(2)}, \gamma^{(1)}\right) & \dots & cov\left(\gamma^{(2)}, \gamma^{(n-1)}\right) & cov\left(\gamma^{(2)}, \varphi_2\right) \\ \dots & \dots & \dots & \dots \\ cov\left(\gamma^{(n-1)}, \gamma^{(1)}\right) & \dots & cov\left(\gamma^{(n-1)}, \gamma^{(n-1)}\right) & cov\left(\gamma^{(n-1)}, \varphi_2\right) \\ cov\left(\psi, \gamma^{(1)}\right) & \dots & cov\left(\psi, \gamma^{(n-1)}\right) & cov\left(\psi, \varphi_2\right) \end{vmatrix}$$

$$+ c_1\,(\varphi_1, \psi)_{\Gamma^{(n)}} + c_2\,(\varphi_2, \psi)_{\Gamma^{(n)}}\,.$$

3. Functional (13.1) defining the inner product operation, $(\varphi, \psi)_{\Gamma^{(n)}}$, is positive definite:

$$(\varphi, \varphi)_{\Gamma^{(n)}} \geq 0, \tag{13.6}$$

and

$$(\varphi, \varphi)_{\Gamma^{(n)}} = 0, \text{ if and only if } \varphi \doteq 0. \tag{13.7}$$

The symbol "dot" above the equality sign in formula (13.7) means that random variable φ is a linear combination of the random variables $\gamma^{(1)}, \gamma^{(2)}, \gamma^{(3)}, ..., \gamma^{(n-1)}$, forming the core of set $\Gamma^{(n)}$:

$$\varphi = \sum_{i=1}^{n-1} b_i^{(i)} \gamma + c, \tag{13.8}$$

where b_i $(i = 1, 2, ..., n-1)$ and c are some constant coefficients.

Inequality (13.6) holds because the determinant of the covariance matrix is always positive for independent random variables (see Chap. 2). This determinant is equal to zero if and only if the random variables, $\gamma^{(1)}, \gamma^{(2)}, \gamma^{(3)}, ..., \gamma^{(n-1)}$, and φ are linearly dependent

According to the theory of Hilbert spaces (Chap. 3), the inner product of some element φ by itself defines the norm square of this element:

$$(\varphi, \varphi)_{\Gamma^{(n)}} = \|\varphi\|^2_{\Gamma^{(n)}} . \tag{13.9}$$

Thus, we can see that the zero value of the Gramian norm of some random variable φ means that this variable is linearly related to the elements of the core, $\gamma^{(1)}, \gamma^{(2)}, \gamma^{(3)}, ..., \gamma^{(n-1)}$. In other words, any linear combination of the elements of the core of the set $\Gamma^{(n)}$ has a zero norm.

We will call the set $\Gamma^{(n)}$ of random variables with the metric (inner product) defined by formula (13.1), the *Gramian space of random variables.*

Similar to the Gramian norm of functions, Eq. (12.11), the Gramian norm of random variables has the following properties:

$$\left\|\gamma^{(i)}\right\|^2_{\Gamma^{(i)}} = \left\|\gamma^{(j)}\right\|^2_{\Gamma^{(j)}}, \text{ for } i, \ j = 1, 2, ..., n. \tag{13.10}$$

The last formula demonstrates that all the random variables, $\gamma^{(1)}, \gamma^{(2)}, ..., \gamma^{(n)}$, have the same norm in the corresponding Gramian spaces $\Gamma^{(j)}$, $j = 1, 2, ..., n$.

This property of the introduced metric allows us to use this metric to select the solutions to the inverse problem, which would correlate the best with the preselected model parameters forming the core of Gramian space $\Gamma^{(n)}$.

In conclusion of this section, we should note that, in Gramian space $\Gamma^{(1)}$, the inner product of two random variables is equal to their covariance:

$$(\varphi, \psi)_{\Gamma^{(1)}} = cov\,(\varphi, \psi)\,, \tag{13.11}$$

and the norm square of φ is simply its variance:

$$\|(\varphi, \psi)\|^2_{\Gamma^{(1)}} = \sigma^2(\varphi). \tag{13.12}$$

13.2 The Maximum Likelihood Method in the Joint Inversion

We can now apply the maximum likelihood method, introduced in Chap. 6, to the solution of the multiphysics inverse problem.

Let us consider geophysical inverse problems for multiple geophysical data sets. It was shown in Chap. 8 that the following operator relationships can describe these problems:

$$d^{(i)} = A^{(i)}(m^{(i)}), \quad i = 1, 2, ..., n, \tag{13.13}$$

where, in a general case, $A^{(i)}$ are nonlinear operators; $d^{(i)}$ $(i = 1, 2, 3, ..., n)$ are the random variables representing different observed data sets; and $m^{(i)}$ $(i = 1, 2, 3, ..., n)$ are the unknown random variables representing different model parameters.

In the framework of the probabilistic approach to solving the inverse problem, we consider the observed data and the model parameters as realizations of some random variables. Assuming that these variables have Gaussian probability distribution, we can apply the maximum likelihood method to solve inverse problems (13.13) (see Chap. 6). According to this method, the optimum values for the model parameters are those that maximize the probability, $P(d^{(i)})$, that the observed data, $d^{(i)}$, are, in fact, observed:

$$P(d^{(i)}) = \max, \quad i = 1, 2, ..., n. \tag{13.14}$$

It was demonstrated in Chap. 6 that for uncorrelated data with uniform variance maximum of $P(d^{(i)})$ occurs when the misfit between the observed, $d^{(i)}$, and predicted data, $A^{(i)}(m^{(i)})$, reaches its minimum:

$$f^{(i)}(\mathbf{m}^{(i)}) = \left\| A^{(i)}(m^{(i)}) - d^{(i)} \right\|^2_D = \min, \quad i = 1, 2, ..., n. \tag{13.15}$$

Combining all functionals $f^{(i)}(\mathbf{m}^{(i)})$ together, we arrive at the following misfit condition for the solution of the multiphysics inverse problem:

$$\Phi\left(m^{(1)}, m^{(2)}, ..., m^{(n)}\right) = \sum_{i=1}^{n-1} w_i^2 f^{(i)}(m^{(i)}) = \min, \tag{13.16}$$

where $\Phi\left(m^{(1)}, m^{(2)}, ..., m^{(n)}\right)$ is combined misfit functional, and w_i, $i = 1, 2, ..., n$, are some scalar weighting coefficients. These coefficients are usually selected as inverse values of the standard deviation of data, $d^{(i)}$:

$$w_i = 1/\sigma_i, \ i = 1, 2, ..., n. \tag{13.17}$$

There are two problems with Eq. (13.16). First of all, direct minimization of the misfit functional $\Phi\left(m^{(1)}, m^{(2)}, ..., m^{(n)}\right)$ can result in an unstable solution of the ill-posed inverse problems (13.13). Second, combining all functionals $f^{(i)}(\mathbf{m}^{(i)})$ in one minimization condition does not impose any requirement on the relationships between the different model parameters, which is the key to joint inversion. However, as we discussed in Chaps. 8 through 12, the joint inversion requires the enforcement of some relations between different model parameters. In the framework of the probabilistic approach, this can be achieved by adding the term containing the covariance matrix between different model parameters, which serves as an analog of Gramian in the deterministic approach (Zhdanov et al. 2012a, b).

13.3 Covariance Representation of the Probabilistic Gramian Stabilizer

The *probabilistic Gramian stabilizer*, S_{G_σ}, can be introduced as the determinant of the covariance matrix between different model parameters in a probabilistic approach to the inversion theory:

$$S_{G_\sigma}(m^{(1)}, m^{(2)}, ...m^{(n)}) =$$

$$= \begin{vmatrix} cov\left(m^{(1)}, m^{(1)}\right) & ... & cov\left(m^{(1)}, m^{(n-1)}\right) & cov\left(m^{(1)}, m^{(n)}\right) \\ cov\left(m^{(2)}, m^{(1)}\right) & ... & cov\left(m^{(2)}, m^{(n-1)}\right) & cov\left(m^{(2)}, m^{(n)}\right) \\ ... & ... & ... & ... \\ cov\left(m^{(n-1)}, m^{(1)}\right) & ... & cov\left(m^{(n-1)}, m^{(n-1)}\right) & cov\left(m^{(n-1)}, m^{(n)}\right) \\ cov\left(m^{(n)}, m^{(1)}\right) & ... & cov\left(m^{(n)}, m^{(n-1)}\right) & cov\left(m^{(n)}, m^{(n)}\right) \end{vmatrix}. \tag{13.18}$$

It was demonstrated in Chap. 2 that the determinant of the covariance matrix is always nonnegative, and it is equal to zero if and only if the random variables, $m^{(1)}, m^{(2)}, ..., m^{(n)}$, are linearly dependent. This property of the probabilistic Gramian is similar to that of the deterministic Gramian defined by formula (12.6) of Chap. 12. The key difference is in the way how we treat the model parameters. In the framework of the deterministic approach, they are described by specific (though unknown) functions. In the framework of the probabilistic approach, the model parameters are the realizations of some unknown random variables.

We should note also that the direct analogy between expressions (12.21) and (13.18) holds when the random model parameters have zero mean values. Indeed, in the case of discrete and real model parameters, according to Eq. (2.20), the statistical estimate of the covariance is as follows:

$$cov(m^{(i)}, m^{(j)}) = \frac{1}{L-1}\sum_{l=1}^{L}\left(m_l^{(i)} - \langle m_l^{(i)}\rangle\right)\left(m_l^{(j)} - \langle m_l^{(j)}\rangle\right), \tag{13.19}$$

where $\langle m_l^{(i)}\rangle$ indicates the mean.

If the mean values are zero, $\langle m_l^{(i,j)}\rangle = 0$, then according to Eq. (13.19), we have

$$cov(m^{(i)}, m^{(j)}) = \frac{1}{L-1}\sum_{l=1}^{L} m_l^{(i)} m_l^{(j)} = \frac{1}{L-1}(m^{(i)}, m^{(j)}). \tag{13.20}$$

Let us consider the Gramian space $\Gamma^{(n)}$ of random variables, representing model parameters with the core elements formed by model parameters $m^{(1)}, m^{(2)}, ..., m^{(n-1)}$. Then according to the definition of a norm in this space, Eqs. (13.1) and (13.9), we can write probabilistic Gramian as the norm of parameter $m^{(n)}$:

$$S_{G_\sigma}(m^{(1)}, m^{(2)}, ...m^{(n)}) = \left\|m^{(n)}\right\|^2_{\Gamma^{(n)}}. \tag{13.21}$$

Following the general principles of regularization theory, we can now introduce a *probabilistic parametric functional*, P^α_σ, as a linear combination of combined misfit functional, $\Phi\left(m^{(1)}, m^{(2)}, ..., m^{(n)}\right)$, and probabilistic Gramian:

$$P^\alpha_\sigma\left(m^{(1)}, m^{(2)}, ..., m^{(n)}\right) = \Phi\left(m^{(1)}, m^{(2)}, ..., m^{(n)}\right) + \alpha S_{G_\sigma}(m^{(1)}, m^{(2)}, ...m^{(n)}), \tag{13.22}$$

where $\alpha \in [0, \infty)$ is regularization parameter.

Substituting Eqs. (13.15), (13.16) into (13.22), we arrive at the following minimization problem for the multiphysics inversion:

$$P^\alpha_\sigma\left(m^{(1)}, m^{(2)}, ..., m^{(n)}\right) =$$

$$= \sum_{i=1}^{n-1} w_i^2 \left\|A^{(i)}(m^{(i)}) - d^{(i)}\right\|^2_D + \alpha S_{G_\sigma}(m^{(1)}, m^{(2)}, ...m^{(n)}) = \min. \tag{13.23}$$

We should note that expression (13.22) is similar to formula (12.2) if the joint stabilizing functionals, S_J, is selected equal to the probabilistic Gramian:

$$S_J(m^{(1)}, m^{(2)}, ...m^{(n)}) = S_{G_\sigma}(m^{(1)}, m^{(2)}, ...m^{(n)}). \tag{13.24}$$

Several standard gradient-type methods exist that can be used to solve the minimization problem—steepest descent, Newton, and conjugate gradient methods. They all require calculating the steepest ascent direction (gradient) of the corresponding functional. In the following sections, we will consider this problem in detail.

13.4 Minimization of the Probabilistic Parametric Functional Using the Regularized Gradient-Type Methods

13.4.1 Steepest Ascent Direction of the Probabilistic Parametric Functional

The first variation of the parametric functional, P_σ^α, with the probabilistic Gramian stabilizer, can be calculated as follows:

$$\delta P_\sigma^\alpha(m^{(1)}, m^{(2)}, ..., m^{(n)})$$

$$= 2\sum_{i=1}^{n} w_i^2 \left(\delta A^{(i)}(m^{(i)}),\ A^{(i)}(m^{(i)}) - d^{(i)}\right)_{\Gamma_D^{(1)}} +$$

$$+ 2\alpha\delta S_{G_\sigma}(m^{(1)}, m^{(2)}, ...m^{(n)}). \tag{13.25}$$

The variation of the probabilistic Gramian stabilizer, according to (13.21), can be expressed as follows:

$$\delta S_{G_\sigma}(m^{(1)}, m^{(2)}, ...m^{(n)}) = \delta \left\| m^{(n)} \right\|_{\Gamma^{(n)}}^2$$

$$= \sum_{i=1}^{n} \delta_{\mathbf{m}^{(i)}} \left\| m^{(n)} \right\|_{\Gamma^{(n)}}^2 = \sum_{i=1}^{n} \delta_{\mathbf{m}^{(i)}} \left\| m^{(i)} \right\|_{\Gamma^{(i)}}^2, \tag{13.26}$$

where we took into account that, according to (13.10):

$$\left\| m^{(n)} \right\|_{\Gamma^{(n)}}^2 = \left\| m^{(j)} \right\|_{\Gamma^{(j)}}^2, \text{ for } j = 1, 2, ..., n. \tag{13.27}$$

Using the properties of the inner product operation in the Gramian spaces of random variables, $\Gamma^{(i)}$, we can calculate the variation of the norm square of $m^{(i)}$ as follows:

$$\delta_{\mathbf{m}^{(i)}} \left\| m^{(i)} \right\|_{\Gamma^{(i)}}^2 = 2\sum_{i=1}^{n} \left(\delta m^{(i)},\ m^{(i)}\right)_{\Gamma^{(i)}}$$

$$= 2\begin{vmatrix} cov\left(m^{(1)}, m^{(1)}\right) & cov\left(m^{(1)}, m^{(2)}\right) & ... & cov\left(m^{(1)}, m^{(i)}\right) & ...cov\left(m^{(1)}, m^{(n)}\right) \\ ... & ... & ... & ... & ...... \\ cov\left(\delta m^{(i)}, m^{(1)}\right) & cov\left(\delta m^{(i)}, m^{(2)}\right) & ... & cov\left(\delta m^{(i)}, m^{(i)}\right) & ...cov\left(\delta m^{(i)}, m^{(n)}\right) \\ ... & ... & ... & ... & .... \\ cov\left(m^{(n)}, m^{(1)}\right) & cov\left(m^{(n)}, m^{(2)}\right) & ... & cov\left(m^{(n)}, m^{(i)}\right) & ...cov\left(m^{(n)}, m^{(n)}\right) \end{vmatrix}$$

$$= 2cov\left(\delta m^{(i)}, \sum_{j=1}^{n} (-1)^{i+j}\, G_{\sigma ij}^{m} m^{(j)}\right) = 2cov\left(\delta m^{(i)}, l_{G_\sigma}^{(i)}\right), \tag{13.28}$$

where $G^{m}_{\sigma ij}$ is the corresponding minor of Gram matrix $G_\sigma(m^{(1)}, m^{(2)}, ..., m^{(n-1)}, m^{(n)})$ formed by eliminating column i and row j, and elements $l^{(i)}_{G_\sigma}$ are the directions of the steepest ascent for the probabilistic Gramian stabilizing functionals,

$$l^{(i)}_{G_\sigma} = \sum_{j=1}^{n} (-1)^{i+j}\, G^{m-m_{apr}}_{\sigma ij}\, m^{(i)}. \tag{13.29}$$

We now introduce the probabilistic Gramian space of the model parameters, $\Gamma^{(1)}_M$, with the inner product defined by covariance according to formula (13.11). Therefore, Eq. (13.28) can be written as follows:

$$\delta_{\mathbf{m}^{(i)}} \left\| m^{(i)} \right\|^2_{\Gamma^{(i)}} = 2\sum_{i=1}^{n} \left(\delta m^{(i)},\ m^{(i)}\right)_{\Gamma^{(i)}} = 2\left(\delta m^{(i)}, l^{(i)}_{G_\sigma}\right)_{\Gamma^{(1)}_M}. \tag{13.30}$$

Substituting the last formula in Eq. (13.26), we arrive at the following expression for the first variation of the probabilistic Gramian:

$$\delta S_{G_\sigma}(m^{(1)}, m^{(2)}, ...m^{(n)}) = \delta \left\| m^{(n)} \right\|^2_{\Gamma^{(n)}}$$

$$= 2\sum_{i=1}^{n} \delta_{\mathbf{m}^{(i)}} \left\| m^{(i)} \right\|^2_{\Gamma^{(i)}} = 2\sum_{i=1}^{n} \left(\delta m^{(i)}, l^{(i)}_{G_\sigma}\right)_{\Gamma^{(1)}_M}. \tag{13.31}$$

Let us examine again expression (13.25) for the variation of the parametric functional:

$$\delta P^{\alpha}_{\sigma}(m^{(1)}, m^{(2)}, ..., m^{(n)})$$

$$= 2\sum_{i=1}^{n} w_i^2 \left(\delta A^{(i)}(m^{(i)}),\ A^{(i)}(m^{(i)}) - d^{(i)}\right)_{\Gamma^{(1)}_D} +$$

$$+ \alpha\delta S_{G_\sigma}(m^{(1)}, m^{(2)}, ...m^{(n)}). \tag{13.32}$$

Considering that operators $A^{(i)}$ are differentiable, we can write

$$\delta A^{(i)}(m^{(i)}) = F^{(i)}_m \delta m^{(i)},$$

where $F^{(i)}_m$ is a linear operator of the Fréchet derivative of $A^{(i)}$. Therefore, the inner product in the first term in formula (13.32) can be modified as follows:

$$\left(\delta A^{(i)}(m^{(i)}),\ A^{(i)}(m^{(i)}) - d^{(i)}\right)_{\Gamma^{(1)}_D} =$$

$$= \left(F_m^{(i)} \delta m^{(i)},\ A^{(i)}(m^{(i)}) - d^{(i)}\right)_{\Gamma_D^{(1)}} = \left(\delta m^{(i)},\ F_m^{(i)\star}\left[A^{(i)}(m^{(i)}) - d^{(i)}\right]\right)_{\Gamma_M^{(1)}}, \tag{13.33}$$

where $F_m^{(i)\star}$are the adjoint Fréchet derivative operators.

Substituting Eqs. (13.33) and (13.31) in the first and second terms of formula (13.32), we obtain the following important result:

$$\delta P_\sigma^\alpha(m^{(1)}, m^{(2)}, ..., m^{(n)})$$

$$= 2\sum_{i=1}^{n} w_i^2 \left(\delta m^{(i)},\ F_m^{(i)\star}\left[A^{(i)}(m^{(i)}) - d^{(i)}\right]\right)_{\Gamma_M^{(1)}} +$$

$$+ 2\alpha \sum_{i=1}^{n} \left(\delta m^{(i)}, l_{G_\sigma}^{(i)}\right)_{\Gamma_M^{(1)}}. \tag{13.34}$$

Combining two sums in Eq. (13.34), we finally arrive at the following compact formula:

$$\delta P_\sigma^\alpha(m^{(1)}, m^{(2)}, ..., m^{(n)}) =$$

$$= 2\sum_{i=1}^{n} \left(\delta m^{(i)}, l_\sigma^{\alpha(i)}(m^{(1)}, m^{(2)}, ..., m^{(n)})\right)_{\Gamma_M^{(1)}}, \tag{13.35}$$

where $l_\sigma^{\alpha(i)}(m^{(1)}, m^{(2)}, ...m^{(n)})$ are the directions of the steepest ascent of the probabilistic functional P_σ^α:

$$l_\sigma^{\alpha(i)}(m^{(1)}, m^{(2)}, ..., m^{(n)})$$

$$= w_i^2 F_m^{(i)\star}\left[A^{(i)}(m^{(i)}) - d^{(i)}\right] + \alpha l_{G_\sigma}^{(i)} = F_m^{(i)\star} r^{(i)} + \alpha l_{G_\sigma}^{(i)}, \tag{13.36}$$

and $r^{(i)}$ are the weighted residuals between the predicted and observed data:

$$r^{(i)} = w_i^2\left[A^{(i)}(m^{(i)}) - d^{(i)}\right]. \tag{13.37}$$

13.4.2 Steepest Descent Method of Joint Inversion

The expression for the steepest ascent direction, introduced above, can be used in constructing the computational schemes for the different gradient-type methods of solving the minimization problem (13.23).

We begin with the most simple steepest descent method.

Let us select

$$\delta m^{(i)} = -k^{\alpha} l_{\sigma}^{\alpha(i)}(m^{(1)}, m^{(2)}, ..., m^{(n)}), \tag{13.38}$$

where k^{α} is some positive real number, and $l_{\sigma}^{\alpha(i)}(m^{(1)}, m^{(2)}, ...m^{(n)})$ are the directions of the steepest ascent of the functional P_{σ}^{α} defined by Eq. (13.36). Substituting formula (13.38) into (13.35), we have

$$\delta P_{\sigma}^{\alpha}(m^{(1)}, m^{(2)}, ...m^{(n)}) = -2k^{\alpha} \sum_{i=1}^{n} \left\| l_{\sigma}^{\alpha(i)}(m^{(1)}, m^{(2)}, ..., m^{(n)}) \right\|_{\Gamma_M^{(1)}}^2 < 0, \tag{13.39}$$

so, the vector,

$$\mathbf{l}_{\sigma}^{\alpha} = \left(l_{\sigma}^{\alpha(1)}, l_{\sigma}^{\alpha(2)}, ..., l_{\sigma}^{\alpha(n)}\right), \tag{13.40}$$

describes the "direction" of increasing (ascent) of the functional P_{σ}^{α}, in other words, the direction of "climbing" on the hill. We can represent vector $\mathbf{l}^{\alpha}$ as a superposition of the steepest ascent (gradient), $\mathbf{l}_{\sigma}$, of misfit functional $\Phi\left(m^{(1)}, m^{(2)}, ..., m^{(n)}\right)$, and the direction of the steepest ascent, $\mathbf{l}_{G_{\sigma}}$, of the probabilistic Gramian, $S_{G_{\sigma}}$:

$$\mathbf{l}_{\sigma}^{\alpha} = \mathbf{l}_{\sigma} + \alpha \mathbf{l}_{G_{\sigma}}, \tag{13.41}$$

where

$$\mathbf{l}_{\sigma} = \left(F_m^{(1)\star} r^{(1)}, F_m^{(2)\star} r^{(2)}, ..., F_m^{(n)\star} r^{(n)}\right), \quad \mathbf{l}_{G_{\sigma}} = \left(l_{G_{\sigma}}^{(1)}, l_{G_{\sigma}}^{(2)}, ..., l_{G_{\sigma}}^{(n)}\right). \tag{13.42}$$

To simplify the notations, we also introduce vector $\mathbf{m}$ formed by different model parameters:

$$\mathbf{m} = \left(m^{(1)}, m^{(2)}, ..., m^{(n)}\right), \tag{13.43}$$

and vector $\mathbf{F}_m$ formed by the corresponding Fréchet derivative operators:

$$\mathbf{F}_m = \left(F_m^{(1)}, F_m^{(2)}, ..., F_m^{(n)}\right). \tag{13.44}$$

We can construct an iterative process for the regularized steepest descent as follows:

$$\mathbf{m}_{n+1} = \mathbf{m}_n + \delta \mathbf{m}_n = \mathbf{m}_n - k_n^{\alpha} \mathbf{l}_{\sigma}^{\alpha}(\mathbf{m}_n), \tag{13.45}$$

where the coefficient k_n^{α} is found by using the minimization of the parametric functional with respect to k_n^{α}:

$$P_{\sigma}^{\alpha}(\mathbf{m}_{n+1}) = P_{\sigma}^{\alpha}(\mathbf{m}_n - k_n^{\alpha} \mathbf{l}_{\sigma}^{\alpha}(\mathbf{m}_n)) = f(k_n^{\alpha}) = \min. \tag{13.46}$$

In particular, applying the linear line search, we find that the minimum of the probabilistic parametric functional is reached if k_n^{α} is determined by the following formula:

$$k_n^\alpha = \frac{\left\| \mathbf{l}_\sigma^\alpha(\mathbf{m}_n) \right\|^2}{\left\| \mathbf{F}_{m_n} \mathbf{l}_\sigma^\alpha(\mathbf{m}_n) \right\|^2 + \alpha \left\| \mathbf{l}_\sigma^\alpha(\mathbf{m}_n) \right\|^2}.$$

The iterative process (13.45) is terminated at $n = N$ when the combined misfit functional reaches the given level ε_0:

$$\Phi(\mathbf{m}_N) \leq \varepsilon_0.$$

13.4.3 Conjugate Gradient Method of Joint Inversion

We have established in Chap. 7 that the conjugate gradient method is based on the same ideas as the steepest descent, and the iterative process is very similar to the last one:

$$\mathbf{m}_{n+1} = \mathbf{m}_n + \delta\mathbf{m}_n = \mathbf{m}_n - \tilde{k}_n^\alpha \widetilde{\mathbf{l}}_\sigma^\alpha(\mathbf{m}_n), \tag{13.47}$$

where

$$\delta\mathbf{m}_n = -\tilde{k}_n^\alpha \widetilde{\mathbf{l}}_\sigma^\alpha(\mathbf{m}_n).$$

However, the "directions" of ascent $\widetilde{\mathbf{l}}_\sigma^\alpha(\mathbf{m}_n)$ are selected differently. In the first step, we use the "direction" of the steepest ascent:

$$\widetilde{\mathbf{l}}_\sigma^\alpha(\mathbf{m}_0) = \mathbf{l}_\sigma^\alpha(\mathbf{m}_0).$$

At the next step, the "direction" of ascent is a linear combination of the steepest ascent at this step and the "direction" of ascent $\widetilde{\mathbf{l}}_\sigma^\alpha(\mathbf{m}_0)$ on the previous step:

$$\widetilde{\mathbf{l}}_\sigma^\alpha(\mathbf{m}_1) = \mathbf{l}_\sigma^\alpha(\mathbf{m}_1) + \beta_1 \widetilde{\mathbf{l}}_\sigma^\alpha(\mathbf{m}_0).$$

At the n-th step

$$\widetilde{\mathbf{l}}_\sigma^\alpha(\mathbf{m}_{n+1}) = \mathbf{l}_\sigma^\alpha(\mathbf{m}_{n+1}) + \beta_{n+1}^\alpha \widetilde{\mathbf{l}}_\sigma^\alpha(\mathbf{m}_n). \tag{13.48}$$

The regularized steepest descent directions are determined according to formulas (13.41) and (13.42).

Determination of the length of the iteration step, coefficient $\widetilde{k}_n^\alpha$, can be based on the linear or parabolic line search:

$$P_\sigma^\alpha(\mathbf{m}_{n+1}) = P_\sigma^\alpha(\mathbf{m}_n - \widetilde{k}_n^\alpha \widetilde{\mathbf{l}}_\sigma^\alpha(\mathbf{m}_n)) = f(\widetilde{k}_n^\alpha) = \min.$$

The solution of this minimization problem gives the following best estimation for the length of the step using a linear line search:

$$\widetilde{k}_n^{\alpha} = \frac{\widetilde{\mathbf{l}}_{\sigma}^{\alpha T}(\mathbf{m}_n)\mathbf{l}^{\alpha}(\mathbf{m}_n)}{\left\|\mathbf{F}_{m_n}\widetilde{\mathbf{l}}_{\sigma}^{\alpha}(\mathbf{m}_n)\right\|^2 + \alpha\left\|\widetilde{\mathbf{l}}_{\sigma}^{\alpha}(\mathbf{m}_n)\right\|^2}. \quad (13.49)$$

One can also use a parabolic line search (Fletcher 1985) to improve the convergence rate of the RCG method.

The CG method requires that the vectors $\widetilde{\mathbf{l}}_{\sigma}^{\alpha}(\mathbf{m}_n)$ introduced above will be mutually conjugate. This requirement is fulfilled if the coefficients β_n are determined by the following formula (Chap. 7):

$$\beta_{n+1}^{\alpha} = \frac{\|\mathbf{l}_{\sigma}^{\alpha}(\mathbf{m}_{n+1})\|^2}{\|\mathbf{l}_{\sigma}^{\alpha}(\mathbf{m}_n)\|^2}. \quad (13.50)$$

Using Eqs. (13.47), (13.48), (13.49), and (13.50), we can obtain $\mathbf{m}$ iteratively.

The iterative process (13.47) is terminated when the combined misfit functional reaches the given level ε_0:

$$\Phi(\mathbf{m}_N) = \|\mathbf{r}_N\|^2 \leq \varepsilon_0.$$

This concludes our description of the probabilistic approach to Gramian inversion. We have established that Gramian is an analog of the determinant of the covariance matrix between the different physical properties representing model parameters. This helps understand better the role of Gramian in enforcing the relationships between different physical models. It also presents an alternative numerical implementation of the Gramian-type constraints by using the statistical estimates of the components of the covariance matrix.

References and Recommended Reading to This Chapter

Fletcher R (1995) Practical methods of optimization. Willey, Chichester-New-York, 436 pp

Foster M (1961) An application of the Wiener-Kolmogorov smoothing theory to matrix inversion. J Soc Ind Appl Math 9:387–392

Franklin JN (1970) Well-posed stochastic extensions of ill-posed linear problems. J Math Anal Appl 31:682–716

Jackson (1972) Interpretation of inaccurate, insufficient and inconsistent data. Geophys J R Astronom Soc 28:97–110

Tarantola A (1987) Inverse problem theory. Elsevier

Tarantola A (2005) Inverse problem theory and methods for model parameter estimation. SIAM, 344 pp

Tarantola A, Valette B (1982) Generalized nonlinear inverse problem solved using the least squares criterion. Rev Geophys Space Phys 20:219–232

Zhdanov MS, Gribenko AV, Wilson G (2012a) Generalized joint inversion of multimodal geophysical data using Gramian constraints. Geophys Res Lett 39, L09301:1–7

Zhdanov MS, Gribenko AV, Wilson G, Funk C (2012b) 3D joint inversion of geophysical data with Gramian constraints: a case study from the Carrapateena IOCG deposit, South Australia. The Leading Edge, November, 1382–1388

Zhdanov MS (2002) Geophysical inverse theory and regularization problems. Elsevier

Zhdanov MS (2015) Inverse theory and applications in geophysics. Elsevier

Zhdanov MS, Jorgensen M, Cox L (2021) Advanced methods of joint inversion of multiphysics data for mineral exploration. Geosciences 11:262

Zhdanov MS, Jorgensen M, Tao M (2023) Probabilistic approach to Gramian inversion of multiphysics data. Front Earth Sci 11:1127597

Chapter 14
Simultaneous Processing and Fusion of Multiphysics Data and Images

Abstract This chapter considers the problem of restoration of blurred images. The image restoration and deblurring problem arises in biomedical, geophysical, astronomical, high-definition television, remote sensing, and other applications. This problem is formulated as the solution of the corresponding ill-posed inverse problem, which can be effectively solved by applying the family of focusing stabilizers. The solution of the joint image deblurring problem is illustrated by using the joint minimum entropy approach. We consider, as an example, the problem of reconstructing blurred images of the brain produced by the Magnetic Resonance Imaging (MRI) method.

Keywords Blurred image · Blurring operator · Image deblurring · Joint image deblurring

14.1 Digital Restoration of the Blurred Images

We begin our discussion with the general concept of image restoration. This concept has been applied for image restoration and deblurring in biomedical, geophysical, astronomical, high-definition television, remote sensing, and other applications. The comprehensive coverage of the image restoration and recovery problem can be found, for example, in Pratt (2007), Stark (2013), Gonzalez (2009), Ekstrom (2012), and many other publications.

The concept of image restoration can be described in compact form as the solution of the following operator equation (e.g., Portniaguine and Zhdanov 2005; Oliveira et al. 2009; Wang and Tao 2014):

$$\mathbf{d} = \mathbf{Bm}, \tag{14.1}$$

where $\mathbf{d}$ is the degraded (blurred) image, $\mathbf{m}$ is the original (ideal) image, and $\mathbf{B}$ is the *blurring linear operator* of the imaging system. Note that the original image, as

M. S. Zhdanov, *Advanced Methods of Joint Inversion and Fusion of Multiphysics Data*, Advances in Geological Science,
https://doi.org/10.1007/978-981-99-6722-3_14

well as the blurred image, can be defined in a plane (2D image: $\mathbf{m} = m(x, y)$, $\mathbf{d} = d(x, y)$) or in a volume (3D image: $\mathbf{m} = m(x, y, z)$, $\mathbf{d} = d(x, y, z)$).

A wide variety of medical, geophysical, radiophysical, and astronomical blurred images can also be described by Eq. (14.1) with different blurring operators (see, for example, Pratt 2007; Ekstrom 2012; Stark 2013). Thus, the problem of image restoration and deblurring, in a general case, is formulated as the solution of the inverse problem (14.1), which is a typical ill-posed problem, as discussed in Chap. 4. Following the principles of regularization theory, a stable solution to the inverse problem (14.1) is based on the minimization of the Tikhonov parametric functional:

$$P^{\alpha}(\mathbf{m}) = \phi(\mathbf{m}) + \alpha s(\mathbf{m}), \tag{14.2}$$

where $\phi(\mathbf{m})$ is a misfit functional determined as a norm of the difference between observed and predicted (theoretical) degraded images:

$$\varphi(\mathbf{m}) = \|\mathbf{Bm} - \mathbf{d}\|^2. \tag{14.3}$$

Functional $s(\mathbf{m})$ is a stabilizing functional (a stabilizer).

In Chap. 4, we discussed several typical stabilizing functionals. One is based on the least-squares criterion, or, in other words, on L_2 norm of the functions describing the image:

$$s_{L_2}(\mathbf{m}) = \|\mathbf{m}\|^2 = (\mathbf{m}, \mathbf{m}) = \left\{ \begin{array}{l} \int_S m^2(x, y)\, ds \\ \int_V m^2(x, y, z)\, dv \end{array} \right\} = \min, \tag{14.4}$$

where S and V are the area of 2D image or volume of 3D image definition, respectively, and (..., ...) denotes the L_2 inner product operation.

Another stabilizer uses a minimum norm of the difference between the selected image and some a priori (reference) image $\mathbf{m}_{apr}$:

$$s_{L_2apr}(\mathbf{m}) = \left\|\mathbf{m} - \mathbf{m}_{apr}\right\|^2 = \min. \tag{14.5}$$

This criterion, as applied to the gradient of image parameters ∇m, brings us to a maximum smoothness stabilizing functional:

$$s_{\max sm}(\mathbf{m}) = \|\nabla\mathbf{m}\|^2_{L_2} = \left\{ \begin{array}{l} \int_S |\nabla m(x, y)|^2\, ds \\ \int_V |\nabla m(x, y, z)|^2\, dv \end{array} \right\} = \min. \tag{14.6}$$

This stabilizer produces smooth images, which in many practical situations don't properly describe the original (ideal) image. It also can result in spurious oscillations when m is discontinuous.

To mitigate this problem, (Rudin et al., 1992) proposed a total variation (TV)-based approach to reconstruct noisy, blurred images. This approach uses a total variation stabilizing functional, which is essentially L_1 norm of the gradient:

$$s_{TV}(\mathbf{m}) = \|\nabla \mathbf{m}\|_{L_1} = \left\{ \begin{array}{l} \int_S |\nabla m(x, y)|\, ds \\ \int_V |\nabla m(x, y, z)|\, dv \end{array} \right\} = \min . \tag{14.7}$$

If the functions $m(x, y)$ or $m(x, y, z)$ representing the images are discontinuous, the calculation of the total variation functional $s_{TV}(\mathbf{m})$ may produce singularities. Vogel and Oman (1998), Vogel (2002) modified the TV stabilizing functional to avoid these singularities as follows (see Chap. 4):

$$s_{\beta TV}(\mathbf{m}) = \|\nabla \mathbf{m}\|_{L_1} = \left\{ \begin{array}{l} \int_S \sqrt{|\nabla m(x, y)|^2 + \beta^2} ds \\ \int_V \sqrt{|\nabla m(x, y, z)|^2 + \beta^2} dv \end{array} \right\} = \min, \tag{14.8}$$

where β is a small number.

We have also considered in Chap. 4 the family of focusing stabilizers introduced by Portniaguine and Zhdanov (1999, 2005), Zhdanov (2002, 2015). To simplify the notations, I review these stabilizers for the case of the 3D (volume) images only.

I present first an L_p-norm minimum support functional (MSL_p), which provides the image with the minimum area of the anomalous image parameter distribution.

$$s_{MSL_p}(\mathbf{m}) = s_{\beta L_p}(\mathbf{m}) = \int_V \frac{|m|^p}{|m|^p + \beta^p} dv, \; 0 \le p < \infty. \tag{14.9}$$

We can also introduce an L_p-norm minimum gradient support functional (MGSL_p), which provides the image with the smallest areas of image variations. This stabilizer is defined as follows:

$$s_{MGSL_p}(\mathbf{m}) = s_{\beta L_p}[\nabla \mathbf{m}] = \int_V \frac{|\nabla m|^p}{|\nabla m|^p + \beta^p} dv, \; 0 \le p < \infty. \tag{14.10}$$

The MGSL_p functional helps produce the sharp reconstructed image. Indeed, this stabilizer forces the areas with rapid changes in the parameters of the image (e.g., brightness, color, etc.) to be reduced, thus increasing the sharpness of the image.

In order to illustrate this property, I consider a simple numerical experiment of image enhancement and sharpening conducted for a seismic image. Figure 14.1 shows an example of the original seismic image of a geological cross section.

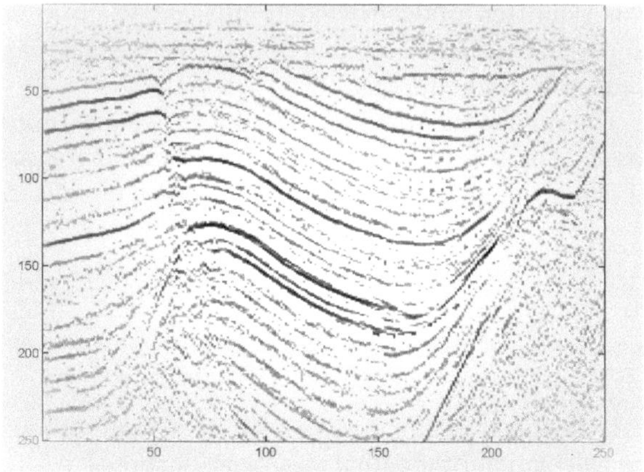

Fig. 14.1 Original seismic depth migration image of a geological cross section

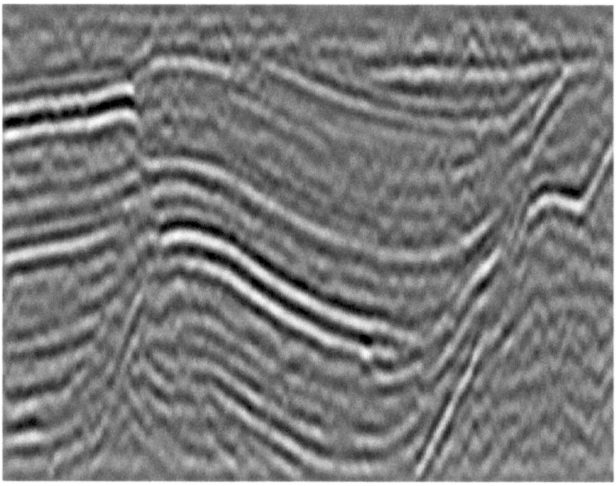

Fig. 14.2 Typical blurred seismic image of the geological cross section

Figure 14.2 presents the typical blurred seismic image of the same cross section, which is generated as the result of the conventional seismic data processing technique of depth migration.

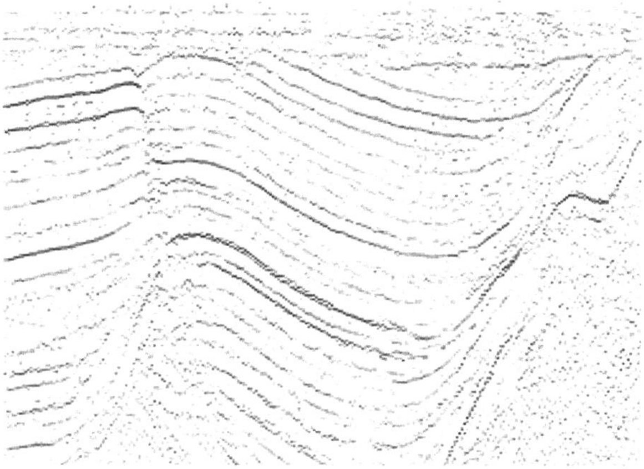

Fig. 14.3 Reconstructed seismic depth migration image of the geological cross section

This image has been processed by the digital image deblurring method with the minimum gradient support functional outlined above (Portniaguine and Zhdanov 2005). The reconstructed image is shown in Fig. 14.3. One can see that this image is practically identical to the original image presented in Fig. 14.1.

14.2 Formulation of the Joint Image Deblurring Problem

Considering multiple images of the same target collected by different physical and/or electron-optical devices with different blurring operators, we can describe the process of joint image deblurring by the following operator relationships:

$$\mathbf{d}^{(i)} = \mathbf{B}^{(i)}(\mathbf{m}^{(i)}), \quad i = 1, 2, 3, ..., N, \tag{14.11}$$

where $\mathbf{d}^{(i)}$ $(i = 1, 2, 3, ..., N)$ are different degraded (blurred) images (which may correspond to different types of sensors); $\mathbf{m}^{(i)}$ $(i = 1, 2, 3, ..., N)$ are the original (ideal) images of physical properties of the target; and $\mathbf{B}^{(i)}$ are the linear blurring operators of the multisensor imaging system, corresponding to different sensors, respectively.

The restoration of the deblurred images, $\mathbf{m}^{(i)}$, from the blurred recorded images, $\mathbf{d}^{(i)}$, can be treated as a solution of the system of equations (14.11).

To this end, we introduce the following parametric functional:

$$P^{\alpha}(\mathbf{m}^{(1)}, \mathbf{m}^{(2)}, ..., \mathbf{m}^{(n)}) = \sum_{i=1}^{N} \left\| \mathbf{B}^{(i)}(\mathbf{m}^{(i)}) - \mathbf{d}^{(i)} \right\|_D^2 + \alpha S_J, \tag{14.12}$$

where α is the regularization parameter and S_J is a joint stabilizing functional responsible for enforcing some relationships between multiple images, $\mathbf{m}^{(i)}$, $(i = 1, 2, 3, ..., N)$.

The choice of the joint stabilizing functional depends on the type of relationships or correlation we want to impose on the multiple images. For example, we can impose the joint constraints by using the joint focusing, joint minimum entropy, or Gramian stabilizers discussed in Chaps. 10, 11, and 12, respectively. To illustrate the application of the joint stabilizers for a fusion of multiple images, let us solve the joint image deblurring problem using the minimum entropy approach, as an example.

We have learned in Chap. 11 that the joint stabilizing functionals based on minimum entropy, S_{JME}, and minimum entropy gradient constraints, S_{JMEG}, are defined according to the following formulas.

A joint minimum entropy stabilizer is introduced as follows:

$$S_{JME} = -\int_V \frac{\sum_{i=1}^N \left|m^{(i)} - m_{apr}^{(i)}\right| + \beta}{Q_J} \log \frac{\sum_{i=1}^N \left|m^{(i)} - m_{apr}^{(i)}\right| + \beta}{Q_J} dv, \tag{14.13}$$

where

$$Q_J = \int_V \left(\sum_{i=1}^N \left|m^{(i)} - m_{apr}^{(i)}\right| + \beta \right) dv.$$

Similarly, we can introduce a joint minimum entropy gradient functional, S_{JMEG}:

$$S_{JMEG} =$$

$$-\int_V \frac{\sum_{i=1}^N \sqrt{\left(\nabla m^{(i)} \cdot \nabla m^{(i)}\right)} + \beta}{Q'_J} \log \frac{\sum_{i=1}^N \sqrt{\left(\nabla m^{(i)} \cdot \nabla m^{(i)}\right)} + \beta}{Q'_J} dv, \tag{14.14}$$

where

$$Q'_J = \int_V \left(\sum_{i=1}^N \sqrt{\left(\nabla m^{(i)} \cdot \nabla m^{(i)}\right)} + \beta \right) dv.$$

According to the basic principles of the regularization method, we have to find the models $\mathbf{m}_\alpha^{(1)}$,.... $\mathbf{m}_\alpha^{(2)}$, ...$\mathbf{m}_\alpha^{(N)}$, a quasi-solution of the inverse problem, which minimizes the parametric functional:

$$P^\alpha(\mathbf{m}^{(1)}, \mathbf{m}^{(2)}, ..., \mathbf{m}^{(n)}) =$$

$$= \sum_{i=1}^N \left\| \mathbf{B}^{(i)}(\mathbf{m}^{(i)}) - \mathbf{d}^{(i)} \right\|_D^2 + \alpha S_{JME,\ JMEG} = \min. \tag{14.15}$$

In order to solve this minimization problem, we calculate the first variation of the parametric functional with joint minimum entropy stabilizers:

$$\delta P^{\alpha}(\mathbf{m}^{(1)}, \mathbf{m}^{(2)}, ...\mathbf{m}^{(n)}) = 2\sum_{i=1}^{N} \left(\delta \mathbf{B}^{(i)}(\mathbf{m}^{(i)}),\ \ \mathbf{B}^{(i)}(\mathbf{m}^{(i)}) - \mathbf{d}^{(i)}\right)_D + \\ + 2\alpha\delta S_{JME,\ JMEG}. \tag{14.16}$$

Taking into consideration that operators $\mathbf{B}^{(i)}$ are linear, we can write

$$\delta \mathbf{B}^{(i)}(\mathbf{m}^{(i)}) = \mathbf{B}^{(i)}\delta \mathbf{m}^{(i)}, \tag{14.17}$$

and

$$\delta S_{JME,\ JMEG} = 2\left(\delta \mathbf{m}^{(i)},\ \mathbf{l}_{JME,\ JMEG}\right), \tag{14.18}$$

where vectors $\mathbf{l}_{JME,\ JMEG}$ are the directions of the steepest ascent for the stabilizing functionals, based on joint minimum entropy and minimum entropy gradient constraints, described by formulas (14.13) and (14.14), respectively.

Substituting expressions (14.17) and (14.18) into formula (14.16), we obtain:

$$\delta P^{\alpha}(\mathbf{m}^{(1)}, \mathbf{m}^{(2)}, ...\mathbf{m}^{(n)})$$

$$= 2\sum_{i=1}^{N} \left(\delta \mathbf{m}^{(i)},\ \left[\mathbf{B}^{(i)\star}\left(\mathbf{A}^{(i)}(\mathbf{m}^{(i)}) - \mathbf{d}^{(i)}\right) + \alpha \mathbf{l}_{JME,\ JMEG}\right]\right), \tag{14.19}$$

where $\mathbf{B}^{(i)\star}$ are the adjoint blurring operators.

Let us select

$$\delta \mathbf{m}^{(i)} = -k^{\alpha}\mathbf{l}^{\alpha(i)}(\mathbf{m}^{(1)}, \mathbf{m}^{(2)}, ...\mathbf{m}^{(n)}), \tag{14.20}$$

where k^{α} is some positive real number, and $\mathbf{l}^{\alpha(i)}(\mathbf{m}^{(1)}, \mathbf{m}^{(2)}, ...\mathbf{m}^{(n)})$ is the direction of the steepest ascent of the functional P^{α}:

$$\mathbf{l}^{\alpha(i)} = \mathbf{B}^{(i)\star}\left(\mathbf{B}^{(i)}(\mathbf{m}^{(i)}) - \mathbf{d}^{(i)}\right) + \alpha \mathbf{l}_{JME,\ JMEG}. \tag{14.21}$$

Then

$$\delta P^{\alpha}(\mathbf{m}^{(1)}, \mathbf{m}^{(2)}, ...\mathbf{m}^{(n)}) = -2k^{\alpha}\sum_{i=1}^{N} \left\|\mathbf{l}^{\alpha(i)}(\mathbf{m}^{(1)}, \mathbf{m}^{(2)}, ...\mathbf{m}^{(n)})\right\|^2. \tag{14.22}$$

The last expression confirms that selection of the perturbations of the images based on formula (14.20) ensures a decrease of the parametric functional.

We can construct an iterative process for the regularized conjugate gradient (RCG) algorithm for solving the minimization problem (14.15), which can be summarized as follows:

$$
\begin{aligned}
&\mathbf{r}_k = \mathbf{B}(\mathbf{m}_k) - \mathbf{d}, \quad \mathbf{l}_k^{\alpha} = \mathbf{l}^{\alpha}(\mathbf{m}_k) && (a)\\
&\beta_k^{\alpha} = \left\| \mathbf{l}_k^{\alpha} \right\|^2 / \left\| \mathbf{l}_{k-1}^{\alpha} \right\|^2, \quad \tilde{\boldsymbol{l}}_k^{\alpha} = \mathbf{l}_k^{\alpha} + \beta_k^{\alpha} \tilde{\boldsymbol{l}}_{k-1}^{\alpha}, \ \tilde{\boldsymbol{l}}_0^{\alpha} = \mathbf{l}_0^{\alpha}, && (b)\\
&s_k^{\alpha} = \left(\tilde{\boldsymbol{l}}_k^{\alpha}, \mathbf{l}_k^{\alpha}\right) / \left\{ \left\| \mathbf{B}\tilde{\mathbf{l}}_k^{\alpha} \right\|^2 + \alpha \left\| \tilde{\mathbf{l}}_k^{\alpha} \right\|^2 \right\}, && (c)\\
&\mathbf{m}_{k+1} = \mathbf{m}_k - s_k^{\alpha} \tilde{\boldsymbol{l}}_k^{\alpha}. && (d)
\end{aligned}
\tag{14.23}
$$

In the last formula, we used the following notations:
$\mathbf{d}$ is a vector of the observed blurred images

$$
\mathbf{d} = \left(\mathbf{d}^{(1)}, \mathbf{d}^{(2)}, \ldots.\mathbf{d}^{(N)}\right)^T; \tag{14.24}
$$

$\mathbf{m}_k$ is a vector of images computed at iteration number k,

$$
\mathbf{m}_k = \left(\mathbf{m}_k^{(1)}, \mathbf{m}_k^{(2)}, \ldots.\mathbf{m}_k^{(N)}\right)^T; \tag{14.25}
$$

$A(\mathbf{m}_k)$ is a vector of the predicted (deblurred) images computed at iteration number k;

and $\mathbf{l}_k^{\alpha}$ is a vector of the direction of the steepest ascent computed at iteration number k,

$$
\mathbf{l}_k^{\alpha} = \left(\mathbf{l}_{Ck}^{\alpha(1)}, \mathbf{l}_{Ck}^{\alpha(2)}, \ldots.\mathbf{l}_{Ck}^{\alpha(N)}\right)^T. \tag{14.26}
$$

The expressions for the steepest ascent directions are shown above in formula (14.21).

The iterative process (14.23) is terminated when the misfit reaches the required level:

$$
\varphi\left(\tilde{\mathbf{m}}_{k+1}\right) = \left\|\tilde{\mathbf{r}}_{k+1}\right\|_D^2 = \delta_d. \tag{14.27}
$$

Note that the focusing stabilizing functionals introduced above can be expressed as pseudo-quadratic functionals as it was discussed in Chap. 11:

$$
\begin{aligned}
s(\mathbf{m}) &= \left(\mathbf{W}_e\left(\mathbf{m} - \mathbf{m}_{apr}\right), \mathbf{W}_e\left(\mathbf{m} - \mathbf{m}_{apr}\right)\right)_{L_2}\\
&= \int_V \left| w_e(\mathbf{r})\left(m(\mathbf{r}) - m_{apr}(\mathbf{r})\right)\right|^2 dv,
\end{aligned}
\tag{14.28}
$$

where $\mathbf{W}_e$ is a linear operator of the multiplication of the images $m(\mathbf{r})$ by the function $w_e(\mathbf{r})$,which may depend on m. If the operator $\mathbf{W}_e$ is independent of $m(\mathbf{r})$, we obtain a quadratic functional, like the minimum norm or the maximum smoothness stabilizing functionals. In general cases, the function w_e may even be a nonlinear function of m, like the minimum entropy (14.13) or minimum entropy gradient (14.14) functionals. It was shown in Chap. 4 that presenting a stabilizing functional in a pseudo-quadratic form simplifies the solution of the regularization problem and makes it possible to develop a unified approach to regularization with different stabilizers.

14.3 Re-weighted Steepest Descent Method of Joint Image Deblurring

In the case of the image deblurring problem, the corresponding parametric functional (14.12) can be written as follows:

$$P^{\alpha}(\mathbf{m}^{(1)}, \mathbf{m}^{(2)}, ...\mathbf{m}^{(N)}) =$$

$$= \sum_{i=1}^{N} (\mathbf{B}^{(i)}(\mathbf{m}^{(i)}) - \mathbf{d}^{(i)})^T (\mathbf{B}^{(i)}(\mathbf{m}^{(i)}) - \mathbf{d}^{(i)}) +$$
$$+ \sum_{i=1}^{N} \alpha^{(i)} (\mathbf{W}_e^{(i)} \mathbf{m}^{(i)} - \mathbf{W}_e^{(i)} \mathbf{m}_{apr}^{(i)})^T (\mathbf{W}_e^{(i)} \mathbf{m}^{(i)} - \mathbf{W}_e^{(i)} \mathbf{m}_{apr}^{(i)}), \tag{14.29}$$

where matrix $\mathbf{W}_e^{(i)}$ is a variable matrix of the minimum entropy (gradient) stabilizer, which depends on $\mathbf{m}^{(1)}, \mathbf{m}^{(2)}, ...\mathbf{m}^{(N)}$.

Therefore, the problem of minimizing the parametric functional, given by Eq. (14.12), can be treated in a similar way to the minimization of the conventional Tikhonov functional. The only difference is that now we introduce some variable weighting matrices $\mathbf{W}_e^{(i)}$ for the image functions. The minimization problem for the parametric functional introduced by Eq. (14.29) can be solved using the ideas of traditional gradient-type methods.

Following Chap. 10, we will use the re-weighted gradient method. In the framework of this approach, the variable weighting matrices $\mathbf{W}_e^{(i)}$, $i = 1, 2, ..N$, are precomputed on each iteration, $\mathbf{W}_e^{(i)} = \mathbf{W}_{en}^{(i)} = \mathbf{W}_e^{(i)}\left(\mathbf{m}_n^{(1)}, \mathbf{m}_n^{(2)}, ...\mathbf{m}_n^{(N)}\right)$ based on the values $\mathbf{m}_n^{(1)}, \mathbf{m}_n^{(2)}, ...\mathbf{m}_n^{(N)}$, obtained on the previous iteration. As a result, they are treated as fixed matrices on each iteration. Under this assumption, we calculate the first variation of the parametric functional (14.29) as follows:

$$\delta P^{\alpha}(\mathbf{m}^{(1)}, \mathbf{m}^{(2)}, ...\mathbf{m}^{(N)}) =$$

$$= 2 \sum_{i=1}^{N} \delta \mathbf{m}^{(i)T} \mathbf{B}^{(i)T} \left(\mathbf{B}^{(i)}(\mathbf{m}^{(i)}) - \mathbf{d}^{(i)}\right) +$$
$$+ 2 \sum_{i=1}^{N} \alpha^{(i)} \delta \mathbf{m}^{(i)T} \mathbf{W}_e^{(i)2} \left(\mathbf{m}^{(i)} - \mathbf{m}_{apr}^{(i)}\right). \tag{14.30}$$

Finally, we obtain

$$\delta P^{\alpha} =$$

$$2\sum_{i=1}^{N} \delta \mathbf{m}^{(i)T}\left[\mathbf{B}^{(i)T}\left(\mathbf{B}^{(i)}(\mathbf{m}^{(i)}) - \mathbf{d}^{(i)}\right) + \alpha^{(i)}\mathbf{W}_{e}^{(i)2}\left(\mathbf{m}^{(i)} - \mathbf{m}_{apr}^{(i)}\right)\right]. \tag{14.31}$$

Following the general scheme of the steepest descent method, we can again select

$$\delta \mathbf{m}^{(i)} = -k^{\alpha(i)}\mathbf{l}^{\alpha(i)}(\mathbf{m}^{(i)}), \tag{14.32}$$

where $k^{\alpha(i)}$ is some positive real number (length of a step) and $\mathbf{l}^{\alpha(i)}(\mathbf{m}^{(i)})$ is a column matrix defining the direction of the steepest ascent:

$$\mathbf{l}^{\alpha(i)}(\mathbf{m}^{(i)}) =$$

$$\mathbf{B}_{m}^{(i)T}\left(\mathbf{B}^{(i)}(\mathbf{m}^{(i)}) - \mathbf{d}^{(i)}\right) + \alpha^{(i)}\mathbf{W}_{e}^{(i)2}\left(\mathbf{m}^{(i)} - \mathbf{m}_{apr}^{(i)}\right). \tag{14.33}$$

Thus, the regularized re-weighted steepest descent method is based on the successive line search in the gradient direction $\mathbf{l}^{\alpha(i)}(\mathbf{m}_{(n)}^{(i)})$:

$$\mathbf{m}_{n+1}^{(i)} = \mathbf{m}_{n}^{(i)} + \mathbf{m}^{(i)} = \mathbf{m}_{n}^{(i)} - k^{\alpha(i)}\mathbf{l}^{\alpha(i)}(\mathbf{m}_{n}^{(i)}), \tag{14.34}$$

where

$$\mathbf{l}^{\alpha(i)}(\mathbf{m}_{n}^{(i)}) = \mathbf{B}^{(i)T}\left(\mathbf{B}^{(i)}(\mathbf{m}_{n}^{(i)}) - \mathbf{d}^{(i)}\right) + \alpha^{(i)}\mathbf{W}_{en}^{(i)2}\left(\mathbf{m}_{n}^{(i)} - \mathbf{m}_{apr}^{(i)}\right). \tag{14.35}$$

14.4 Re-weighted Conjugate Gradient Method of Joint Image Deblurring

The regularized re-weighted conjugate gradient (RRCG) method can be developed in the same way as the steepest descent method; however, the model parameters are updated based on the successive line search in the conjugate gradient direction $\tilde{l}^{\alpha(i)}(\mathbf{m}_{n}^{(i)})$:

$$\mathbf{m}_{n+1}^{(i)} = \mathbf{m}_{n}^{(i)} + \mathbf{m}^{(i)} = \mathbf{m}_{n}^{(i)} - k^{\alpha(i)}\tilde{l}^{\alpha(i)}(\mathbf{m}_{n}^{(i)}). \tag{14.36}$$

The conjugate gradient directions $\tilde{l}^{\alpha(i)}(\mathbf{m}_{n}^{(i)})$ are selected as follows. In the initial step, we use the "direction" of regularized steepest ascent for the initial model $\mathbf{m}_0$:

$$\tilde{\mathbf{l}}_{0}^{\alpha(i)} = \tilde{\mathbf{l}}^{\alpha(i)}(\mathbf{m}_{0}^{(i)}) = \mathbf{l}^{\alpha(i)}(\mathbf{m}_{0}^{(i)}) =$$

$$= \mathbf{B}^{(i)T}\left(\mathbf{B}^{(i)}(\mathbf{m}_0^{(i)}) - \mathbf{d}^{(i)}\right) + \alpha^{(i)}\mathbf{W}_{e0}^{(i)2}\left(\mathbf{m}_0^{(i)} - \mathbf{m}_{apr}^{(i)}\right), \tag{14.37}$$

where $\mathbf{F}_{m_0}^{(i)}$ is the Fréchet derivative matrix for the initial model and $\mathbf{W}_{e0}^{(i)2} = \mathbf{W}_{e0}^{(i)2}\left(\mathbf{m}_0^{(i)}\right)$.

In the next step, the "direction" of ascent is a linear combination of the regularized steepest ascent on this step and the "direction" of ascent $\widetilde{\mathbf{l}}_0^{\alpha(i)}$ on the previous step:

$$\widetilde{\mathbf{l}}_1^{\alpha(i)} = \mathbf{l}_1^{\alpha(i)} + \beta_1^{\alpha(i)}\widetilde{\mathbf{l}}_0^{\alpha(i)}. \tag{14.38}$$

In the $(n+1)$th step

$$\widetilde{\mathbf{l}}_{n+1}^{\alpha(i)} = \mathbf{l}_{n+1}^{\alpha(i)} + \beta_{n+1}^{\alpha(i)}\widetilde{\mathbf{l}}_n^{\alpha(i)}, \tag{14.39}$$

where the regularized steepest ascent directions are determined now according to formula (14.37), and

$$\widetilde{\mathbf{l}}_n^{\alpha(i)} = \widetilde{\mathbf{l}}^{\alpha(i)}(\mathbf{m}_n^{(i)});\ \mathbf{l}_n^{\alpha(i)} = \mathbf{l}^{\alpha(i)}(\mathbf{m}_n^{(i)}). \tag{14.40}$$

The length of each iteration step, the coefficients $k_n^{\alpha(i)}$, can be determined with a linear or parabolic line search:

$$P^{\alpha}(\mathbf{m}_{n+1}^{(1)}, \mathbf{m}_{n+1}^{(2)}, ...\mathbf{m}_{n+1}^{(N)}) =$$

$$P^{\alpha}(\mathbf{m}_n^{(1)} - k_n^{\alpha(1)}\widetilde{\mathbf{l}}_n^{\alpha(1)}, ..., \mathbf{m}_n^{(N)} - k_n^{\alpha(N)}\widetilde{\mathbf{l}}_n^{\alpha(N)}) = \min. \tag{14.41}$$

Solution of this minimization problem gives the following best estimate for the lengths of the step using a linear line search:

$$k_n^{\alpha(i)} = \frac{\widetilde{\mathbf{l}}_n^{\alpha(i)T}\mathbf{l}_n^{\alpha(i)}}{\widetilde{\mathbf{l}}_n^{\alpha(i)T}\left(\mathbf{B}^{(i)T}\mathbf{B}^{(i)} + \alpha\mathbf{W}_{en}^{(i)2}\right)\widetilde{\mathbf{l}}_n^{\alpha(i)}}. \tag{14.42}$$

One can use a parabolic line search also (Zhdanov, 2002) to improve the convergence rate of the RRCG method.

The CG method requires that the vectors $\widetilde{\mathbf{l}}_n^{\alpha(i)}$ introduced above will be mutually conjugate. This requirement is fulfilled if the coefficients $\beta_n^{\alpha(i)}$ are determined by the formula

$$\beta_{n+1}^{\alpha(i)} = \frac{\|\mathbf{l}_{n+1}^{\alpha(i)}\|^2}{\|\mathbf{l}_n^{\alpha(i)}\|^2}. \tag{14.43}$$

Using equations (14.34), (14.37), and (14.42), we can obtain $\mathbf{m}^{(i)}$ iteratively.

Note that due to re-weighting, the stabilizing functional can change, and even increase from iteration to iteration,

$$s\left(\mathbf{m}_{n+1}^{(i)}\right) =$$

$$(\mathbf{m}_{n+1}^{(i)} - \mathbf{m}_{apr}^{(i)})^T \mathbf{W}_{e(n+1)}^{(i)2} (\mathbf{m}_{n+1}^{(i)} - \mathbf{m}_{apr}^{(i)}) = \gamma_n^{(i)} s\left(\mathbf{m}_n^{(i)}\right), \tag{14.44}$$

where

$$\gamma_n^{(i)} = \frac{s\left(\mathbf{m}_{n+1}^{(i)}\right)}{s\left(\mathbf{m}_n^{(i)}\right)} =$$

$$\frac{(\mathbf{m}_{n+1}^{(i)} - \mathbf{m}_{apr}^{(i)})^T \mathbf{W}_{e(n+1)}^{(i)2} (\mathbf{m}_{n+1}^{(i)} - \mathbf{m}_{apr}^{(i)})}{(\mathbf{m}_n^{(i)} - \mathbf{m}_{apr}^{(i)})^T \mathbf{W}_{en}^{(i)2} (\mathbf{m}_n^{(i)} - \mathbf{m}_{apr}^{(i)})}. \tag{14.45}$$

In order to ensure the convergence of the parametric functional to the global minimum, we use adaptive regularization and decrease the α_{n+1}, if $\gamma_n^{(i)} > 1$:

$$\alpha_{n+1}^{(i)} = \begin{cases} \alpha_n^{(i)}, \text{ if } \gamma_n^{(i)} \le 1, \\ \alpha_n^{(i)}/\gamma_n^{(i)}, \text{ if } \gamma_n^{(i)} > 1. \end{cases} \tag{14.46}$$

So, the product of the regularization parameter $\alpha_{n+1}^{(i)}$ and the stabilizer $s\left(\mathbf{m}_{n+1}\right)$ decreases or does not change:

$$\alpha_{n+1}^{(i)} s\left(\mathbf{m}_{n+1}^{(i)}\right) = \begin{cases} \alpha_n^{(i)} s\left(\mathbf{m}_{n+1}^{(i)}\right) = \alpha_n^{(i)} \gamma_n^{(i)} s\left(\mathbf{m}_n^{(i)}\right), & \text{if } \gamma_n^{(i)} \le 1, \\ \alpha_n^{(i)} s\left(\mathbf{m}_{n+1}^{(i)}\right) / \gamma_n^{(i)} = \alpha_n^{(i)} s\left(\mathbf{m}_n^{(i)}\right), & \text{if } \gamma_n^{(i)} > 1. \end{cases} \tag{14.47}$$

We also decrease the regularization parameter α_{n+1},

$$\alpha_{n+1}^{(i)\prime} = q\alpha_{n+1}^{(i)}, \; q < 1, \tag{14.48}$$

if the total misfit for all data does not decrease fast enough:

$$\sum_{i=1}^{N} \left\| \mathbf{B}^{(i)}(\mathbf{m}_n^{(i)}) - \mathbf{d}^{(i)} \right\|^2 - \sum_{i=1}^{N} \left\| \mathbf{B}^{(i)}(\mathbf{m}_{n+1}^{(i)}) - \mathbf{d}^{(i)} \right\|^2 \tag{14.49}$$

$$< 0.01 \sum_{i=1}^{N} \left\| \mathbf{B}^{(i)}(\mathbf{m}_n^{(i)}) - \mathbf{d}^{(i)} \right\|^2. \tag{14.50}$$

Numerical experiments demonstrate that the recommended choice of the empirical coefficient q is within an interval (0.5; 0.9).

The algorithm of the RRCG method can be summarized as follows:

$$\mathbf{r}_n^{(i)} = \mathbf{B}^{(i)}(\mathbf{m}_n^{(i)}) - \mathbf{d}^{(i)}, \qquad \mathbf{g}_n^{(i)} = \mathbf{W}_{en}^{(i)}(\mathbf{m}_n^{(i)} - \mathbf{m}_{apr}^{(i)}), \qquad (a)$$

$$\mathbf{l}_n^{\alpha_n(i)} = \mathbf{B}^{(i)T}\mathbf{r}_n^{(i)} + \alpha_n \mathbf{W}_{en}^{(i)}\mathbf{g}_n^{(i)}, \qquad (b)$$

$$\beta_n^{\alpha_n(i)} = \left\|\mathbf{l}_n^{\alpha_n(i)}\right\|^2 / \left\|\mathbf{l}_{n-1}^{\alpha_{n-1}(i)}\right\|^2, \qquad \tilde{\boldsymbol{l}}_n^{\alpha_n(i)} = \mathbf{l}_n^{\alpha_n(i)} + \beta_n^{\alpha_n(i)}\tilde{\boldsymbol{l}}_{n-1}^{\alpha_{n-1}(i)}, \; \tilde{\boldsymbol{l}}_0^{\alpha_0(i)} = \mathbf{l}_0^{\alpha_0(i)}, \; (c)$$

$$\tilde{k}_n^{\alpha_n(i)} = \left(\tilde{\boldsymbol{l}}_n^{\alpha_n(i)T}\, \mathbf{l}_n^{\alpha_n(i)}\right) / \left[\tilde{\boldsymbol{l}}_n^{\alpha_n(i)T}\left(\mathbf{B}^{(i)T}\mathbf{B}^{(i)} + \alpha_n \mathbf{W}_{en}^{(i)2}\right)\tilde{\boldsymbol{l}}_n^{\alpha_n(i)}\right], \; (d)$$

$$\mathbf{m}_{n+1}^{(i)} = \mathbf{m}_n^{(i)} - \tilde{k}_n^{\alpha_n(i)}\,\tilde{\boldsymbol{l}}_n^{\alpha_n(i)}, \qquad \gamma_n^{(i)} = \|\mathbf{g}_{n+1}\|^2 / \|\mathbf{g}_n\|^2, \qquad (e)$$

$$\alpha_{n+1}^{(i)} = \alpha_n^{(i)}, \text{ if } \gamma_n^{(i)} \le 1, \text{ and } \alpha_{n+1}^{(i)} = \alpha_n^{(i)}/\gamma_n^{(i)}, \text{ if } \gamma_n^{(i)} > 1, \qquad (f)$$

$$\alpha_{n+1}^{(i)\prime} = q\alpha_{n+1}^{(i)}, \; q < 1, \text{ if } \left\|\mathbf{W}_d^{(i)}\mathbf{r}_n^{(i)}\right\|^2 - \left\|\mathbf{W}_d^{(i)}\mathbf{r}_{n+1}^{(i)}\right\|^2 < 0.01\left\|\mathbf{W}_d^{(i)}\mathbf{r}_n^{(i)}\right\|^2, \; (g)$$
(14.51)

where $\alpha_n^{(i)}$ are the subsequent values of the regularization parameter. The iterative process (14.51) is terminated when the misfit reaches the given level ε_0:

$$\phi(\mathbf{m}_N^{(i)}) = \left\|\mathbf{r}_N^{(i)}\right\|^2 \le \varepsilon_0. \qquad (14.52)$$

14.5 Numerical Examples of Reconstructing the Blurred MRI Images

Let us consider that we have two independent original images of the same target, $\mathbf{m}^{(1)}$and $\mathbf{m}^{(2)}$, collected by different physical and/or electron-optical devices with different blurring operators, $\mathbf{B}^{(i)}$:

$$\mathbf{d}^{(i)} = \mathbf{B}^{(i)}(\mathbf{m}^{(i)}), \; i = 1, 2. \qquad (14.53)$$

We also assume that the original images, as well as the blurred images, can be defined in a plane:

$$m^{(i)} = m^{(i)}(x, y), \; d^{(i)} = d^{(i)}(x, y), \; i = 1, 2;$$

or in a discrete form:

$$m_{kl}^{(i)} = m^{(i)}(x_k, y_l), \; d_{kl}^{(i)} = d^{(i)}(x_k, y_k), \text{ where } -K \le k \le K, \; -L \le l \le L, \text{ and } i = 1, 2.$$

We can represent a blurred 2D image d_{kn} as a convolution of the original image $m_{k'n'}$ with the kernel $b\,(k, n)$ of the blurring operator:

$$d_{kn} = \sum_{k-W}^{k+W} \sum_{n-W}^{n+W} b\left(k - k', n - n'\right) m_{k'n'}, \text{ where } -K \leq k, k' \leq K, \; -L \leq l, l' \leq L. \tag{14.54}$$

In the last formula, $2W$ is the width of the kernel $b\,(k, n)$.

In this section, we consider, as an example, the problem of reconstructing blurred images produced by the *Magnetic Resonance Imaging (MRI)* method. It is based on nuclear magnetic resonance (NMR) principles, a spectroscopic technique used to obtain microscopic physical data about molecules. Magnetic resonance imaging is performed through the strong pulse of the magnetic field in the radio frequency (RF) range of the electromagnetic spectrum. All the atoms, including those that compose the human body, have a property known as spin (a fundamental property of all atoms in nature like mass or charge). The human body is mainly composed of fat and water, which makes the human body content of about 63% hydrogen. The MRI machine applies a very strong magnetic field of up to 20,000 gauss in the radio frequency (RF) range specific to hydrogen. The pulse directed to a specific body area causes the protons to absorb energy and spin in different directions, known as resonance. When the RF pulse is turned off, the hydrogen protons slowly return to their natural alignment within the magnetic field and release excess stored energy. This generates the secondary RF electromagnetic field, which is recorded by the corresponding receivers (induction coils). The recorded signal is transformed into a 3D image of the specific part of the body using the specific computer visualization software.

The following is the example of the reconstruction of brain MRI images using the joint minimum entropy method outlined above (Fig. 14.4).[1]

The original (ideal) MRI images are shown in panels (a) and (e) of this figure. In real medical applications, these images could be taken on two different occasions and can be variously degraded by deviating conditions of MRI testing.

In order to computer simulate this situation, we have applied the Gaussian digital filter to the original image (a) of Fig. 14.4 using MATLAB. The coefficients of the Gaussian filter are as follows:

$$b_G\,(k, l) = \frac{a_G\,(k, l)}{\sum_{-W}^{+W} \sum_{-W}^{+W} a_G\,(k, l)}, \tag{14.55}$$

where

$$a_G\,(k, l) = \exp\left[\frac{-\left(k^2 + l^2\right)}{2\sigma^2}\right], \; -W \leq k, l \leq W.$$

We have also applied the Pillbox filter to the original image (e) of Fig. 14.4 using MATLAB. The Pillbox filter is defined by the following equations:

$$b_P\,(k, l) = \frac{a_P\,(k, l)}{\sum_{-W}^{+W} \sum_{-W}^{+W} a_P\,(k, l)}, \tag{14.56}$$

[1] These images were computer simulated by graduate student Xiaolei Tu.

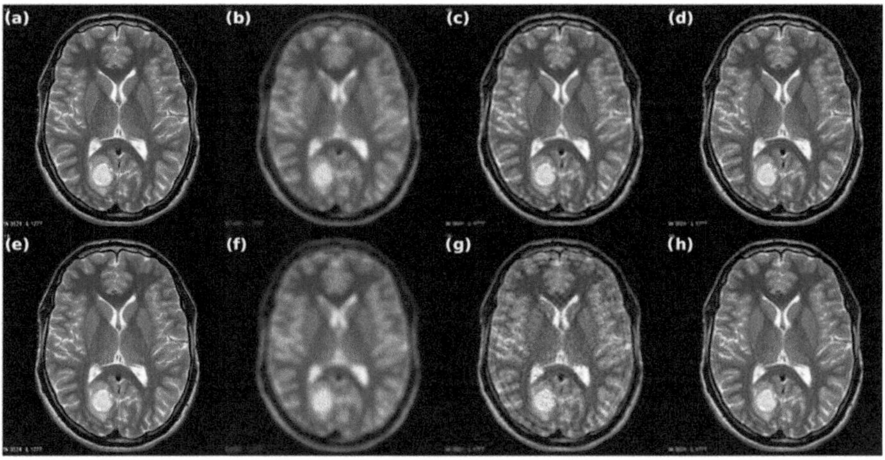

Fig. 14.4 Reconstruction of MRI images of the brain using minimum gradient entropy stabilizer: panels **a** and **e** are original (actual) MRI images; **b** is the blurred image by a Gaussian filter with a standard derivation of 7, contaminated with 2% random Gaussian noise; **f** is the blurred image by a Pillbox filter with a radius of 7, contaminated with 7% random Gaussian noise; panels **c** and **g** present the separately reconstructed images with minimum gradient entropy stabilizer; panels **d** and **h** show the jointly reconstructed images with joint minimum gradient entropy stabilizer

where

$$a_P(k,l) = \begin{cases} 1, & k^2 + l^2 \leq R^2 \\ 0, & \text{otherwise} \end{cases}.$$

In this numerical experiment, we use the Gaussian filter with $W = 21$, and $\sigma = 7$, and the Pillbox filter with R = 15. We then contaminated the images blurred by Gaussian and Pillbox filters with 2% and 7% Gaussian noise, respectively. Note that the Gaussian noise added to the two images is uncorrelated. The noise-contaminated blurred images are shown in panels (b) and (f) of Fig. 14.4.

We first reconstructed the images using the deblurring method outlined above with separate minimum gradient entropy stabilizers calculated individually for each image. The reconstructed images are shown in panels (c) and (g) of Fig. 14.4. We can see an improvement in these images compared to the blurred images (b) and (f); however, some brain structure details are still diffused and unfocused.

The degraded images (b) and (f) were also jointly reconstructed using the joint minimum gradient entropy stabilizer. The results of joint image enhancement and deblurring, (d) and (h), show significant improvement in resolution and quality of the deblurred images, which become almost indistinguishable from the original MRI images (a) and (e) of Fig. 14.4. In summary, the joint image reconstruction recovers the details of the deblurred images better than the separate reconstructions.

References and Recommended Reading

Ekstrom MP (2012) Image recovery: theory and applications. Academic

Gonzalez RC (2009) Digital image processing. Pearson Education, India

Oliveira JP, Bioucas-Dias JM, Figueiredo MA (2009) Adaptive total variation image deblurring: a majorization–minimization approach. Signal Process 89(9):1683–1693

Portniaguine O, Zhdanov MS (1999) Focusing geophysical inversion images. Geophysics 64:874–887

Portniaguine ON, Zhdanov MS (2005) Method of digital image enhancement and sharpening: US Patent No. 6,879,735

Pratt WK (2007) Digital image processing: PIKS scientific inside, vol 4. Wiley-Interscience, Hoboken

Rudin LI, S. Osher, Fatemi E (1992) Nonlinear total variation based noise removal algorithms. Phys D 60:259–268

Stark H (2013) Image recovery: theory and applications. Academic

Vogel CR, Oman ME (1998) Fast, robust total variation-based reconstruction of noisy, blurred images. IEEE Trans Image Process 7(6):813–824

Vogel CR (2002) Computational methods for inverse problems. Soc Indust Appl Math

Wang R, Tao D (2014) Recent progress in image deblurring: arXiv preprint arXiv:1409.68308

Zhdanov MS (2002) Geophysical inverse theory and regularization problems. Elsevier

Zhdanov MS (2015) Inverse theory and applications in geophysics. Elsevier

Part IV
AI-Aided Inversion

Chapter 15
Machine Learning in the Context of Inversion Theory

Abstract This chapter reviews the basic ideas of Artificial Intelligence (AI) and Machine Learning (ML) methods. The similarities and differences between the machine learning approach and inverse problem solution are discussed in detail. We introduce the concept of multilayer perceptron, which forms an artificial neural network, and discuss the importance of the Universal Approximation Theorem for neural networks. It is shown that network training can be treated as the inverse problem solution. The backpropagation algorithm of deep neural network training is presented as well. We also demonstrate that training the deep neural network (DNN) or finding the optimal weights of the DNN can be reduced to the application of the standard methods of regularized inversion.

Keywords Artificial intelligence (AI) · Machine learning (ML) · Perceptron · Activation function · Artificial neural network

15.1 Machine Learning Versus Inversion

The term *machine learning* was initially introduced for solving pattern recognition problems (see, for example, Bishop 2006; Mohri et al. 2012). Machine learning has been developed for solving complex problems for which there was no clear specification of how the observed data were related to the object or image under investigation. In this case, the machine learning approach attempts to simulate how human beings learn from our experience and practical experiments rather than following specific physical or other natural science laws.

Over the years, several mathematical techniques have been developed to describe in numerical form the learning process. These algorithms are usually subdivided into two classes—unsupervised learning and supervised learning (Mehlig 2021). Unsupervised learning involves such techniques as data and model clustering, principal component analysis (PCA), singular value decomposition (SVD), etc. These methods can recover the concealed patterns of images or data. The importance of

M. S. Zhdanov, *Advanced Methods of Joint Inversion and Fusion of Multiphysics Data*, Advances in Geological Science,
https://doi.org/10.1007/978-981-99-6722-3_15

unsupervised learning is in its ability to discover similarities and differences in the observed data or models without using a priori information.

Another class—supervised learning—includes linear and nonlinear regressions, classification, ranking, and neural network simulation, among others (Mohri et al. 2012; Schuster 2023). Supervised learning is based on training the algorithms on the known input and output parameters, measuring the effectiveness of this training through the corresponding error functions, and continuing the training until the desired level of errors is reached. This process is similar to an inverse problem solution when we iteratively update the model parameters until the misfit between the observed and predicted data reaches the desired level of errors. However, there are some important differences.

For example, in formulating the inverse problem, which is the subject of this book, we always assume that we know the specific laws or mathematical rules which relate the observed data, $\mathbf{d}$, to the corresponding model, $\mathbf{m}$. We express these rules in the form of the following operator equation:

$$\mathbf{d} = A(\mathbf{m}), \tag{15.1}$$

where A is the forward modeling operator. We always suppose that the forward modeling operator is known because it represents specific physical laws or mathematical rules.

In the framework of the supervised machine learning approach, in a general case, there is no need to know the specific form of this operator. Instead, the algorithm is based on the assumption that multiple pairs, $\{\mathbf{m}_i, \mathbf{d}_i;\ i = 1, 2, ...N\}$, of the observed data and the models generating these data are known. The "machine" or computer operates with these sets to "learn" the relationships between the data and models through numerical experiments. It is important to note that for the machine learning algorithm, there is no difference if we want to find the actions of the forward operator, A, or of the inverse operator, A^{-1}, transforming the data into the models:

$$\mathbf{m} = A^{-1}(\mathbf{d}). \tag{15.2}$$

The algorithm simulates the actions of the inverse operator based on the known pairs of the input/output parameters.

In the following sections, we will show that the machine learning algorithm can provide an accurate numerical approximation of the inverse operator. Indeed, the majority of machine learning methods use the concept of an *artificial neural network (ANN)* as a learning engine. According to the *Universal Approximation Theorem* for neural networks, which we will discuss in detail below, practically any continuous function can be approximated by a corresponding neural network with any given accuracy, ε (Hornik et al. 1989).

From the theory of ill-posed inverse problems (Chap. 4), we know that, in a general case, the inverse operator, A^{-1}, may not necessarily be continuous and bounded. Therefore, the Universal Approximation Theorem would not work to solve the ill-posed problems. However, we demonstrated in Chap. 4 that we can replace the solution of one ill-posed inverse problem (15.1) with the solutions of the family of well-posed problems,

$$\mathbf{d} = A_{\alpha}(\mathbf{m}), \tag{15.3}$$

assuming that these solutions,

$$\mathbf{m}_{\alpha} = A_{\alpha}^{-1}(\mathbf{d}), \tag{15.4}$$

asymptotically go to the true solution, $\mathbf{m}_t$, as α tends to zero:

$$\mathbf{m}_{\alpha} = A_{\alpha}^{-1}(\mathbf{d}) \rightarrow \mathbf{m}_t, \text{ if } \alpha \rightarrow 0, \tag{15.5}$$

where $\alpha \geq 0$ is a regularization parameter; and A_{α}^{-1} are continuous and bounded inverse operators for the well-posed problems (15.3). The operators A_{α}^{-1} are called the regularizing operatorsfor Eq. (15.1)), $R(\mathbf{d}, \alpha)$:

$$R(\mathbf{d}, \alpha) = A_{\alpha}^{-1}(\mathbf{d}). \tag{15.6}$$

We can now apply the Universal Approximation Theorem to produce the required approximation of the regularizing operators:

$$\|M_{\alpha}(\mathbf{d}) - R(\mathbf{d}, \alpha)\| < \varepsilon_{\alpha}, \tag{15.7}$$

where $M_{\alpha}(\mathbf{d})$ denotes the results obtained by the corresponding machine learning algorithm, and $\varepsilon_{\alpha} > 0$ are small positive values of accuracy levels of these approximations, which can vary with α.

Thus, we can see that both machine learning inversion and the classical inversion methods are used to solve the same problem of finding the corresponding regularizing operators (15.6). The main difference between these two approaches is in the way how they construct $R(\mathbf{d}, \alpha)$. The classical inversion methods discussed in the previous chapters of this book rely on the a priori knowledge of the forward modeling operator and the properties of the inverse models to reduce the computer memory requirements and speed up the calculations. In other words, classical methods use the laws of physics or other related sciences to design the inversion algorithms. On the contrary, the machine learning approach does not use any specific physical laws. Instead, it develops the approximations of these laws by training the algorithms on the known input and output parameters. The major limitation of this approach is related to the fact that the volume of the known parameters and the computer power required to process this volume increases significantly with the complexity of the inverse problem. However, with the recent dramatic increase in computer power, this limitation has become less critical, which resulted in the renewed interest in artificial neural networks and machine learning algorithms.

In the following chapter, we will show that by synthesizing the ideas from both approaches to inversion—machine learning and classical methods of regularization—we can arrive at a new powerful technique for solving general inverse problems.

15.2 Artificial Neural Networks

The concept of "*neural network*" was first introduced as a model of biological neurons (McCulloch and Pitts 1943). This concept provided the basis for building the first prototypes of machines possessing "*artificial intelligence*" (AI) by Rosenblatt (1957, 1962) and others. However, those efforts did not receive much traction because of the limited resources of computers at that time. The modern dramatic increase in the interest in AI and machine learning can be attributed first and foremost to the rapid scaling up of computing power during the last decade. At the same time, the concept of a neural network is still a fundamental building block of machine learning algorithms.

15.2.1 Perceptron

In the center of the classical neural network is the *perceptron* proposed by Rosenblatt (1962), which describes the nonlinear input and output process mimicking the operations of the human brain's neurons. To explain the concept of the perceptron, we consider a neural network containing one input layer, one hidden layer, and one output layer of neurons, as shown in Fig. 15.1.

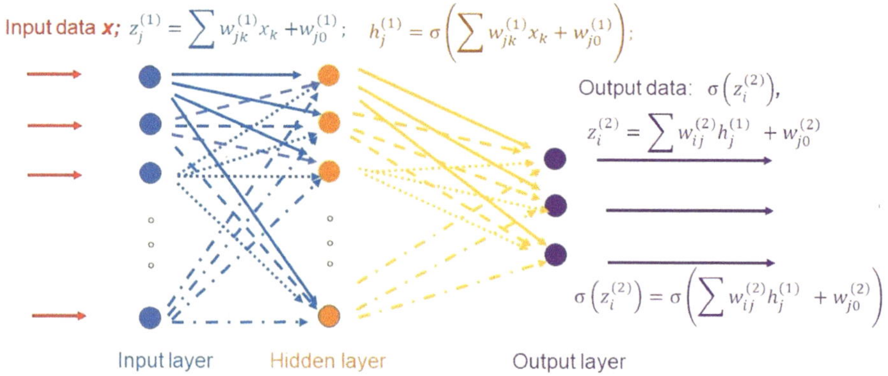

Fig. 15.1 Schematic diagram of the perceptron

The actions of this basic neural network model can be described as a set of linear and nonlinear transformations. The first transformation is given by a linear combination, $z_j^{(1)}$, of the input variables, $x_i, x_2, ...x_{N_0}$, with some weights:

$$z_j^{(1)} = \sum_{k=1}^{N_0} w_{jk}^{(1)} x_k + w_{j0}^{(1)}, \ j = 1, 2, ..., N_1, \tag{15.8}$$

where the weight of the link from the kth input neuron to the jth output neuron is denoted as $w_{jk}^{(1)}$ (where the second index always refers to the link's "beginning"; the first, to its "end"), and N_0 is the number of neurons in the input layer. Thus, each link in Fig. 15.1 for neuron from one layer to neuron in the other layer is associated with the weights. The parameters $w_{0j}^{(1)}$ are called *biases*.

The neurons forming the second layer are often called *hidden neurons*. There are N_1 hidden neurons in our model. The actions of the hidden neurons are quite different from that of the first-layer neurons. The hidden neurons transform the input variables by nonlinear differentiable *activation functions*. The most widely used activation function is a sigmoid (S-shaped) function, $\sigma(z)$, which is defined as follows:

$$\sigma(z) = \frac{1}{1 + e^{-bz}} = \left(1 + e^{-bz}\right)^{-1}, \tag{15.9}$$

where b is a parameter of the sigmoid.

The sigmoidal function is a continuous, monotonically increasing function with a characteristic S-like curve, as shown in Fig. 15.2.

In this case, the jth hidden neuron from the first hidden layer receives as input the weighted sum, $z_j^{(1)}$ of input data, x_k, and subjects this sum to the neuron activation function, $\sigma(z)$:

$$h_j^{(1)} = \sigma\left(z_j^{(1)}\right) = \sigma\left(\sum_{k=1}^{N_0} w_{jk}^{(1)} x_k + w_{j0}^{(1)}\right), \ j = 1, 2, ..., N_1. \tag{15.10}$$

There are N_2 output neurons. The output neuron generates a linear combination of the inputs from the first hidden layer $h_j^{(1)}$, $(j = 1, 2, ..., N_1)$:

$$z_i^{(2)} = \sum_{j=1}^{N_1} w_{ij}^{(2)} h_j^{(1)} + w_{i0}^{(2)} = \sum_{j=1}^{N_1} w_{ij}^{(2)} \sigma\left(z_j^{(1)}\right) + w_{i0}^{(2)}, \ i = 1, 2, ..., N_2, \tag{15.11}$$

and subject it the same activation function, $\sigma(z)$, as shown in Fig. 15.1:

$$h_i^{(2)} = \sigma\left(z_i^{(2)}\right) = \sigma\left(\sum_{j=1}^{N_1} w_{ij}^{(2)} h_j^{(1)} + w_{i0}^{(2)}\right), \ i = 1, 2, ..., N_2. \tag{15.12}$$

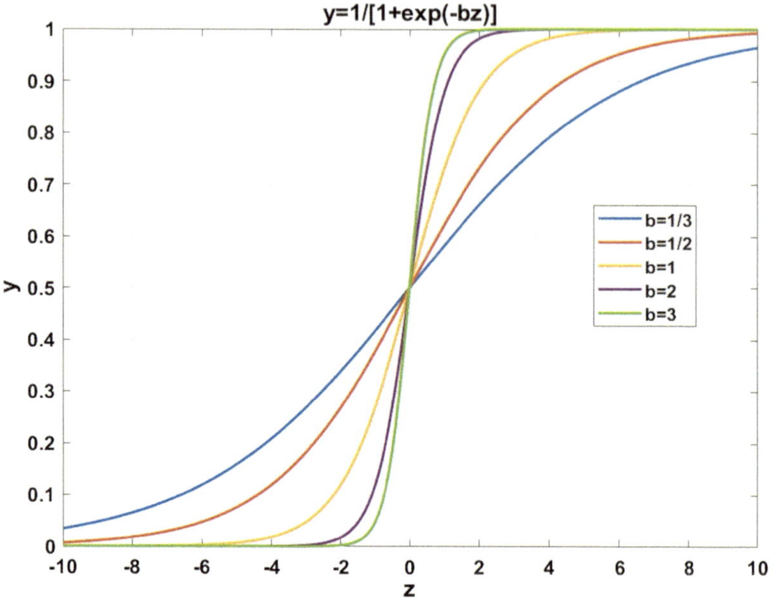

Fig. 15.2 Examples of the sigmoidal function

Thus, the output data, $h_i^{(2)}$, $i = 1, 2, ..., N_2$, resulting from input data propagation from the input layer through two neuron layers with activation function $\sigma(z)$, can be expressed as follows:

$$h_i^{(2)} = \sigma\left(\sum_{j=1}^{N_1} w_{ij}^{(2)} \sigma\left(\sum_{k=1}^{N_0} w_{jk}^{(1)} x_k + w_{0j}^{(1)}\right) + w_{i0}^{(2)}\right), \quad i = 1, 2, ..., N_2, \qquad (15.13)$$

where N_2 is the total number of the output neurons.

Expression (15.13) shows that perceptron transforms the input data by a combination of linear and nonlinear operations. As a result, according to the Universal Approximation Theorem, the perceptron operation can approximate any continuous operator applied to the input data (with the proper selection of the weights). The key is to find these weights. I will discuss a solution to this problem using a more general concept of multilayer perceptron.

15.2.2 *Multilayer Perceptron*

The term *multilayer perceptron* is usually applied to a fully connected class of artificial neural network, where the data flow from the input neuron layer to the output layer is strictly feedforward.

Let us revisit the neural network formed by the perceptron (Fig. 15.1). The ith neuron of the output layer receives the weighted sum, $z_i^{(2)}$, of the values, $h_j^{(1)}$, $j = 1, 2, \ldots N_1$, coming from the neurons of the first hidden layer and, again, subjects it to the same activation function, $\sigma(z)$, as shown in formula (15.12). This is how the ith output is obtained.

If we have two hidden layers, the process will continue as follows. The ith output from the output layer is obtained using the following formula:

$$h_i^{(3)} = \sigma\left(z_i^{(3)}\right) = \sigma\left(\sum_{j=1}^{N_2} w_{ij}^{(3)} h_j^{(2)} + w_{i0}^{(3)}\right), \quad i = 1, 2, \ldots N_3. \tag{15.14}$$

This process can be extended to any number of hidden layers as we discuss in the following section.

It is convenient to describe the transformations performed by the neural network using the vector and matrix notations to simplify the mathematical analysis of this process. For example, we can introduce the vector-columns,

$$\mathbf{x} = (x_1, x_2, \ldots., x_{N_0})^T \text{ and } \mathbf{z}^{(q)} = (z_1^{(q)}, z_2^{(q)}, \ldots., z_{N_q}^{(q)})^T, \; q = 1, 2, 3; \tag{15.15}$$

formed by the input variables (where the upper subscript "T" denotes a transposition), and the matrices, $\mathbf{w}^{(1)}$, $\mathbf{w}^{(2)}$, and $\mathbf{w}^{(3)}$, formed by the weights, $w_{jk}^{(1)}$, $w_{ij}^{(2)}$ and $w_{ij}^{(3)}$, with $N_1 \times N_0$, $N_2 \times N_1$, and $N_3 \times N_2$ components, respectively. We also introduce the bias vectors as follows:

$$\mathbf{b}^{(1)} = (w_{10}^{(1)}, w_{20}^{(1)}, \ldots., w_{N_1 0}^{(1)})^T; \;\; \mathbf{b}^{(2)} = (w_{10}^{(2)}, w_{20}^{(2)}, \ldots., w_{N_2 0}^{(2)})^T;$$

$$\mathbf{b}^{(3)} = (w_{10}^{(3)}, w_{20}^{(3)}, \ldots., w_{N_2 0}^{(3)})^T.$$

Using these notations, we have the output of the first neuron layer, formula (15.8), equal to:

$$\mathbf{z}^{(1)} = \mathbf{w}^{(1)}\mathbf{h}^{(0)} + \mathbf{b}^{(1)}, \tag{15.16}$$

where for convenience, we use the following notations:

$$\mathbf{x} = \mathbf{h}^{(0)}.$$

The output of the first hidden layer, Eq. (15.10) can be rewritten as follows:

$$\mathbf{h}^{(1)} = \boldsymbol{\sigma}\left(\mathbf{z}^{(1)}\right) = \boldsymbol{\sigma}\left(\mathbf{w}^{(1)}\mathbf{h}^{(0)} + \mathbf{b}^{(1)}\right), \tag{15.17}$$

where

$$\mathbf{h}^{(1)} = \left[h_1^{(1)}, h_2^{(1)}, \ldots., h_{N_1}^{(1)}\right]^T,$$

and $\boldsymbol{\sigma}\left(\mathbf{z}^{(1)}\right)$ denotes the vector obtained by application of the sigmoidal activation function, $\sigma(z)$ to every component of vector $\mathbf{z}^{(1)}$.

The output of the neuron of the second hidden layer, formula (15.12), takes the form:

$$\mathbf{h}^{(2)} = \boldsymbol{\sigma}\left(\mathbf{w}^{(2)}\mathbf{h}^{(1)} + \mathbf{b}^{(2)}\right), \tag{15.18}$$

where

$$\mathbf{h}^{(2)} = \left[h_1^{(2)}, h_2^{(2)}, \; ...., h_{N_2}^{(2)}\right]^T.$$

The output of the neuron of the output layer, formula (15.14), can be written as follows:

$$\mathbf{h}^{(3)} = \boldsymbol{\sigma}\left(\mathbf{w}^{(3)}\mathbf{h}^{(2)} + \mathbf{b}^{(3)}\right), \tag{15.19}$$

where

$$\mathbf{h}^{(3)} = \left[h_1^{(3)}, h_2^{(3)}, \; ...., h_{N_3}^{(3)}\right]^T.$$

Therefore, Eq. (15.14) can be presented as the following recursive formula:

$$\begin{aligned}
\mathbf{M}(\mathbf{x}; \mathbf{W}, \; \mathbf{B}) = \mathbf{h}^{(3)} &= \boldsymbol{\sigma}\left(\mathbf{w}^{(3)}\mathbf{h}^{(2)} + \mathbf{b}^{(3)}\right) \\
&= \boldsymbol{\sigma}\left(\mathbf{w}^{(3)}\boldsymbol{\sigma}\left(\mathbf{w}^{(2)}\mathbf{h}^{(1)} + \mathbf{b}^{(2)}\right) + \mathbf{b}^{(3)}\right) \\
&= \boldsymbol{\sigma}\left(\mathbf{w}^{(3)}\boldsymbol{\sigma}\left(\mathbf{w}^{(2)}\boldsymbol{\sigma}\left(\mathbf{w}^{(1)}\mathbf{h}^{(0)} + \mathbf{b}^{(1)}\right) + \mathbf{b}^{(2)}\right) + \mathbf{b}^{(3)}\right),
\end{aligned} \tag{15.20}$$

where $\mathbf{M}(\mathbf{x}; \mathbf{W}, \; \mathbf{B})$ is called *the neural network operator*. Symbols $\mathbf{W} = \left(\mathbf{w}^{(1)}, \; \mathbf{w}^{(2)}, \mathbf{w}^{(3)}\right)$ and $\mathbf{B} = \left(\mathbf{b}^{(1)}, \; \mathbf{b}^{(2)}, \mathbf{b}^{(3)}\right)$ denote the groups of weights and biases, respectively.

Thus, the neural network can be mathematically represented as a nonlinear operator, $\mathbf{M}(\mathbf{x}; \mathbf{W}, \; \mathbf{B})$, acting on a vector of input variables, $\mathbf{x} = \mathbf{h}^{(\mathbf{0})}$, which is controlled by the matrices representing the weights, $\left(\mathbf{w}^{(1)}, \; \mathbf{w}^{(2)}, \mathbf{w}^{(3)}\right)$, and the biases $\left(\mathbf{b}^{(1)}, \; \mathbf{b}^{(2)}, \mathbf{b}^{(3)}\right)$. The process of supervised learning of the neural network is equivalent to adjusting these weights and biases to achieve the most accurate representation of the true output parameters by the network. Below we will consider this process for a more general case of a neural network with $(Q - 1)$ hidden layers.

15.2.3 Deep Neural Networks

We have presented in Fig. 15.1 the most commonly used architecture of the artificial neural network. At the same time, this architecture can be easily expanded by including an additional number of hidden layers. For example, if we consider a multilayer network with $(Q - 1)$ hidden layers (Fig. 15.3), the corresponding neural network operator can be expressed by the following recursive formulas:

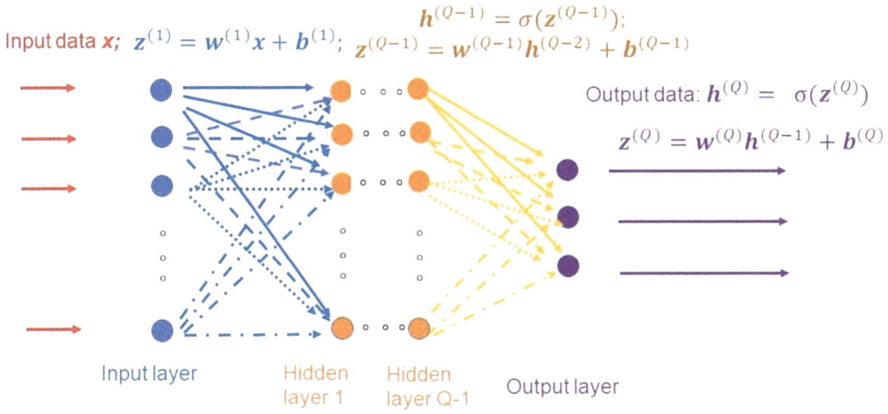

Fig. 15.3 Schematic diagram of the multilayer perceptron

$$\mathbf{M}\,(\mathbf{x};\,\mathbf{W},\ \mathbf{B}) = \mathbf{h}^{(Q)},$$

$$\mathbf{h}^{(Q)} = \sigma(z^{(Q)}),\ \ \mathbf{z}^{(Q)} = \mathbf{w}^{(Q)}\mathbf{h}^{(Q-1)} + \mathbf{b}^{(Q)},$$

$$\mathbf{h}^{(Q-1)} = \boldsymbol{\sigma}\left(\mathbf{z}^{(Q-1)}\right),\ \ \mathbf{z}^{(Q-1)} = \mathbf{w}^{(Q-1)}\mathbf{h}^{(Q-2)} + \mathbf{b}^{(Q-1)},$$

$$::,$$

$$\mathbf{h}^{(3)} = \boldsymbol{\sigma}\left(\mathbf{z}^{(3)}\right),\ \ \mathbf{z}^{(3)} = \mathbf{w}^{(3)}\mathbf{h}^{(2)} + \mathbf{b}^{(3)},$$

$$\mathbf{h}^{(2)} = \boldsymbol{\sigma}\left(\mathbf{z}^{(2)}\right),\ \ \mathbf{z}^{(2)} = \mathbf{w}^{(2)}\mathbf{h}^{(1)} + \mathbf{b}^{(2)},$$

$$\mathbf{h}^{(1)} = \boldsymbol{\sigma}\left(\mathbf{z}^{(1)}\right),\ \ \mathbf{z}^{(1)} = \mathbf{w}^{(1)}\mathbf{h}^{(0)} + \mathbf{b}^{(1)},$$

$$\mathbf{h}^{(0)} = \mathbf{x}, \tag{15.21}$$

where for compactness of the notations, symbols $\mathbf{W} = \left(\mathbf{w}^{(1)},\ \mathbf{w}^{(2)},\ \ldots,\ \mathbf{w}^{(Q)}\right)$ and $\mathbf{B} = \left(\mathbf{b}^{(1)},\ \mathbf{b}^{(2)},\ \ldots,\ \mathbf{b}^{(Q)}\right)$ denote the groups of weights and biases, respectively.

The artificial neural network with an increased number of hidden layers between the input and the output layers is usually called a *deep neural network (DNN).*

According to Eq. (15.21), the transition from the $(q-1)$th hidden neuron layer to the qth hidden layer can be described as follows:

$$\mathbf{h}^{(q)} = \boldsymbol{\sigma}\left(\mathbf{z}^{(q)}\right),\ \mathbf{z}^{(q)} = \mathbf{w}^{(q)}\mathbf{h}^{(q-1)} + \mathbf{b}^{(q)}. \tag{15.22}$$

Note that in the last equation, $\mathbf{b}^{(q)}$, $\mathbf{z}^{(q)}$ and $\mathbf{h}^{(q)}$ are vector-columns formed by biases, inputs, and outputs of the hidden layer q,

$$\mathbf{b}^{(q)} = \left[b_1^{(q)}, b_2^{(q)}, \ \ldots., b_{N_q}^{(q)}\right]^T,$$

$$\mathbf{z}^{(q)} = \left[z_1^{(q)}, z_2^{(q)}, \ \ldots., z_{N_q}^{(q)}\right]^T,$$

$$\mathbf{h}^{(q)} = \left[h_1^{(q)}, h_2^{(q)}, \ \ldots., h_{N_q}^{(q)}\right]^T, \tag{15.23}$$

where N_q is the number of neurons in layer q.

Matrix $\mathbf{w}^{(q)}$ is $\left[N_q \times N_{q-1}\right]$ matrix formed by the corresponding weights:

$$\begin{bmatrix} w_{11}^{(q)} & w_{12}^{(q)} & w_{1N_{q-1}}^{(q)} \\ w_{21}^{(q)} & w_{22}^{(q)} & w_{2N_{q-1}}^{(q)} \\ & & \\ w_{N_q1}^{(q)} & w_{N_q2}^{(q)} & w_{N_qN_{q-1}}^{(q)} \end{bmatrix}. \tag{15.24}$$

Using notations of equations (15.21), we can rewrite the key equation of the machine learning algorithm for inverse problem solution based on inequality (15.7) in the following form:

$$\|\mathbf{M}\left(\mathbf{x}; \mathbf{W}, \ \mathbf{B}\right) - R(\mathbf{d}, \alpha)\| = \min. \tag{15.25}$$

Therefore, the machine learning process is reduced to solving the minimization problem (15.25) for weights, $\mathbf{W} = \left(\mathbf{w}^{(1)}, \ \mathbf{w}^{(2)}, \ldots, \mathbf{w}^{(Q)}\right)$ and biases, $\mathbf{B} = \left(\mathbf{b}^{(1)}, \ \mathbf{b}^{(2)}, \ldots, \mathbf{b}^{(Q)}\right)$. This is a very challenging problem considering the huge number of unknown weights and biases. In order to develop an effective method of solving this problem, we first reformulate the recursive equations (15.21) describing the action of the neural network operator in a more convenient form of matrix-vector multiplications, which simplifies the mathematical analysis.

To this end, we introduce a new column vector $\mathbf{W}^{(q)}$ obtained by concatenating the rows of matrix $\mathbf{w}^{(q)}$, Eq. (15.24), as follows:

$$\mathbf{W}^{(q)} = \left[w_{11}^{(q)} \ldots w_{1N_{q-1}}^{(q)} \, w_{21}^{(q)} \ldots w_{2N_{q-1}}^{(q)} \, w_{N_q1}^{(q)} \ldots w_{N_qN_{q-1}}^{(q)}\right]^T. \tag{15.26}$$

We also introduce $\left[N_q \times \left(N_{q-1}N_q\right)\right]$ matrix $\mathbf{H}^{(q-1)}$ composed of vector-columns (15.23) of the hidden layers outputs, $\mathbf{h}^{(q-1)T}$,

$$\mathbf{H}^{(q-1)} = \begin{bmatrix} \mathbf{h}^{(q-1)T} & 0 & \ldots & 0 \\ 0 & \mathbf{h}^{(q-1)T} & \ldots & 0 \\ \vdots & \vdots & \ddots & \vdots \\ 0 & 0 & \ldots & \mathbf{h}^{(q-1)T} \end{bmatrix}, \tag{15.27}$$

where according to definition (15.23), $\mathbf{h}^{(q-1)T}$ is a vector-row of length N_{q-1}.

Then, it is easy to show the following matrix identity:

$$\mathbf{w}^{(q)}\mathbf{h}^{(q-1)} = \mathbf{H}^{(q-1)}\mathbf{W}^{(q)}. \qquad (15.28)$$

Indeed, according to Eq. (15.24), $\mathbf{w}^{(q)}$ is a rectangular $[N_q \times N_{q-1}]$ matrix, and vector-column $\mathbf{h}^{(q-1)}$ has the length of N_{q-1}. Therefore, their product is a vector-column of the length N_q with the ith component, $a_i^{(q)}$, equal to

$$a_i^{(q)} = \sum_{j=1}^{N_q} w_{ij}^{(q)} h_j^{(q-1)}, \quad i = 1, 2, ..., N_q. \qquad (15.29)$$

At the same time, according to Eq. (15.27), matrix $\mathbf{H}^{(q-1)}$ is a rectangular $[N_q \times (N_{q-1}N_q)]$ matrix, and vector-column $\mathbf{W}^{(q)}$ has the length of $(N_{q-1}N_q)$. The product of matrix $\mathbf{H}^{(q-1)}$ and vector-column $\mathbf{W}^{(q)}$ results in vector-column of the length N_q with the same ith component as one defined by expression (15.29) above:

$$\left[h_1^{(q-1)}, h_2^{(q-1)}, \ ...., h_{N_q}^{(q-1)}\right] \begin{bmatrix} w_{i1}^{(q)} \\ w_{i2}^{(q)} \\ \vdots \\ w_{iN_q}^{(q)} \end{bmatrix} =$$

$$= \sum_{j=1}^{N_q} w_{ij}^{(q)} h_j^{(q-1)} = a_j^{(q)}, \quad i = 1, 2, ..., N_q. \qquad (15.30)$$

This proves the vector identity (15.28).

Therefore, expression (15.22) for a transition from the $(q-1)$th hidden neuron layer to the qth hidden layer takes the following form:

$$\mathbf{h}^{(q)} = \sigma\left(\mathbf{z}^{(q)}\right), \quad \mathbf{z}^{(q)} = \mathbf{H}^{(q-1)}\mathbf{W}^{(q)} + \mathbf{b}^{(q)}, \qquad (15.31)$$

Formula (15.31) plays an important role in the derivation of the backpropagation method of neural network training.

15.2.4 Universal Approximation Theorem

It is important that we consider the networks with feedforward architecture, where the data flow from input to output neurons is strictly feedforward. This means that the data processing can extend over multiple layers of neurons. Still, no closed cycles are present, that is, connections extending from outputs of units to inputs of units in the same layer or previous layers.

This is important due to the remarkable property of the feedforward architecture arising from the Universal Approximation Theorem for neural networks (Cybenko 1989; Hornik et al. 1989).

This theorem states, in general terms, that a superposition of sigmoidal functions can approximate any continuous operator $\mathbf{F}(\mathbf{x})$ with arbitrary accuracy, ε (provided a sufficiently large number of sigmoidal functions used for this approximation). In terms of the neural network approach, the Universal Approximation Theorem proves that the continuous function can be approximated to arbitrary accuracy by a network with a single hidden layer, if there is a sufficient number of neurons in the hidden layer. Therefore, considering that operator $\mathbf{M}\,(\mathbf{x};\,\mathbf{W},\ \mathbf{B})$ according to Eq. (15.20) is a linear combination of sigmoidal functions, we can state that there exist the weights, $\mathbf{W}$, and biases, $\mathbf{B}$, with the property that

$$\|\mathbf{M}\,(\mathbf{x};\,\mathbf{W},\ \mathbf{B}) - \mathbf{F}\,(\mathbf{x})\|_\infty < \varepsilon, \tag{15.32}$$

where in a general case, $\|...\|_\infty$ is a uniform norm (see Chap. 3).

The question arises of how to find the optimal weights and biases delivering condition (15.32).

15.3 Network Training as the Inverse Problem Solution

15.3.1 Formulation of the Regularized Inverse Problem for Training the Deep Neural Network

The main goal of machine learning algorithms is to find the optimal weights of a neural network so it can correctly predict the output values from the input variables. This process is called *neural network training*. In order to train the network, we should assume that we know a training set comprising of the pairs of vectors, $\mathbf{x}^{(k)}$ and $\mathbf{y}^{(k)}$, representing the input and output data, respectively:

$$\left\{\mathbf{x}^{(k)}\right\} \text{ and } \left\{\mathbf{y}^{(k)}\right\},\ k = 1, 2, \ldots\ldots, K, \tag{15.33}$$

where input vectors form a subset E_x of the Euclidean space E_{N_0}, $\mathbf{x} \in E_x$, and output vectors form a subset E_y of the Euclidean space E_{N_Q}, $\mathbf{y} \in E_y$.

For example, vectors $\mathbf{y}^{(k)}$ can comprise different physical properties (e.g., density, magnetization, conductivity, seismic attributes, etc.) at specific locations in the subsurface, and vectors $\mathbf{x}^{(k)}$ can include various physical fields (e.g., gravity, magnetic, electromagnetic, seismic) measured in the observation points. Training consists in calculating the optimal weights and biases, which ensures that the neural network predicts the correct values of output data, $\mathbf{y}^{(k)}$, for a given input data, $\mathbf{x}^{(k)}$ (the "training sample").

In mathematical dressing, this problem is equivalent to the solution of the. following operator equation with respect to the weights, $\mathbf{w}^{(1)}$, $\mathbf{w}^{(2)}$, ..., $\mathbf{w}^{(Q)}$, and biases, $\mathbf{b}^{(1)}$, $\mathbf{b}^{(2)}$, ..., $\mathbf{b}^{(Q)}$:

$$\mathbf{y}^{(k)} = \mathbf{M}\left(\mathbf{x}^{(k)};\ \mathbf{W},\ \mathbf{B}\right),\ k = 1, 2, \ldots.., K. \tag{15.34}$$

Thus, we have arrived at the classical inverse problem similar to the one introduced in Chap. 1 of this book. The solution to this inverse problem—*the training process*—can be achieved by applying any of the large variety of inversion methods discussed in this book. In addition, one can use the probabilistic or deterministic approaches to solving this problem. As an illustration of a typical training method, we consider below the solution based on the deterministic approach.

It is important to note that inverse problem (15.34) is, in a general case, an ill-posed problem. The regularized solution of this problem can be obtained by minimization of Tikhonov parametric functional, $P^{\alpha}(\mathbf{W},\ \mathbf{B})$, as follows:

$$P^{\alpha}(\mathbf{W},\ \mathbf{B}) = \varphi(\mathbf{W},\ \mathbf{B}) + \alpha S(\mathbf{W},\ \mathbf{B}) = \min, \tag{15.35}$$

where φ and S are misfit and stabilizing functionals, respectively, and α is a regularization parameter.

The misfit functional is defined as a sum of the squares of the corresponding least-square norms of the difference between the predicted output, $\mathbf{M}\left(\mathbf{x}^{(k)};\ \mathbf{W},\ \mathbf{B}\right)$, and the learning sample output, $\mathbf{y}^{(k)}$, calculated over all training set pairs

$$\varphi(\mathbf{W},\ \mathbf{B}) = \frac{1}{K}\sum_{k=1}^{K}\left\|\mathbf{M}\left(\mathbf{x}^{(k)};\ \mathbf{W},\ \mathbf{B}\right) - \mathbf{y}^{(k)}\right\|_{E_{N_Q}}^{2}, \tag{15.36}$$

where $\|\ldots\|_{E_{N_Q}}$ is the least-square norm defined in the Euclidean space E_{N_Q}.

A conventional way of selecting the stabilizing functional is in the form of the square norm of the difference between the current weights and biases, $\mathbf{w}^{(q)}$, $\mathbf{b}^{(q)}$, and a priori selected parameters, $\mathbf{w}_{apr}^{(q)}$, $\mathbf{b}_{apr}^{(q)}$:

$$S(\mathbf{W},\ \mathbf{B}) = \sum_{q=1}^{Q}\left[\|\mathbf{w}^{(q)} - \mathbf{w}_{apr}^{q}\|_F^2 + \|\mathbf{b}^{(q)} - \mathbf{b}_{apr}^{q}\|_{E_{N_Q}}^2\right], \tag{15.37}$$

where $\|\ldots\|_F$ is the Frobenius norm of the matrices. If no a priori (initial) weights are known, a standard minimum norm stabilizer can be used:

$$S_{MN}(\mathbf{W},\ \mathbf{B}) = \sum_{q=1}^{Q}\left[\|\mathbf{w}^{(q)}\|_F^2 + \|\mathbf{b}^{(q)}\|_{E_{N_Q}}^2\right]. \tag{15.38}$$

The problem of parametric functional minimization can be solved using the gradient-type methods described in Chap. 7. We will discuss the implementation of some of these techniques below.

We call the parameters, $\mathbf{W}, \mathbf{B}$, delivering the minimum of the parametric functional, $P^{\alpha}(\mathbf{W}, \mathbf{B})$, the trained parameters, $\mathbf{W}^{tr}, \mathbf{B}^{tr}$.

After the trained weights and biases are determined, one can use the trained ANN in order to predict the output (predicted) values, $\mathbf{y}^{pr(k)}$, for a new set of the input variables, $\mathbf{x}^{inp(k)}$, as follows:

$$\mathbf{y}^{pr(k)} = \mathbf{M}\left(\mathbf{x}^{inp(k)}; \mathbf{W}^{tr}, \mathbf{B}^{tr}\right), \; k = 1, 2, \ldots\ldots, K. \tag{15.39}$$

15.3.2 Network Training by the Steepest Descent Method

We have demonstrated in Chap. 7 that the steepest descent method represents the simplest but still powerful technique for functional minimization. The critical step of this method involves computing the steepest descent direction of the corresponding parametric functional.

Indeed, in the framework of the conventional steepest descent method, we can update the weights and biases iteratively as follows:

$$\mathbf{w}_{n+1}^{(q)} = \mathbf{w}_n^{(q)} + \delta\mathbf{w}^{(q)} = \mathbf{w}_n^{(q)} - k_{\mathbf{w}_n}^{(q)}\mathbf{l}^{\alpha}(\mathbf{w}_n^{(q)}),$$

$$\mathbf{b}_{n+1}^{(q)} = \mathbf{b}_n^{(q)} + \delta\mathbf{b}^{(q)} = \mathbf{b}_n^{(q)} - k_{\mathbf{b}_n}^{(q)}\mathbf{l}^{\alpha}(\mathbf{b}_n^{(q)}), \tag{15.40}$$

where $k_{\mathbf{w}_n}^{(q)} > 0$ and $k_{\mathbf{b}_n}^{(q)} > 0$ are positive coefficients, and $\mathbf{l}^{\alpha}(\mathbf{w}_n^{(q)})$ and $\mathbf{l}^{\alpha}(\mathbf{b}_n^{(q)})$ are the steepest ascent directions satisfying the following conditions:

$$\delta_{\mathbf{w}^{(q)}} P^{\alpha}(\mathbf{W}, \mathbf{B}) = -2k_{\mathbf{w}_n}^{(q)}(\mathbf{l}(\mathbf{w}_n^{(q)}), \mathbf{l}(\mathbf{w}_n^{(q)})) < 0,$$

$$\delta_{\mathbf{b}^{(q)}} P^{\alpha}(\mathbf{W}, \mathbf{B}) = -2k_{\mathbf{b}_n}^{(q)}(\mathbf{l}(\mathbf{b}_n^{(q)}), \mathbf{l}(\mathbf{b}_n^{(q)})) < 0, \tag{15.41}$$

and symbols $\delta_{\mathbf{w}^{(q)}}$ and $\delta_{\mathbf{b}^{(q)}}$ denote the variations of the parametric functional with respect to weights, $\mathbf{w}^{(q)}$, and biases, $\mathbf{b}^{(q)}$, respectively

Thus, $\mathbf{l}^{\alpha}(\mathbf{w}_n^{(q)})$ and $\mathbf{l}^{\alpha}(\mathbf{b}_n^{(q)})$ describe the "directions" of increasing (ascent) of the functional P^{α}, because they are opposite to the descent directions,

$$\delta\mathbf{w}^{(q)} = -k_{\mathbf{w}_n}^{(q)}\mathbf{l}^{\alpha}\left(\mathbf{w}_n^{(q)}\right); \; \delta\mathbf{b}^{(q)} = -k_{\mathbf{b}_n}^{(q)}\mathbf{l}^{\alpha}\left(\mathbf{b}_n^{(q)}\right). \tag{15.42}$$

The positive coefficients $k_{\mathbf{w}_n}^{(q)} > 0$ and $k_{\mathbf{b}_n}^{(q)} > 0$ are the steepest descent step lengths. In machine learning theory, these parameters are known as *the learning rates*. They are defined by a line search according to the following conditions:

$$P^{\alpha}(\mathbf{w}_{n+1}^{(q)}) = P^{\alpha}\left(\mathbf{w}_{n}^{(q)} - k_{\mathbf{w}_n}^{(q)}\mathbf{l}^{\alpha}(\mathbf{w}_{n}^{(q)})\right) = \Phi_{\mathbf{w}_n}^{\alpha}(k_{\mathbf{w}_n}^{(q)}) = \min,$$

$$P^{\alpha}(\mathbf{b}_{n+1}^{(q)}) = P^{\alpha}\left(\mathbf{b}_{n}^{(q)} - k_{\mathbf{b}_n}^{(q)}\mathbf{l}^{\alpha}(\mathbf{b}_{n}^{(q)})\right) = \Phi_{\mathbf{b}_n}^{\alpha}(k_{\mathbf{b}_n}^{(q)}) = \min . \quad (15.43)$$

Thus, the proposed algorithm to conduct the network training can be described by iterative process (15.40) if we know how to compute the steepest ascent directions (gradients), $\mathbf{l}^{\alpha}(\mathbf{w}_n^{(q)})$ and $\mathbf{l}^{\alpha}(\mathbf{b}_n^{(q)})$, of the parametric functional. This problem can be solved by using the backpropagation method which we discuss below.

15.3.3 Backpropagation Method for Steepest Descent Calculation

We have learned above that the action of the neural network operator can be described as a recursive process summarized in Eq. (15.21). This specific structure of the network operator makes it possible to develop a very efficient method of computing the steepest ascent directions (gradients) of the misfit and parametric functionals.

We introduce the basic principles of the *backpropagation method* considering the problem of misfit functional minimization first. After that, we will expand this method to a general case of parametric functional minimization.

According to Eq. (15.36), the misfit functional, $\varphi(\mathbf{W},\ \mathbf{B})$, can be treated as an average over the misfit functionals, $\varphi_k(\mathbf{W},\ \mathbf{B})$, calculated separately for each pair of the training input and output data $\{\mathbf{x}^{(k)};\ \mathbf{y}^{(k)}\}$:

$$\varphi(\mathbf{W},\ \mathbf{B}) = \frac{1}{K}\sum_{k=1}^{K}\varphi_k(\mathbf{W},\ \mathbf{B}), \quad (15.44)$$

where

$$\varphi_k(\mathbf{W},\ \mathbf{B}) = \left\|\mathbf{M}\left(\mathbf{x}^{(k)};\ \mathbf{W},\ \mathbf{B}\right) - \mathbf{y}^{(k)}\right\|_{E_{N_Q}}^2 . \quad (15.45)$$

Therefore, we can simplify the explanation of the backpropagation method by focusing on the minimization of the individual misfit, $\varphi_k(\mathbf{W},\ \mathbf{B})$, only. This is equivalent to the assumption that we have just one training set $\{\mathbf{x}^{(k)} = \mathbf{x}^{tr};\ \mathbf{y}^{(k)} = \mathbf{y}^{tr}\}$. After developing the minimization method for one training set, we can easily apply the result to the arbitrary number of the training sets by averaging over all available training samples. With this assumption in mind, we present expression (15.45) for the misfit functional under consideration as follows:

$$\varphi_k(\mathbf{W},\ \mathbf{B}) = \left\|\mathbf{M}\left(\mathbf{x}^{tr};\ \mathbf{W},\ \mathbf{B}\right) - \mathbf{y}^{tr}\right\|_{E_{N_Q}}^2 . \quad (15.46)$$

We can write the norm square of the difference between the predicted output values $\mathbf{M}\left(\mathbf{x}^{tr}; \mathbf{W}, \mathbf{B}\right)$ and the training sample of the output data, $\mathbf{y}^{tr}$, using the matrix multiplication:

$$\left\|\mathbf{M}\left(\mathbf{x}^{tr}; \mathbf{W}, \mathbf{B}\right) - \mathbf{y}^{tr}\right\|_{E_{N_Q}}^{2}$$

$$= \left(\mathbf{M}\left(\mathbf{x}^{tr}; \mathbf{W}, \mathbf{B}\right) - \mathbf{y}^{tr}\right)^{T} \left(\mathbf{M}\left(\mathbf{x}^{tr}; \mathbf{W}, \mathbf{B}\right) - \mathbf{y}^{tr}\right). \tag{15.47}$$

We can also recall that, according to Eq. (15.21) the predicted output values are generated by the output neuron layer Q :

$$\mathbf{M}\left(\mathbf{x}^{tr}; \mathbf{W}, \mathbf{B}\right) = \mathbf{h}^{(Q)} = \boldsymbol{\sigma}\left(\mathbf{z}^{(Q)}\right), \tag{15.48}$$

where $\boldsymbol{\sigma}\left(\mathbf{z}^{(Q)}\right)$ is the corresponding activation function.

Therefore, Eq. (15.47) for the misfit functional takes the following form:

$$\varphi_k\left(\mathbf{W}, \mathbf{B}\right) = \left(\boldsymbol{\sigma}\left(\mathbf{z}^{(Q)}\right) - \mathbf{y}^{tr}\right)^{T} \left(\boldsymbol{\sigma}\left(\mathbf{z}^{(Q)}\right) - \mathbf{y}^{tr}\right)$$

$$= \left(\mathbf{h}^{(Q)} - \mathbf{y}^{tr}\right)^{T} \left(\mathbf{h}^{(Q)} - \mathbf{y}^{tr}\right). \tag{15.49}$$

To simplify the future notations, we will drop the subscript "k" from the symbol of the misfit functional, denoting φ_k as φ. Following the standard logic of the steepest descent method, summarized in Chap. 7, the steepest ascent direction can be found by calculating the variations of the misfit functional and imposing the descent conditions:

$$\delta_{\mathbf{w}^{(q)}}\varphi\left(\mathbf{W}, \mathbf{B}\right) < 0; \quad q = 1, 2, ..., Q,$$

$$\delta_{\mathbf{b}^{(q)}}\varphi\left(\mathbf{W}, \mathbf{B}\right) < 0; \quad q = 1, 2, ..., Q. \tag{15.50}$$

In the last formulas, $\delta_{\mathbf{w}^{(q)}}\varphi$ represents the variations of the misfit corresponding to perturbations (changes) in the weights, $\mathbf{w}^{(q)}$. The term $\delta_{\mathbf{b}^{(q)}}\varphi$ describes the variations of the misfit corresponding to perturbations (changes) in the biases, $\mathbf{b}^{(q)}$.

The descent conditions (15.50) serve as the basis for finding the ascent directions (gradients) of the misfit functional with respect to the weights, $\mathbf{w}^{(q)}$, and biases, $\mathbf{b}^{(q)}$. However, as the first step in solving this problem, it is convenient to consider the variations of the misfit with respect to input in the qth neuron layer, $\mathbf{z}^{(q)}$.

To this end, let us recall some rules of differentiation of vector and scalar functions of vector argument. First of all, if we are given a vector function $\mathbf{g}\left(\mathbf{z}\right)$ of the vector argument $\mathbf{z}$ with $\mathbf{g} \in E_M$ and $\mathbf{z} \in E_N$, we can write it in the form of vector-column:

$$\mathbf{g}(\mathbf{z}) = \begin{bmatrix} g_1(\mathbf{z}) \\ g_2(\mathbf{z}) \\ \vdots \\ g_M(\mathbf{z}) \end{bmatrix} = \begin{bmatrix} g_1(z_1, z_2, ..z_N) \\ g_2(z_1, z_2, ..z_N) \\ \vdots \\ g_M(z_1, z_2, ..z_N) \end{bmatrix}.$$

The Fréchet derivative (Jacobian) matrix for this function is $[M \times N]$ matrix defined as follows:

$$\mathbf{F}_{\mathbf{z}}(\mathbf{g}) = \begin{bmatrix} \frac{\partial g_1}{\partial z_1} & \frac{\partial g_1}{\partial z_2} & \cdots & \frac{\partial g_1}{\partial z_N} \\ \frac{\partial g_2}{\partial z_1} & \frac{\partial g_2}{\partial z_2} & \cdots & \frac{\partial g_2}{\partial z_N} \\ \vdots & \vdots & \ddots & \vdots \\ \frac{\partial g_M}{\partial z_1} & \frac{\partial g_M}{\partial zx_2} & \cdots & \frac{\partial g_M}{\partial z_N} \end{bmatrix}. \tag{15.51}$$

Therefore, the first variation of $\mathbf{g}(\mathbf{z})$ can be calculated as follows:

$$\delta \mathbf{g}(\mathbf{z}) = \mathbf{F}_z(\mathbf{g})\, \delta \mathbf{z}, \tag{15.52}$$

where $\delta \mathbf{z}$ is $[N \times 1]$ matrix-column:

$$\delta \mathbf{z} = \begin{bmatrix} \delta z_1 \\ \delta z_2 \\ \vdots \\ \delta z_N \end{bmatrix}.$$

In particular, if we have a function $f(\mathbf{z})$, its Fréchet derivative is $[1 \times N]$ matrix-row:

$$\mathbf{F}_{\mathbf{z}}(f) = \begin{bmatrix} \frac{\partial f}{\partial z_1} & \frac{\partial f}{\partial z_2} & \cdots & \frac{\partial f}{\partial z_N} \end{bmatrix}. \tag{15.53}$$

We can use formula (15.53) to derive the variation of the misfit with respect to the input in the last Qth layer, $\mathbf{z}^{(Q)}$:

$$\delta_{\mathbf{z}^{(Q)}} \varphi = \mathbf{F}_{\mathbf{z}^{(Q)}}(\varphi)\; \delta \mathbf{z}^{(Q)}, \tag{15.54}$$

where symbol $\mathbf{F}_{\mathbf{z}^{(Q)}}(\varphi)$ denotes the Fréchet derivative (Jacobian) matrix of misfit functional φ with respect to $\mathbf{z}^{(Q)}$. According to (15.53), this is a vector-row $\left[1 \times N_Q\right]$ matrix defined by the following formula

$$\mathbf{F}_{\mathbf{z}^{(Q)}}(\varphi) = \begin{bmatrix} \frac{\partial \varphi}{\partial z_1^{(Q)}} & \frac{\partial \varphi}{\partial z_2^{(Q)}} & \cdots & \frac{\partial \varphi}{\partial z_{N_Q}^{(Q)}} \end{bmatrix}. \tag{15.55}$$

An efficient way to calculate $\mathbf{F}_{\mathbf{z}^{(Q)}}(\varphi)$ is based on the chain rule of differentiation, which we summarize below. Let us consider a composite vector function,

$$\Phi(\mathbf{z}) = \mathbf{f}[\mathbf{h}(\mathbf{z})], \tag{15.56}$$

where $\mathbf{z} \in E_{N_q}$. Vector function $\mathbf{h}(\mathbf{z}) \in E_{N_l}$ transforms E_{N_q} into E_{N_l}, vector function $\mathbf{f}[\mathbf{h}] \in E_{N_q}$ transforms E_{N_l} into E_{N_m}, and $\Phi(\mathbf{z})$ transforms E_{N_q} into E_{N_m}.

Then, the Fréchet derivative matrix (Jacobian) of Φ with respect to $\mathbf{z}$ is $\left[N_m \times N_q\right]$ matrix defined according to the following chain rule:

$$\mathbf{F_z}(\Phi) = \mathbf{F_h}(\mathbf{f})\,\mathbf{F_z}(\mathbf{h}), \tag{15.57}$$

where $\mathbf{F_h}(\mathbf{f})$ and $\mathbf{F_z}(\mathbf{h})$ are $[N_m \times N_l]$ and $\left[N_l \times N_q\right]$ Fréchet derivative matrices of functions $\mathbf{f}[\mathbf{h}]$ and $\mathbf{h}(\mathbf{z})$, respectively.

We apply now the above rules of Fréchet derivative calculations to find the variation of the misfit functional (15.49). According to Eq. (15.54), we need to find $\mathbf{F}_{\mathbf{z}^{(Q)}}(\varphi)$.

We can use the chain rule of differentiation and present Fréchet derivative $\mathbf{F}_{\mathbf{z}^{(Q)}}(\varphi)$ as the product of $\mathbf{F}_{\mathbf{h}^{(Q)}}(\varphi)$, Fréchet derivative of φ with regard $\mathbf{h}^{(Q)}$, and $\mathbf{F}_{\mathbf{z}^{(Q)}}\left(\mathbf{h}^{(Q)}\right)$, Fréchet derivative of $\mathbf{h}^{(Q)}$ with regard to $\mathbf{z}^{(Q)}$:

$$\mathbf{F}_{\mathbf{z}^{(Q)}}(\varphi) = \mathbf{F}_{\mathbf{h}^{(Q)}}(\varphi)\,\mathbf{F}_{\mathbf{z}^{(Q)}}\left(\mathbf{h}^{(Q)}\right). \tag{15.58}$$

In the last formula, $\mathbf{F}_{\mathbf{h}^{(Q)}}(\varphi)$ is $[1 \times N]$ matrix-row, defined by the following equation, similar to (15.55):

$$\mathbf{F}_{\mathbf{h}^{(Q)}}(\varphi) = \left[\frac{\partial\varphi}{\partial h_1^{(Q)}}\ \frac{\partial\varphi}{\partial h_2^{(Q)}}\ \cdots\ \frac{\partial\varphi}{\partial h_{N_Q}^{(Q)}}\right]. \tag{15.59}$$

Using compact matrix notations given by formula (15.49), $\mathbf{F}_{\mathbf{h}^{(Q)}}(\varphi)$ can be written as follows:

$$\mathbf{F}_{\mathbf{h}^{(Q)}}(\varphi) = \frac{\partial\varphi_k}{\partial\mathbf{h}^{(Q)}}$$

$$= \frac{\partial}{\partial\mathbf{h}^{(Q)}}\left[\left(\mathbf{h}^{(Q)} - \mathbf{y}^{tr}\right)^T \left(\mathbf{h}^{(Q)} - \mathbf{y}^{tr}\right)\right] = 2\left(\mathbf{h}^{(Q)} - \mathbf{y}^{tr}\right)^T. \tag{15.60}$$

We can also recall that, according to Eq. (15.21) the predicted output values are generated by the output neuron layer Q:

$$\mathbf{h}^{(Q)} = \sigma\left(\mathbf{z}^{(Q)}\right).$$

Therefore

$$\mathbf{F}_{\mathbf{z}^{(Q)}}\left(\mathbf{h}^{(Q)}\right) = \sigma'\left(\mathbf{z}^{(Q)}\right), \tag{15.61}$$

where $\sigma'\left(\mathbf{z}^{(Q)}\right)$ is the matrix of the Fréchet derivative of the activation function $\sigma\left(\mathbf{z}^{(Q)}\right)$.

It can be demonstrated that the Fréchet derivative of the vector function can be calculated based on the following equations:

$$\delta\boldsymbol{\sigma}\left(\mathbf{z}^{(Q)}\right)=\delta\begin{bmatrix}\sigma\left(z_1^{(Q)}\right)\\ \sigma\left(z_2^{(Q)}\right)\\ \sigma\left(z_3^{(Q)}\right)\\ \vdots\\ \sigma\left(z_{N_q}^{(Q)}\right)\end{bmatrix}=\begin{bmatrix}\delta\sigma\left(z_1^{(Q)}\right)\\ \delta\sigma\left(z_2^{(Q)}\right)\\ \delta\sigma\left(z_3^{(Q)}\right)\\ \vdots\\ \delta\sigma\left(z_{N_q}^{(Q)}\right)\end{bmatrix}=\begin{bmatrix}\sigma'\left(z_1^{(Q)}\right)\delta z_1^{(Q)}\\ \sigma'\left(z_2^{(Q)}\right)\delta z_2^{(Q)}\\ \sigma'\left(z_3^{(Q)}\right)\delta z_3^{(Q)}\\ \vdots\\ \sigma'\left(z_{N_q}^{(Q)}\right)\delta z_{N_q}^{(Q)}\end{bmatrix}$$

$$=\begin{bmatrix}\sigma'\left(z_1^{(Q)}\right) & 0 & 0\ \ldots & 0\\ 0 & \sigma'\left(z_2^{(Q)}\right) & 0\ \ldots & 0\\ \vdots & \vdots & \vdots\ \ddots & \vdots\\ 0 & 0 & 0\ \ldots & \sigma'\left(z_{N_q}^{(Q)}\right)\end{bmatrix}\begin{bmatrix}\delta z_1^{(Q)}\\ \delta z_2^{(Q)}\\ \vdots\\ \delta z_{N_q}^{(Q)}\end{bmatrix}$$

$$=\boldsymbol{\sigma}'\left(\mathbf{z}^{(Q)}\right)\delta\mathbf{z}^{(Q)}=\mathbf{F}_{\mathbf{z}^{(Q)}}\left(\mathbf{h}^{(Q)}\right)\delta\mathbf{z}^{(Q)}, \tag{15.62}$$

where $\boldsymbol{\sigma}'\left(\mathbf{z}^{(Q)}\right)$ is $\left[N_Q \times N_Q\right]$ diagonal matrix formed by the derivatives of function $\sigma(z)$:

$$\boldsymbol{\sigma}'\left(\mathbf{z}^{(Q)}\right)=\mathbf{diag}\left[\sigma'\left(z_i^{(Q)}\right)\right]. \tag{15.63}$$

Note that, for a sigmoidal function, its derivative is equal to:

$$\sigma'(z)=b\left(1-\sigma(z)\right)\sigma(z). \tag{15.64}$$

Thus, from Eq. (15.63) we have

$$\boldsymbol{\sigma}'\left(\mathbf{z}^{(Q)}\right)=b\mathbf{diag}\left[\left(1-\sigma\left(z_i^{(Q)}\right)\right)\sigma\left(z_i^{(Q)}\right)\right]. \tag{15.65}$$

Substituting Eqs. (15.60) and (15.61) back into formula (15.58), we arrive at the following equation:

$$\mathbf{F}_{\mathbf{z}^{(Q)}}(\varphi)=\mathbf{F}_{\mathbf{h}^{(Q)}}(\varphi)\,\mathbf{F}_{\mathbf{z}^{(Q)}}\left(\mathbf{h}^{(Q)}\right)=2\left(\mathbf{h}^{(Q)}-\mathbf{y}^{tr}\right)\boldsymbol{\sigma}'\left(\mathbf{z}^{(Q)}\right), \tag{15.66}$$

where $\mathbf{F}_{\mathbf{z}^{(Q)}}(\varphi)$ is $[1 \times N]$ matrix-row of the Fréchet derivative of functional φ with respect to the input of the Qth neuron layer, $\mathbf{z}^{(Q)}$.

However, we need to find the Fréchet derivatives $\mathbf{F}_{\mathbf{z}^{(q)}}(\varphi)$ with respect to the inputs to every hidden layer q. In order to solve this problem, we again use the chain rule according to formula (15.58):

$$\mathbf{F}_{\mathbf{z}^{(q)}}(\varphi) = \mathbf{F}_{\mathbf{z}^{(q+1)}}(\varphi)\ \mathbf{F}_{\mathbf{z}^{(q)}}\left(\mathbf{z}^{(q+1)}\right), \tag{15.67}$$

where $\mathbf{F}_{\mathbf{z}^{(q+1)}}(\varphi)$ is $\left[1 \times N_{q+1}\right]$ matrix of derivatives of the misfit functional with respect to $\mathbf{z}^{(q+1)}$, and $\mathbf{F}_{\mathbf{z}^{(q)}}\left(\mathbf{z}^{(q+1)}\right)$ denotes $\left[N_{q+1} \times N_q\right]$ matrix of Fréchet derivative of $\mathbf{z}^{(q+1)}$ with respect to $\mathbf{z}^{(q)}$.

At the same time, according to (15.22)

$$\mathbf{z}^{(q+1)} = \mathbf{w}^{(q+1)}\boldsymbol{\sigma}\left(\mathbf{z}^{(q)}\right) + \mathbf{b}^{(q+1)}. \tag{15.68}$$

Therefore, using (15.61), we can write:

$$\mathbf{F}_{\mathbf{z}^{(q)}}\left(\mathbf{z}^{(q+1)}\right) = \mathbf{w}^{(q+1)}\mathbf{F}_{\mathbf{z}^{(q)}}\left[\boldsymbol{\sigma}\left(\mathbf{z}^{(q)}\right)\right] = \mathbf{w}^{(q+1)}\boldsymbol{\sigma}'\left(\mathbf{z}^{(q)}\right), \tag{15.69}$$

where $\boldsymbol{\sigma}'\left(\mathbf{z}^{(q)}\right)$ is the diagonal matrix of the Fréchet derivatives of the activation function $\boldsymbol{\sigma}\left(\mathbf{z}^{(Q)}\right)$, defined in Eq. (15.63) above.

Substituting (15.69) into (15.67) we have:

$$\mathbf{F}_{\mathbf{z}^{(q)}}(\varphi) = \mathbf{F}_{\mathbf{z}^{(q+1)}}(\varphi)\,\mathbf{w}^{(q+1)}\boldsymbol{\sigma}'\left(\mathbf{z}^{(q)}\right). \tag{15.70}$$

Formula (15.70) shows that Fréchet derivative matrix of the misfit functional with respect to the input variables at the qth neuron layer, an $\left[1 \times N_q\right]$ matrix-row $\mathbf{F}_{\mathbf{z}^{(q)}}(\varphi)$, is related to similar matrix at the $(q+1)$th neuron layer, an $\left[1 \times N_{q+1}\right]$ vector-row $\mathbf{F}_{\mathbf{z}^{(q+1)}}(\varphi)$, by simple multiplication with the corresponding $\left[N_{q+1} \times N_q\right]$ matrix of weights $\mathbf{w}^{(q+1)}$, and diagonal $\left[N_q \times N_q\right]$ matrix of the derivative of the activation function, $\boldsymbol{\sigma}'\left(\mathbf{z}^{(q)}\right)$. One can quickly check that the dimensions of all matrices and their product in formula (15.70) are consistent:

$$\left[1 \times N_q\right] = \left[1 \times N_{q+1}\right] \cdot \left[N_{q+1} \times N_q\right] \cdot \left[N_q \times N_q\right].$$

Note that we have already found the expression of the Fréchet derivative at the output neuron layer:

$$\mathbf{F}_{\mathbf{z}^{(Q)}}(\varphi) = 2\left(\mathbf{h}^{(Q)} - \mathbf{y}^{tr}\right)\boldsymbol{\sigma}'\left(\mathbf{z}^{(Q)}\right). \tag{15.71}$$

Thus, Eqs. (15.70) and (15.71) provide a recursive method to find $\mathbf{F}_{\mathbf{z}^{(q)}}$ for any q by propagating the solutions backward from the last neuron layer of the network: $q = Q,\ Q-1, Q-2,\ \ldots., 2,\ 1.$

Descent conditions (15.50), however, require that the variations of the misfit functional, φ, with respect to weights, $\mathbf{w}^{(q)}$, and biases, $\mathbf{b}^{(q)}$, must be negative. We will demonstrate, using the chain rule, that these variations can be calculated based on the known Fréchet derivatives with respect to inputs, $\mathbf{z}^{(q)}$.

Indeed, we can write the expressions for the variations of the misfit functional with respect to the weights and biases as a product of $\left[1 \times N_q\right]$ vector-row $\mathbf{F}_{\mathbf{z}^{(q)}}(\varphi)$ and vector–columns $\delta_{\mathbf{w}^{(q)}}\mathbf{z}^{(q)}$ and $\delta_{\mathbf{b}^{(q)}}\mathbf{z}^{(q)}$ of the variations of the inputs:

$$\delta_{\mathbf{w}^{(q)}}\varphi = \mathbf{F}_{\mathbf{z}^{(q)}}(\varphi)\ \delta_{\mathbf{w}^{(q)}}\mathbf{z}^{(q)},$$

$$\delta_{\mathbf{b}^{(q)}}\varphi = \mathbf{F}_{\mathbf{z}^{(q)}}(\varphi)\ \delta_{\mathbf{b}^{(q)}}\mathbf{z}^{(q)}, \tag{15.72}$$

where, according to (15.22):

$$\mathbf{z}^{(q)} = \mathbf{w}^{(q)}\mathbf{h}^{(q-1)} + \mathbf{b}^{(q)}. \tag{15.73}$$

Applying the variation operators $\delta_{\mathbf{w}^{(q)}}$ and $\delta_{\mathbf{b}^{(q)}}$ to Eq. (15.73), we obtain at once that

$$\delta_{\mathbf{w}^{(q)}}\mathbf{z}^{(q)} = \delta\mathbf{w}^{(q)}\mathbf{h}^{(q-1)},$$

$$\delta_{\mathbf{b}^{(q)}}\mathbf{z}^{(q)} = \delta\mathbf{b}^{(q)}. \tag{15.74}$$

Substituting Eq. (15.74) into (15.72), we have:

$$\delta_{\mathbf{w}^{(q)}}\varphi = \mathbf{F}_{\mathbf{z}^{(q)}}(\varphi)\,\delta\mathbf{w}^{(q)}\mathbf{h}^{(q-1)}, \tag{15.75}$$

$$\delta_{\mathbf{b}^{(q)}}\varphi = \mathbf{F}_{\mathbf{z}^{(q)}}(\varphi)\,\delta\mathbf{b}^{(q)}. \tag{15.76}$$

In the last formula, $\delta\mathbf{w}^{(q)}$ is $\left[N_q \times N_{q-1}\right]$ matrix formed by the variations of the corresponding weights:

$$\delta\mathbf{w}^{(q)} = \begin{bmatrix} \delta w_{11}^{(q)} & \delta w_{12}^{(q)} & \ldots & \delta w_{1N_{q-1}}^{(q)} \\ \delta w_{21}^{(q)} & \delta w_{22}^{(q)} & \ldots & \delta w_{2N_{q-1}}^{(q)} \\ \vdots & \vdots & \ddots & \vdots \\ \delta w_{N_q 1}^{(q)} & \delta w_{N_q 2}^{(q)} & \ldots & \delta w_{N_q N_{q-1}}^{(q)} \end{bmatrix}. \tag{15.77}$$

Vector column $\mathbf{h}^{(q-1)}$ has a dimension of $\left[N_{q-1} \times 1\right]$. Therefore, the misfit functional variation, $\delta_{\mathbf{w}^{(q)}}\varphi$, is a scalar (matrix with one cell):

$$[1 \times 1] = \left[1 \times N_q\right] \cdot \left[N_q \times N_{q-1}\right] \cdot \left[N_{q-1} \times 1\right].$$

Our goal is to find the variations of the weights, $\delta\mathbf{w}^{(q)}$, which would guarantee the descent conditions (15.50).

$$\delta_{\mathbf{w}^{(q)}}\varphi\left(\mathbf{w}^{(1)}, .., \mathbf{w}^{(Q)}\right) < 0;\ \ q = 1, 2, ..., Q. \tag{15.78}$$

With this goal in mind, we can rewrite formula (15.76) using the identity (15.28):

$$\delta_{\mathbf{w}^{(q)}}\varphi = \mathbf{F}_{\mathbf{z}^{(q)}}(\varphi)\,\delta\mathbf{w}^{(q)}\mathbf{h}^{(q-1)} = \mathbf{F}_{\mathbf{z}^{(q)}}(\varphi)\,\mathbf{H}^{(q-1)}\delta\mathbf{W}^{(q)}, \tag{15.79}$$

where matrix $\mathbf{H}^{(q-1)}$ is a rectangular $\left[N_q \times \left(N_{q-1}N_q\right)\right]$ matrix,

$$\mathbf{H}^{(q-1)} = \begin{bmatrix} \mathbf{h}^{(q-1)T} & 0 & \dots & 0 \\ 0 & \mathbf{h}^{(q-1)T} & \dots & 0 \\ \vdots & \vdots & \ddots & \vdots \\ 0 & 0 & \dots & \mathbf{h}^{(q-1)T} \end{bmatrix}, \tag{15.80}$$

and vector-column $\delta\mathbf{W}^{(q)}$ has the length of $\left(N_{q-1}N_q\right)$:

$$\mathbf{W}^{(q)} = \left[\, w_{11}^{(q)} \dots w_{1N_{q-1}}^{(q)}\; w_{21}^{(q)} \dots w_{2N_{q-1}}^{(q)}\; w_{N_q1}^{(q)} \dots w_{N_qN_{q-1}}^{(q)} \,\right]^T. \tag{15.81}$$

Following the standard logic of the steepest descent method (Chap. 7), in order to satisfy the descent condition (15.78), we select the perturbation of weights as follows:

$$\delta\mathbf{W}^{(q)} = -k_{\mathbf{W}^{(q)}}\mathbf{l}_{\mathbf{W}^{(q)}} = -k_{\mathbf{W}^{(q)}}\left(\mathbf{F}_{\mathbf{z}^{(q)}}(\varphi)\,\mathbf{H}^{(q-1)}\right)^T, \tag{15.82}$$

where $k_{\mathbf{W}^{(q)}}$ is some positive real number (length of a step) and $\mathbf{l}_{\mathbf{W}^{(q)}}$ is a $\left[N_q \times 1\right]$ vector-column defining the direction of the steepest ascent of the misfit functional:

$$\mathbf{l}_{\mathbf{W}^{(q)}} = \left(\mathbf{F}_{\mathbf{z}^{(q)}}(\varphi)\,\mathbf{H}^{(q-1)}\right)^T. \tag{15.83}$$

Using notations (15.83), Eq. (15.79) for the variation of the misfit functional can be written as follows:

$$\delta_{\mathbf{w}^{(q)}}\varphi = \mathbf{F}_{\mathbf{z}^{(q)}}(\varphi)\,\delta\mathbf{w}^{(q)}\mathbf{h}^{(q-1)} = \mathbf{l}_{\mathbf{W}^{(q)}}^T\delta\mathbf{W}^{(q)}, \tag{15.84}$$

Finally, substituting Eq. (15.82) for $\delta\mathbf{W}^{(q)}$ into (15.84), we have

$$\delta_{\mathbf{w}^{(q)}}\varphi = -k_{\mathbf{W}^{(q)}}\mathbf{l}_{\mathbf{W}^{(q)}}^T\mathbf{l}_{\mathbf{W}^{(q)}} = -k_{\mathbf{W}^{(q)}}\,\|\mathbf{l}_{\mathbf{W}^{(q)}}\|^2 < 0. \tag{15.85}$$

We can see that $\mathbf{l}_{\mathbf{W}^{(q)}}$ describes the "direction" of increasing (ascent) of the functional φ, because it is opposite to the descent direction, $\delta\mathbf{W}^{(q)}$.

We can also find the directions of steepest ascent with respect to the biases. Indeed, the first variation of the misfit functional with respect to biases is equal:

$$\delta_{\mathbf{b}^{(q)}}\varphi = \mathbf{F}_{\mathbf{z}^{(q)}}(\varphi)\,\delta\mathbf{b}^{(q)}. \tag{15.86}$$

According to descent condition (15.50) for biases, we select the perturbation of $\mathbf{b}^{(q)}$ as follows:

$$\delta\mathbf{b}^{(q)} = -k_{\mathbf{b}^{(q)}}\mathbf{l}_{\mathbf{b}^{(q)}} = -k_{\mathbf{b}^{(q)}}\left(\mathbf{F}_{\mathbf{z}^{(q)}}(\varphi)\right)^T, \tag{15.87}$$

where $k_{\mathbf{b}^{(q)}}$ is some positive real number (length of a step) and $\mathbf{F}_{\mathbf{z}^{(q)}}(\varphi)$ is $[1 \times N]$ matrix-row of the Fréchet derivative of functional φ with respect to the input of the qth hidden neuron layer, $\mathbf{z}^{(q)}$.

$\mathbf{l}_{\mathbf{b}^{(q)}}$ is a vector-column of the steepest ascent of the misfit functional, equal to the transposed Fréchet derivative matrix with respect to the inputs to the qth hidden layer of the network:

$$\mathbf{l}_{\mathbf{b}^{(q)}} = \left(\mathbf{F}_{\mathbf{z}^{(q)}}(\varphi)\right)^T. \tag{15.88}$$

Using notations (15.88), Eq. (15.86) for the variation of the misfit functional can be written as follows:

$$\delta_{\mathbf{b}^{(q)}}\varphi = \mathbf{F}_{\mathbf{z}^{(q)}}(\varphi)\,\delta\mathbf{b}^{(q)} = \mathbf{l}_{\mathbf{b}^{(q)}}^T\delta\mathbf{b}^{(q)}, \tag{15.89}$$

By substituting Eq. (15.87) into (15.89), we have

$$\delta_{\mathbf{b}^{(q)}}\varphi = -k_{\mathbf{b}^{(q)}}\mathbf{l}_{\mathbf{b}^{(q)}}^T\mathbf{l}_{\mathbf{b}^{(q)}} = -k_{\mathbf{W}^{(q)}}\left\|\mathbf{l}_{\mathbf{b}^{(q)}}\right\|^2 < 0. \tag{15.90}$$

Thus, $\mathbf{l}_{\mathbf{b}^{(q)}}$ describes the "direction" of increasing (ascent) of the functional φ, because it is opposite to the descent direction, $\delta\mathbf{b}^{(q)}$.

15.3.4 Backpropagation Algorithm

Let us summarize the main operations applied on every iteration step, n, of the steepest descent method based on the *backpropagation algorithm*.

1. **Feedforward propagation.** Apply the training set $\mathbf{x}^{tr}$ to the network and feed-forward through all neuron layers to find the activations of all hidden neuron layers using formulas (15.21):

$$\mathbf{h}_n^{(0)} = \mathbf{x}^{tr};$$

$$\mathbf{h}_n^{(1)} = \sigma\left(\mathbf{z}_n^{(1)}\right), \quad \mathbf{z}_n^{(1)} = \mathbf{w}_n^{(1)}\mathbf{h}_n^{(0)} + \mathbf{b}_n^{(1)},$$

$$\mathbf{h}_n^{(2)} = \sigma\left(\mathbf{z}_n^{(2)}\right), \quad \mathbf{z}_n^{(2)} = \mathbf{w}_n^{(2)}\mathbf{h}_n^{(1)} + \mathbf{b}_n^{(2)},$$

$$\vdots$$

$$\mathbf{h}_n^{(Q)} = \sigma(z_n^{(Q)}), \quad \mathbf{z}_n^{(Q)} = \mathbf{w}_n^{(Q)}\mathbf{h}_n^{(Q-1)} + \mathbf{b}_n^{(Q)},$$

$$\mathbf{M}\left(\mathbf{x}^{tr}; \mathbf{W}_n, \mathbf{B}_n\right) = \mathbf{h}_n^{(Q)}, \tag{15.91}$$

where index n denotes the parameters of the network (inputs, outputs, weights and biases) at the current iteration of the steepest descent method.

2. **Output residual**. Calculate the residuals between the output of the network and the training sample, $\mathbf{y}^{tr}$:

$$\mathbf{r}_n = \mathbf{M}\left(\mathbf{x}^{tr};\, \mathbf{W}_n,\ \mathbf{B}_n\right) - \mathbf{y}^{tr} = \mathbf{h}_n^{(Q)} - \mathbf{y}^{tr}. \tag{15.92}$$

3. **Output Fréchet derivative calculations**. Calculate the Fréchet derivative $\mathbf{F}_{\mathbf{z}_n^{(Q)}}(\varphi)$ of the misfit functional with respect to the input to the last hidden layer Q:

$$\mathbf{F}_{\mathbf{z}_n^{(Q)}}(\varphi) = 2\left(\mathbf{h}_n^{(Q)} - \mathbf{y}^{tr}\right)\sigma'\left(\mathbf{z}_n^{(Q)}\right) = 2\mathbf{r}_n\sigma'\left(\mathbf{z}_n^{(Q)}\right). \tag{15.93}$$

4. **Backpropagation of the Fréchet derivatives.** Calculate the Fréchet derivatives $\mathbf{F}_{\mathbf{z}_n^{(q)}}(\varphi)$ of the misfit functional with respect to the input to each hidden layer q by backpropagation formula:

$$\mathbf{F}_{\mathbf{z}_n^{(q)}}(\varphi) = \mathbf{F}_{\mathbf{z}_n^{(q+1)}}(\varphi)\,\mathbf{w}_n^{(q+1)}\sigma'\left(\mathbf{z}_n^{(q)}\right). \tag{15.94}$$

5. **Evaluation of the steepest descent directions (gradients)**. The steepest descent directions of the misfit functional are given by the following formulas:

$$\mathbf{l}_{\mathbf{W}_n^{(q)}} = \left(\mathbf{F}_{\mathbf{z}_n^{(q)}}(\varphi)\,\mathbf{H}_n^{(q-1)}\right)^T, \tag{15.95}$$

$$\mathbf{l}_{\mathbf{b}_n^{(q)}} = \left(\mathbf{F}_{\mathbf{z}_n^{(q)}}(\varphi)\right)^T. \tag{15.96}$$

6. **Updating the weights and biases using the steepest descent method**. The parameters of the neural network are updated according to the following formulas:

$$\mathbf{W}_{n+1}^{(q)} = \mathbf{W}_n^{(q)} + \delta\mathbf{W}_n^{(q)} = \mathbf{W}_n^{(q)} - k_{\mathbf{W}_n^{(q)}}\mathbf{l}_{\mathbf{W}_n^{(q)}},$$

$$\mathbf{b}_{n+1}^{(q)} = \mathbf{b}_n^{(q)} + \delta\mathbf{b}_n^{(q)} = \mathbf{b}_n^{(q)} - k_{\mathbf{b}_n^{(q)}}\mathbf{l}_{\mathbf{b}_n^{(q)}}, \tag{15.97}$$

where the iteration steps (learning rates), $k_{\mathbf{W}_n^{(q)}}$ and $k_{\mathbf{b}_n^{(q)}}$, are defined by a line search according to the conditions:

$$\varphi(\mathbf{w}_{n+1}^{(q)}) = \varphi\left(\mathbf{w}_n^{(q)} - k_{\mathbf{W}_n^{(q)}}\mathbf{l}^{\alpha}(\mathbf{w}_n^{(q)})\right) = \Phi_{\mathbf{w}_n}^{\alpha}(k_{\mathbf{W}_n^{(q)}}) = \min,$$

$$\varphi(\mathbf{b}_{n+1}^{(q)}) = \varphi\left(\mathbf{b}_n^{(q)} - k_{\mathbf{b}_n^{(q)}}\mathbf{l}^{\alpha}(\mathbf{b}_n^{(q)})\right) = \Phi_{\mathbf{b}_n}^{\alpha}(k_{\mathbf{b}_n^{(q)}}) = \min. \tag{15.98}$$

The iterative process described above is terminated at $n = N$ when the misfit functional reaches the given level of error ε_0 :

$$\varphi((\mathbf{x}^{tr}; \mathbf{W}_n, \mathbf{B}_n)) \leq \varepsilon_0. \tag{15.99}$$

15.3.5 Regularized Steepest Descent Method of Deep Neural Network Training

We can now return to the problem of regularized network training based on minimization of the parametric functional (15.35), which we copy here for convenience:

$$P^{\alpha}(\mathbf{W}, \mathbf{B}) = \varphi(\mathbf{W}, \mathbf{B}) + \alpha S_{MN}(\mathbf{W}, \mathbf{B}) = \min, \tag{15.100}$$

where

$$\varphi(\mathbf{W}, \mathbf{B}) = \left\| \mathbf{M}\left(\mathbf{x}^{tr}; \mathbf{W}, \mathbf{B}\right) - \mathbf{y}^{tr} \right\|^2_{E_{N_Q}}, \tag{15.101}$$

and S_{MN} is the minimum norm stabilizer,

$$S_{MN}(\mathbf{W}, \mathbf{B}) = \frac{1}{2} \sum_{q=1}^{Q} \left[\|\mathbf{w}^{(q)}\|^2_F + \|\mathbf{b}^{(q)}\|^2_{E_q} \right]. \tag{15.102}$$

Using matrix notations, we can present the parametric functional in the form similar to Eq. (15.49) for the misfit functional:

$$P^{\alpha}(\mathbf{W}, \mathbf{B}) = \left(\mathbf{h}^{(Q)} - \mathbf{y}^{tr}\right)^T \left(\mathbf{h}^{(Q)} - \mathbf{y}^{tr}\right) +$$

$$+ \alpha \frac{1}{2} \sum_{q=1}^{Q} \left[\mathbf{W}^{(q)T} \mathbf{W}^{(q)} + \mathbf{b}^{(q)T} \mathbf{b} \right] = \min. \tag{15.103}$$

Following a standard logic of the gradient-type optimization methods, let us calculate the first variation of the parametric functional (15.103):

$$\delta_{\mathbf{W}^{(q)}} P^{\alpha}(\mathbf{W}, \mathbf{B}) = \delta_{\mathbf{W}^{(q)}} \varphi(\mathbf{W}, \mathbf{B}) + \alpha \mathbf{W}^{(q)T} \delta \mathbf{W}^{(q)}, \tag{15.104}$$

$$\delta_{\mathbf{b}^{(q)}} P^{\alpha}(\mathbf{W}, \mathbf{B}) = \delta_{\mathbf{b}^{(q)}} \varphi(\mathbf{W}, \mathbf{B}) + \alpha \mathbf{b}^{(q)T} \delta \mathbf{b}^{(q)} \tag{15.105}$$

We have found already the expressions (15.84) and (15.89) for the variations of the misfit functional:

$$\delta_{\mathbf{W}^{(q)}}\varphi = \mathbf{F}_{\mathbf{z}^{(q)}}(\varphi)\,\mathbf{H}^{(q-1)}\delta\mathbf{W}^{(q)} = \mathbf{l}^{T}_{\mathbf{W}^{(q)}}\delta\mathbf{W}^{(q)} \tag{15.106}$$

$$\delta_{\mathbf{b}^{(q)}}\varphi = \mathbf{F}_{\mathbf{z}^{(q)}}(\varphi)\,\delta\mathbf{b}^{(q)} = \mathbf{l}^{T}_{\mathbf{b}^{(q)}}\delta\mathbf{b}^{(q)}, \tag{15.107}$$

where

$$\mathbf{l}_{\mathbf{W}^{(q)}} = \left(\mathbf{F}_{\mathbf{z}^{(q)}}(\varphi)\,\mathbf{H}^{(q-1)}\right)^{T};\quad \mathbf{l}_{\mathbf{b}^{(q)}} = \left(\mathbf{F}_{\mathbf{z}^{(q)}}(\varphi)\right)^{T} \tag{15.108}$$

Substituting these expressions into (15.104) and (15.105), we obtain:

$$\delta_{\mathbf{W}^{(q)}}P^{\alpha}(\mathbf{W},\ \mathbf{B}) = \mathbf{l}^{T}_{\mathbf{W}^{(q)}}\delta\mathbf{W}^{(q)} + \alpha\mathbf{W}^{(q)T}\delta\mathbf{W}^{(q)} =$$

$$= \left(\mathbf{l}_{\mathbf{W}^{(q)}} + \alpha\mathbf{W}^{(q)}\right)^{T}\delta\mathbf{W}^{(q)}, \tag{15.109}$$

and

$$\delta_{\mathbf{b}^{(q)}}P^{\alpha}(\mathbf{W},\ \mathbf{B}) = \mathbf{l}^{T}_{\mathbf{b}^{(q)}}\delta\mathbf{b}^{(q)} + \alpha\mathbf{b}^{(q)T}\delta\mathbf{b}^{(q)}$$

$$= \left(\mathbf{l}_{\mathbf{b}^{(q)}} + \alpha\mathbf{b}^{(q)}\right)^{T}\delta\mathbf{b}^{(q)} \tag{15.110}$$

From the last formulas we obtain at once the expressions for the regularized steepest ascent directions:

$$\mathbf{l}^{\alpha}_{\mathbf{W}^{(q)}} = \mathbf{l}_{\mathbf{W}^{(q)}} + \alpha\mathbf{W}^{(q)}, \tag{15.111}$$

and

$$\mathbf{l}^{\alpha}_{\mathbf{b}^{(q)}} = \mathbf{l}_{\mathbf{b}^{(q)}} + \alpha\mathbf{b}^{(q)}. \tag{15.112}$$

Substituting formulas (15.111) and (15.112) back into (15.109) and (15.110), we have:

$$\delta_{\mathbf{W}^{(q)}}P^{\alpha}(\mathbf{W}) = \mathbf{l}^{\alpha T}_{\mathbf{W}^{(q)}}\delta\mathbf{W}^{(q)},$$

$$\delta_{\mathbf{b}^{(q)}}P^{\alpha}(\mathbf{W}) = \mathbf{l}^{\alpha T}_{\mathbf{b}^{(q)}}\delta\mathbf{b}^{(q)}. \tag{15.113}$$

All the standard principles of the regularized inversion are applied in this case.

Indeed, following the general scheme of the steepest descent method, we can select

$$\delta\mathbf{W}^{(q)} = -k^{\alpha}_{\mathbf{W}}\mathbf{l}^{\alpha}_{\mathbf{W}^{(q)}},\quad \delta\mathbf{b}^{(q)} = -k^{\alpha}_{\mathbf{b}}\mathbf{l}^{\alpha}_{\mathbf{b}^{(q)}}, \tag{15.114}$$

where $k^{\alpha}_{\mathbf{W}}$ and $k^{\alpha}_{\mathbf{b}}$ are some positive real numbers (step lengths). Substituting (15.114) into (15.113), we can see that

$$\delta_{\mathbf{W}^{(q)}}P^{\alpha}(\mathbf{W}) = -k^{\alpha}_{\mathbf{W}}\mathbf{l}^{\alpha T}_{\mathbf{W}^{(q)}}\mathbf{l}^{\alpha}_{\mathbf{W}^{(q)}} = -k^{\alpha}_{\mathbf{W}}\left\|\mathbf{l}^{\alpha}_{\mathbf{W}^{(q)}}\right\|^{2} < 0,$$

$$\delta_{\mathbf{b}^{(q)}}P^{\alpha}(\mathbf{W}) = -k^{\alpha}_{\mathbf{b}}\mathbf{l}^{\alpha T}_{\mathbf{b}^{(q)}}\mathbf{l}^{\alpha}_{\mathbf{b}^{(q)}} = -k^{\alpha}_{\mathbf{b}}\left\|\mathbf{l}^{\alpha}_{\mathbf{b}^{(q)}}\right\|^{2} < 0. \tag{15.115}$$

The iterative process of the method is constructed according to formulas

$$\mathbf{W}_{n+1}^{(q)} = \mathbf{W}_n^{(q)} + \delta\mathbf{W}^{(q)} = \mathbf{W}_n^{(q)} - k_{\mathbf{W}}^{\alpha}\mathbf{l}_{\mathbf{W}^{(q)}}^{\alpha},$$

$$\mathbf{b}_{n+1}^{(q)} = \mathbf{b}_n^{(q)} + \delta\mathbf{b}^{(q)} = \mathbf{b}_n^{(q)} - k_{\mathbf{b}}^{\alpha}\mathbf{l}_{\mathbf{b}^{(q)}}^{\alpha}, \tag{15.116}$$

where the coefficients $k_{\mathbf{W}}^{\alpha}$ and $k_{\mathbf{b}}^{\alpha}$ are defined by the line search method.

The described method represents a conventional steepest descent algorithm of the parametric functional minimization. As was discussed in the previous chapters of the book, the more efficient Newton and/or conjugate-gradient methods can be used to solve the minimization problem.

Thus, training of the DNN, or finding the optimal weights of the DNN is reduced to the application of the standard methods of the regularized inversion.

15.4 Convolution Neural Network

In some applications, it is convenient to include a convolution operator in the construction of neural network; Schuster (2023). This concept was originally introduced to improve the performance of machine learning algorithms in solving computer vision problems. The idea is that the data representing the visual images are usually behave smoothly, so that there is some sort of similarity in the values of neighboring pixels of the images. The same can be applied to all types of the data used in geophysical, medical and other applications. Therefore, instead of operating with a single-point data, we can use as input signal in the neural layer of the ANN the convolution of the data within a specific digital window. For example we can write formula (15.8) as follows:

$$z_j^{(1)} = \sum_{k=1}^{M} w_{jk}^{(1)} * x_k + w_{j0}^{(1)}, \tag{15.117}$$

where symbol "$*$" denotes a convolution operator:

$$w_{jk}^{(1)} * x_k = \sum_{p=1}^{P-1} w_{jp}^{(1)} x_{k-p}, \tag{15.118}$$

and P is the size of digital window used in the convolution process.

Therefore, the output from the first hidden layer (formula (15.10)) will take the form:

$$h_j^{(1)} = \sigma\left(z_j^{(1)}\right) = \sigma\left(\sum_{k=1}^{M} w_{jk}^{(1)} * x_k + w_{j0}^{(1)}\right), \quad j = 1, 2, \ldots L. \tag{15.119}$$

In a general case of multilayer convolution neural network (CNN) the transition from the $(q-1)$th hidden neuron layer to the qth hidden layer can be described as follows:

$$\mathbf{h}^{(q)} = \sigma\left(\mathbf{z}^{(q)}\right), \ \mathbf{z}^{(q)} = \mathbf{w}^{(q)} * \mathbf{h}^{(q-1)} + \mathbf{b}^{(q)}. \tag{15.120}$$

Thus, all the expressions derived above for a conventional ANN can be easily applied to the CNN by using the convolution operation. Considering that this operation is a linear one, it is easy to show that the training of the CNN can be, in general, described by the same algorithms we discussed above for the conventional ANN.

References and Recommended Reading

Aggarwal C. C (2018) Neural networks and deep learning. Springer

Bishop CM (2006) Pattern recognition and machine learning. Springer

Cybenko (1989) Approximation by superpositions of sigmoidal function. Math Control, Signals Syst 2:304–314

Hornik K, Stinchcombe M, Halbert H (1989) Multilayer feedforward networks are universal approximators. In: Neural networks, vol 2. Pergamon Press, pp 359–366

McCulloch WS, Pitts W (1943) A logical calculus of the ideas immanent in nervous activity. Bull Math Biophys 5(4):115–133

Mehlig B (2021) Machine learning with neural networks: an introduction for scientists and engineers. Cambridge University Press

Minsky ML, Papert SA (1969, 1990) Perceptrons, Expanded edn. MIT Press

Mohri M, Rostamizadeh A, Talwalkar A (2012) Foundations of machine learning. MIT Press

Rosenblatt F (1957) The Perceptron—a perceiving and recognizing automaton. Report 85-460-1. Cornel Aeronautical Laboratory

Rosenblatt F (1962) Principles of neurodynamics. Spartan, Washington DC Books.

Schuster G (2023) Machine learning methods in geoscience. Society of Exploration Geophysicists

Chapter 16
Machine Learning Inversion of Multiphysics Data

Abstract In this Chapter, we discuss the application of machine learning methods to the solution of the inverse problem. Considering that machine learning algorithms do not require the knowledge of the function or operator they approximate, one can apply the neural network operator to determine the regularized solution of the inverse problem. This operator is called a regularizing neural network operator, and the corresponding network is the regularizing neural network (RNN). This chapter presents the methods of constructing the RNN algorithms using different types of stabilizing functionals introduced in the previous chapters of the book. We also introduce the concept of a knowledge-based neural network. It is based on the ability of the neural network to accurately approximate any mathematical or physical law represented by the forward modeling operator. Thus, a knowledge-based neural network provides a solution based on a priori knowledge about the laws governing the relationships between the model and data. The methods of joint inversion of multiphysics data using a regularizing neural network (RNN) are also considered.

Keywords Neural network operator · Regularizing neural network · Knowledge-based neural network

16.1 Approximation of the Regularizing Operator of the Inverse Problem by the Neural Network Operator

We consider now the classical inverse problem introduced in Chap. 1:

$$\mathbf{d} = A(\mathbf{m}), \tag{16.1}$$

where $\mathbf{m}$ represents a model characterizing the structure and properties of the target, and $\mathbf{d}$ denotes the observed data. We also assume, as usual, that linear or nonlinear forward modeling operator, A, is known and defined by physical and/or mathematical laws that relate the given model to the observed data.

M. S. Zhdanov, *Advanced Methods of Joint Inversion and Fusion of Multiphysics Data*, Advances in Geological Science,
https://doi.org/10.1007/978-981-99-6722-3_16

In a general case, inverse problem (16.1) is ill posed, which means that the solution may not exist or be nonunique and/or unstable. The regularized solution, $\mathbf{m}_\alpha$, of the ill-posed inverse problem (16.1) can be found by solving a family of well-posed problems:

$$\mathbf{m}_\alpha = A_\alpha^{-1}(\mathbf{d}) = R(\mathbf{d},\alpha) \rightarrow \mathbf{m}_t, \text{ if } \alpha \rightarrow 0, \tag{16.2}$$

where regularizing operator $R(\mathbf{d},\alpha)$is unknown.

The convenience of machine learning algorithms is that they do not require the knowledge of the function or operator they approximate. Instead, they are based on the known pairs of vectors, representing the data, $\mathbf{d}^{tr} = (d_1^{tr}, d_2^{tr}, ..., d_M^{tr})^T$, and models, $\mathbf{m}^{tr} = (m_1^{tr}, m_2^{tr}, ..., m_N^{tr})^T$, generating these data

According to the Universal Approximation Theorem, there always exists a neural network operator, $\mathbf{N}_\alpha$ ($\mathbf{d}$, $\mathbf{W}$, $\mathbf{B}$), which approximates the regularizing operator with the given accuracy ε:

$$\|\mathbf{N}_\alpha(\mathbf{d}, \mathbf{W}, \mathbf{B}) - R(\mathbf{d},\alpha)\|_\infty \leq \varepsilon. \tag{16.3}$$

We call operator $\mathbf{N}_\alpha$ ($\mathbf{d}$, $\mathbf{W}$, $\mathbf{B}$) a *regularizing neural network operator*, and the corresponding network *the regularizing neural network (RNN)*. The RNN operator provides a regularized solution of the original inverse problem (16.1):

$$\mathbf{m}_\alpha = \mathbf{N}_\alpha(\mathbf{d}, \mathbf{W}, \mathbf{B}). \tag{16.4}$$

The question is how to find the weights, $\mathbf{W}$, and biases, $\mathbf{B}$, of the RNN operator, $\mathbf{N}_\alpha$ ($\mathbf{d}$, $\mathbf{W}$, $\mathbf{B}$).

The general principles of solving this problem are the same as discussed above for the artificial neural network. However, the stabilizing term in Eq. (15.37) is different. Instead of imposing some conditions on the weights and biases, we use the stabilizing functional based on the desired property of the inverse problem solution, $\mathbf{m}$.

We write the corresponding parametric functional, P_{RNN}^α ($\mathbf{W}$, $\mathbf{B}$), in the following form:

$$P_{RNN}^\alpha(\mathbf{W}, \mathbf{B}) = \varphi(\mathbf{W}, \mathbf{B}) + \alpha S_{RNN}(\mathbf{W}, \mathbf{B}), \tag{16.5}$$

where φ ($\mathbf{W}$, $\mathbf{B}$) is the misfit functional defined as the square of the corresponding least-square norm of the difference between the predicted models,

$$\mathbf{m} = \mathbf{N}_\alpha\left(\mathbf{d}^{tr}, \mathbf{W}, \mathbf{B}\right), \tag{16.6}$$

and the training sample model, $\mathbf{m}^{tr}$:

$$\varphi(\mathbf{W}, \mathbf{B}) = \left\|\mathbf{m} - \mathbf{m}^{tr}\right\|_{L_2}^2 = \left\|\mathbf{N}_\alpha\left(\mathbf{d}^{tr}, \mathbf{W}, \mathbf{B}\right) - \mathbf{m}^{tr}\right\|_{L_2}^2, \tag{16.7}$$

where $\|...\|_{L_2}$ is the L_2 norm.

In order to define the stabilizing term $S_{RNN}(\mathbf{W}, \mathbf{B})$, we recall the family of stabilizing functionals for model parameters, introduced in Chap. 4. To simplify the notations, we reproduce here the expressions for these functionals assuming that the model parameters are described by some function, $m(\mathbf{r})$, defined over the target domain, V. This function can always be discretized or parameterized, when we use these stabilizers in the construction of the regularized neural network operator.

The following stabilizers can be used, for example, to impose an additional constraint on the behavior of the model parameters.

1. Stabilizer based on the minimum norm of the difference between the selected model and some a priori model $\mathbf{m}_{apr}$:

$$S_{MN}(\mathbf{m}) = \left\|\mathbf{m} - \mathbf{m}_{apr}\right\|_{L_2}^2 = \min. \tag{16.8}$$

The corresponding RNN stabilizer can be obtained by substituting the neural network operator $\mathbf{N}_\alpha(\mathbf{d}, \mathbf{W}, \mathbf{B})$ into (16.8):

$$S_{MN}(\mathbf{W}, \mathbf{B}) = S_{MN}(\mathbf{m})$$

$$= S_{MN}\left[\mathbf{N}_\alpha(\mathbf{d}, \mathbf{W}, \mathbf{B})\right] = \left\|\mathbf{N}_\alpha(\mathbf{d}, \mathbf{W}, \mathbf{B}) - \mathbf{m}_{apr}\right\|_{L_2}^2. \tag{16.9}$$

Note that, in a general case, we can use all available data, $\mathbf{d}$, in calculations of the stabilizing functional. In contrast, we use only the training set of data, $\mathbf{d}^{tr}$, for misfit functional calculations.

2. Maximum smoothness stabilizing functional:

$$S_{\max sm}(\mathbf{m}) = \|\nabla \mathbf{m}\|_{L_2}^2 = (\nabla \mathbf{m}, \nabla \mathbf{m})_{L_2} =$$

$$\int_V |\nabla m(\mathbf{r})|^2 \, dv = \min. \tag{16.10}$$

We construct the related RNN stabilizer by substituting the neural network operator $\mathbf{N}_\alpha(\mathbf{d}, \mathbf{W}, \mathbf{B})$ into (16.10):

$$S_{RNN,\max sm}(\mathbf{W}, \mathbf{B}) = S_{\max sm}(\mathbf{m})$$

$$= S_{\max sm}\left[\mathbf{N}_\alpha(\mathbf{d}, \mathbf{W}, \mathbf{B})\right] = \|\nabla \mathbf{N}_\alpha(\mathbf{d}, \mathbf{W}, \mathbf{B})\|_{L_2}^2. \tag{16.11}$$

3. By using the L_p norm stabilizer,

$$S_{L_p}(\mathbf{m}) = \|\mathbf{m}\|_{L_p}^p = \int_V |m(\mathbf{r})|^p \, dv, \quad 0 \leq p < \infty, \tag{16.12}$$

we arrive at L_p norm RNN stabilizer:

$$S_{RNN,L_p}(\mathbf{W}, \mathbf{B}) = S_{L_p}(\mathbf{m}) = S_{L_p}[\mathbf{N}_\alpha(\mathbf{d}, \mathbf{W}, \mathbf{B})] = \|\mathbf{N}_\alpha(\mathbf{d}, \mathbf{W}, \mathbf{B})\|_{L_p}^p. \quad (16.13)$$

4. The minimum support RNN stabilizing functional is determined as follows:

$$S_{RNN,MS}(\mathbf{W}, \mathbf{B}) = S_{MS}(\mathbf{m}) = S_{MS}[\mathbf{N}_\alpha(\mathbf{d}, \mathbf{W}, \mathbf{B})], \quad (16.14)$$

where

$$S_{MS}(\mathbf{m}) = \int_V \frac{(m - m_{apr})^2}{(m - m_{apr})^2 + \beta^2} dv. \quad (16.15)$$

5. The minimum gradient support RNN functional is given by the following expressions:

$$S_{RNN,MGS}(\mathbf{W}, \mathbf{B}) = S_{MGS}(\mathbf{m}) = S_{MGS}[\mathbf{N}_\alpha(\mathbf{d}, \mathbf{W}, \mathbf{B})], \quad (16.16)$$

where

$$S_{MGS}(\mathbf{m}) = \int_V \frac{\nabla m \cdot \nabla m}{\nabla m \cdot \nabla m + \beta^2} dv. \quad (16.17)$$

We can extend this list by including all other stabilizing functions discussed in the previous chapters.

Thus, the weights and biases, $\mathbf{W}^{tr}, \mathbf{B}^{tr}$, of the trained RNN operator $\mathbf{N}_\alpha(\mathbf{d}, \mathbf{W}^{tr}, \mathbf{B}^{tr})$ can be found by minimizing the parametric functional (16.5):

$$P_{RNN}^\alpha(\mathbf{W}, \mathbf{B}) = \left\|\mathbf{N}_\alpha(\mathbf{d}^{tr}, \mathbf{W}, \mathbf{B}) - \mathbf{m}^{tr}\right\|^2 + \alpha S_*[\mathbf{N}_\alpha(\mathbf{d}, \mathbf{W}, \mathbf{B})] = \min, \quad (16.18)$$

where S_* stands for any functional from the family of stabilizing functionals considered in this book.

The minimization problem (16.18) can be solved by applying the same gradient-type methods and backpropagation algorithm we introduced before in Chap. 15 for the regularized neural network training. The only difference is that in Sect. 15.3.5 we applied the stabilizing functional to the weights and biases, while in solving inverse problem (16.1) we can apply the stabilizing functionals to the model parameters, produced by the neural network during the training process according to formula (16.6).

After training, the regularizing neural network operator $\mathbf{N}_\alpha(\mathbf{d}, \mathbf{W}^{tr}, \mathbf{B}^{tr})$ can be applied to the observed data, $\mathbf{d}$, to generate the regularized solution of the inverse problem (16.1):

$$\mathbf{m}_\alpha = \mathbf{N}_\alpha(\mathbf{d}, \mathbf{W}^{tr}, \mathbf{B}^{tr}). \quad (16.19)$$

16.2 Knowledge-Based Neural Network

In the previous section, we described the machine learning inversion method based on neural network approximation of the inverse operator $A^{-1}(\mathbf{d})$. This approach does not require the knowledge of the forward operator, $A(\mathbf{m})$. However, in the majority of inverse problems, this operator is known. Therefore, using this knowledge in constructing the corresponding machine learning algorithm for solving the inverse problem can be advantageous. In this section, we present an algorithm based on this approach. We call this approach a *knowledge-based neural network* by analogy with the *physics-informed neural networks* introduced in Raissiet al. (2019), and Karniadakis et al. (2021), see also Schuster (2023).

The cited papers introduced the concept of physics-informed neural networks, which are trained by taking into account the known laws of physics. This is achieved by employing deep neural networks capable of solving the partial differential equations describing the corresponding laws of physics.

I present here an approach to solving the same problem of incorporating the laws of physics and mathematics in the machine learning algorithm based on the ability of the neural network to accurately approximate any mathematical or physical law, represented by the forward modeling operator, A. It is important to note that, in the knowledge-based neural network framework, we explicitly use the corresponding mathematical or physical law, represented by operator A, in the machine learning algorithm.

To illustrate this approach, let us consider the classical inverse problem again,

$$\mathbf{d} = A(\mathbf{m}), \tag{16.20}$$

where A is the forward modeling operator, $\mathbf{d}$ are the observed data, and $\mathbf{m}$ is the distribution of the model parameters on the plane or in the 3D volume, V, described by the following continuous function:

$$\mathbf{m} = \mathbf{m}(\mathbf{r}), \mathbf{r} \in V, \tag{16.21}$$

where $\mathbf{r}$ is a radius-vector of the observation point in some Carthesian coordinate system.

According to the Universal Approximation Theorem discussed above, there exists a feedforward neural network that approximates $\mathbf{m}(\mathbf{r})$ with a given accuracy ε:

$$\|\mathbf{N}_m(\mathbf{r}, \mathbf{W}, \mathbf{B}) - \mathbf{m}(\mathbf{r})\|_\infty < \varepsilon, \tag{16.22}$$

where $\mathbf{N}_m(\mathbf{r}, \mathbf{W}, \mathbf{B})$ is the corresponding neural network operator.

In other words, for the purpose of solving inverse problem (16.20), we can represent the model parameters distribution as follows:

$$\mathbf{m}(\mathbf{r}) = \mathbf{N}_m(\mathbf{r}, \mathbf{W}, \mathbf{B}). \tag{16.23}$$

Following the classical principles of the regularization theory, we reduce the solution of the inverse problem to minimization of the corresponding parametric functional:

$$P^{\alpha}(\mathbf{m}) = \varphi(\mathbf{m}) + \alpha S(\mathbf{m}) = \min, \tag{16.24}$$

where $\varphi(\mathbf{m})$ is a conventional misfit functional between the observed and predicted data,

$$\varphi(\mathbf{m}) = \|A(\mathbf{m}) - \mathbf{d}\|_{L_2}^2 = \|A(\mathbf{N}_m(\mathbf{r}, \mathbf{W}, \mathbf{B})) - \mathbf{d}\|_{L_2}^2, \tag{16.25}$$

and $S(\mathbf{m})$ is a properly selected stabilizing functional.

We can substitute now neural network representation (16.23) into (16.24):

$$P^{\alpha}(\mathbf{N}_m(\mathbf{r}, \mathbf{W}, \mathbf{B})) = \varphi(\mathbf{N}_m(\mathbf{r}, \mathbf{W}, \mathbf{B})) + \alpha S(\mathbf{N}_m(\mathbf{r}, \mathbf{W}, \mathbf{B})) = \min. \tag{16.26}$$

To simplify the further discussion, we introduce the following notations:

$$P_N^{\alpha}(\mathbf{W}, \mathbf{B}) = P^{\alpha}(\mathbf{N}_m(\mathbf{r}, \mathbf{W}, \mathbf{B})),$$

$$\varphi_N(\mathbf{W}, \mathbf{B}) = \varphi(\mathbf{N}_m(\mathbf{r}, \mathbf{W}, \mathbf{B})) = \|A(\mathbf{N}_m(\mathbf{r}, \mathbf{W}, \mathbf{B})) - \mathbf{d}\|_{L_2}^2,$$

$$S_N(\mathbf{W}, \mathbf{B}) = S(\mathbf{N}_m(\mathbf{r}, \mathbf{W}, \mathbf{B})). \tag{16.27}$$

Using these notations, expression (16.26) can be written in the following equivalent form:

$$P_N^{\alpha}(\mathbf{W}, \mathbf{B}) = \varphi_N(\mathbf{W}, \mathbf{B}) + \alpha S_N(\mathbf{W}, \mathbf{B}) = \min. \tag{16.28}$$

Thus, the inverse problem is reduced to finding the optimal weights and biases of the corresponding neural network. This can be done by applying the proper optimization algorithms, introduced in Chaps. 6 and 7, to the solution of Eq. (16.28).

After solving this problem, the corresponding model can be found from Eq. (16.23).

We can see from expressions (16.25) and (16.27) that calculation of the parametric functional $P_N^{\alpha}(\mathbf{W}, \mathbf{B})$ involves the application of the corresponding mathematical or physical laws, represented by operator A, to the model, generated by the neural network $\mathbf{m}(\mathbf{r}) = \mathbf{N}_m(\mathbf{r}, \mathbf{W}, \mathbf{B})$. In other words, in the process of finding the optimal weights and biases of the knowledge-based neural network, we explicitly use our knowledge about related laws.

The advantage of using the knowledge-based neural network over regularizing neural network (RNN) is that the former explicitly depends on the known forward modeling operator, A, while the latter finds the regularizing inverse operator, $R(\mathbf{d},\alpha)$, without any assumption about the mathematical or physical laws governing the solution of the forward problem. Simply stated, a knowledge-based neural network provides *a solution* based on a priori knowledge about the laws governing the relationships between the model and data, while the application of the RNN can be treated as *an intuition-based solution.*

16.3 Joint Inversion of Multiphysics Data Using Regularizing Neural Network (RNN)

We can extend the application of the RNN and machine learning algorithms to joint inversion of multiphysics data.

Let us consider the multiphysics inverse problem described by the following system of operator equations:

$$\mathbf{d}^{(i)} = A^{(i)}(\mathbf{m}^{(i)}), \quad i = 1, 2, ..., n, \tag{16.29}$$

where in a general case, $A^{(i)}$ are nonlinear operators, $\mathbf{d}^{(i)}$ $(i = 1, 2, ..., n)$ are different observed data sets, and $\mathbf{m}^{(i)}$ $(i = 1, 2, ..., n)$ are the unknown sets of model parameters. For simplicity, we assume that both the observed data and the model parameters are dimensionless, which could be easily done by applying the corresponding weights (see Chap. 8).

We consider n regularizing neural networks with the corresponding RNN operators, $\mathbf{N}^{(i)}\left(\mathbf{d}^{(i)}, \mathbf{W}^{(i)}, \mathbf{B}^{(i)}\right)$, which can be used to transform the observed data into the corresponding model parameters:

$$\mathbf{m}^{(i)} = \mathbf{N}^{(i)}\left(\mathbf{d}^{(i)}, \mathbf{W}^{(i)}, \mathbf{B}^{(i)}\right), \quad i = 1, 2, ..., n. \tag{16.30}$$

We also assume that we have n sets of pairs of vectors, representing the data, $\mathbf{d}^{tr(i)} = (d_1^{tr(i)}, d_2^{tr(i)}, ..., d_M^{tr(i)})^T$, and models, $\mathbf{m}^{tr(i)} = (m_1^{tr(i)}, m_2^{tr(i)}, ..., m_N^{tr(i)})^T$, generating these data, which can be used for training the RNN operators $\mathbf{N}^{(i)}\left(\mathbf{d}^{(i)}, \mathbf{W}^{(i)}, \mathbf{B}^{(i)}\right)$, $(i = 1, 2, ...n)$.

The training process can be represented as a minimization of the following joint parametric functional:

$$P^{\alpha}(\mathbf{W}^{(1)}, ..., \mathbf{W}^{(n)}; \mathbf{B}^{(1)}, ..., \mathbf{B}^{(n)}) =$$

$$= \sum_{i=1}^{n} \left\| \mathbf{N}^{(i)}\left(\mathbf{d}^{tr(i)}, \mathbf{W}^{(i)}, \mathbf{B}^{(i)}\right) - \mathbf{m}^{tr(i)} \right\|^2 + \alpha c_1 \sum_{i=1}^{n} S_*^{(i)}\left[\mathbf{N}^{(i)}\left(\mathbf{d}^{(i)}, \mathbf{W}^{(i)}, \mathbf{B}^{(i)}\right)\right]$$

$$+ \alpha c_2 S_{J*}\left[\mathbf{N}^{(1)}\left(\mathbf{d}^{(1)}, \mathbf{W}^{(1)}, \mathbf{B}^{(1)}\right), ..., \mathbf{N}^{(n)}\left(\mathbf{d}^{(n)}, \mathbf{W}^{(n)}, \mathbf{B}^{(n)}\right)\right] = \min, \tag{16.31}$$

where α is the regularization parameter, and c_1 and c_2 are the weighting coefficients determining the weights of the different stabilizers in the parametric functional.

The terms $S_*^{(i)}$ are the stabilizing functionals, based on minimum norm, minimum support, and minimum gradient support constraints, respectively, defined above in Eqs. (16.9), (16.14) and (16.16). The term S_{J*} is the joint stabilizing functional, which can be represented by joint structural, focusing, entropy, or Gramian stabilizers, discussed in Chaps. 9, 10, 11, and 12, respectively. It is also important to note that, in expression (16.31), the misfit functionals between the predicted and known models are calculated for the training sets, $\mathbf{d}^{tr(i)}$ and $\mathbf{m}^{tr(i)}$, only. At the same time, the stabilizing functionals are applied to all models produced by the corresponding neural network operators, $\mathbf{N}^{(i)}\left(\mathbf{d}^{(i)}, \mathbf{W}^{(i)}, \mathbf{B}^{(i)}\right)$, for all available data, $\mathbf{d}^{(i)}$, $i = 1, 2, ...n$.

The training process is now reduced to minimization of the joint parametric functional (16.31). This problem can be solved using the technique discussed above in Chap. 15. After the training of the neural networks $\mathbf{N}^{(i)}\left(\mathbf{d}^{(i)}, \mathbf{W}^{(i)}, \mathbf{B}^{(i)}\right)$ is completed, the solution of the inverse problem can be found by application of these networks to the observed data using formulas (16.30).

16.4 Joint Inversion of Multiphysics Data Using Knowledge-Based Neural Network

In this section, I describe the technique of joint inversion of multiphysics data based on a synthesis of neural network and forward modeling operators. We consider again the multiphysics inverse problem given by the system of operator Eq. (16.29). According to the Universal Approximation Theorem, there exist the corresponding neural network operators, $\mathbf{N}_m^{(i)}(\mathbf{r}, \mathbf{W}, \mathbf{B})$, approximating $\mathbf{m}^{(i)}(\mathbf{r})$ with a given accuracy ε:

$$\left\|\mathbf{N}_m^{(i)}(\mathbf{r}, \mathbf{W}, \mathbf{B}) - \mathbf{m}^{(i)}(\mathbf{r})\right\|_\infty < \varepsilon, \quad i = 1, 2, ..., n; \tag{16.32}$$

or

$$\mathbf{m}^{(i)}(\mathbf{r}) \approx \mathbf{N}_m^{(i)}\left(\mathbf{r}, \mathbf{W}^{(i)}, \mathbf{B}^{(i)}\right), \quad i = 1, 2, ..., n. \tag{16.33}$$

The solution of the multiphysics inverse problem (16.29) is equivalent to the minimization of the corresponding parametric functional:

$$P^\alpha(\mathbf{W}^{(1)}, ..., \mathbf{W}^{(n)}; \mathbf{B}^{(1)}, ..., \mathbf{B}^{(n)}) =$$

$$= \sum_{i=1}^{n} \left\| A\left[\mathbf{N}_m^{(i)}\left(\mathbf{r}, \mathbf{W}^{(i)}, \mathbf{B}^{(i)}\right)\right] - \mathbf{d}^{(i)}\right\|_{L_2}^2 + \alpha c_1 \sum_{i=1}^{n} S_*\left[\mathbf{N}_m^{(i)}\left(\mathbf{r}, \mathbf{W}^{(i)}, \mathbf{B}^{(i)}\right)\right]$$

$$+ \alpha c_2 S_{J*}\left[\mathbf{N}_m^{(1)}\left(\mathbf{r}, \mathbf{W}^{(1)}, \mathbf{B}^{(1)}\right), \mathbf{N}_m^{(2)}\left(\mathbf{r}, \mathbf{W}^{(2)}, \mathbf{B}^{(2)}\right), ...\mathbf{N}_m^{(n)}\left(\mathbf{r}, \mathbf{W}^{(n)}, \mathbf{B}^{(n)}\right)\right] = \min, \tag{16.34}$$

where α is the regularization parameter, and c_1 and c_2 are the weighting coefficients determining the weights of the different stabilizers in the parametric functional. As above, the terms $S_*^{(i)}$ are the stabilizing functionals, based on minimum norm, minimum support, or minimum gradient support constraints, respectively, defined above in Eqs. (16.9), (16.14) and (16.16). The term S_{J*} is the joint stabilizing functional (joint structural, focusing, entropy, or Gramian stabilizers).

One can see that in the framework of this approach, neural network operators, $\mathbf{N}_m^{(i)}(\mathbf{r}, \mathbf{W}, \mathbf{B})$, serve as the parameterization of the corresponding model functions $\mathbf{m}^{(i)}(\mathbf{r})$. Thus, in the case of multiphysics inversion, the inverse problem is also reduced to finding the optimal weights and biases of the corresponding neural networks. This can be achieved by using the backpropagation algorithm described in Chap. 15.

The principle difference between minimization problems (16.34) and (16.31) can be described as follows. In case of joint inversion based on a regularizing neural network (RNN) (Eq. (16.31)), we do not use any knowledge about a specific form of the forward modeling operators. In other words, RNN does not use the laws of physics but finds the inverse problem solution based on the training sets only. This is what I call *an intuition-based solution.* At the same time, joint inversion based on forward modeling operators (Eq. (16.34)) employs the corresponding laws of physics in the framework of the machine learning algorithm *(a knowledge-based solution).* This approach seems more attractive for the following reasons. First, it explicitly uses the known forward modeling operator, while RNN has to approximate the inverse operator, which may require more intense computations. Second, the knowledge of the specific form of the forward modeling operator imposes additional constraints on the solution, making it more robust.

16.5 Approximation of the Joint Stabilizing Functional by the Neural Network Operator

Another approach to joint inversion is based on the approximation of the joint stabilizing functional by the neural network operator. The idea is that we can use the neural network to enforce some relationships between the different model parameters without a priori knowledge about the specific form of these relationships.

Let us consider for simplicity the joint inverse problem for two model parameters, $\mathbf{m}^{(1)}$ and $\mathbf{m}^{(2)}$:

$$\mathbf{d}^{(1)} = \mathbf{A}^{(1)}(\mathbf{m}^{(1)}), \quad \mathbf{d}^{(2)} = \mathbf{A}^{(2)}(\mathbf{m}^{(2)}). \tag{16.35}$$

We assume that there may exist some relationship between $\mathbf{m}^{(1)}$ and $\mathbf{m}^{(2)}$ described by a continuous operator, Φ:

$$\mathbf{m}^{(2)} = \Phi\left(\mathbf{m}^{(1)}\right). \tag{16.36}$$

We can represent Eq. (16.36) in the form similar to Eq. (8.6) of Chap. 8:

$$\mathbf{C}(\mathbf{m}^{(1)}, \mathbf{m}^{(2)}) = \Phi\left(\mathbf{m}^{(1)}\right) - \mathbf{m}^{(2)} = 0, \tag{16.37}$$

where operator $\mathbf{C}(\mathbf{m}^{(1)}, \mathbf{m}^{(2)})$, in a general case, is some unknown operator defined on a set of functions $\mathbf{m}^{(1)}$ and $\mathbf{m}^{(2)}$.

In Chap. 8, we developed a method of joint inversion under the assumption that operator $\mathbf{C}(\mathbf{m}^{(1)}, \mathbf{m}^{(2)})$ was known and could be given in an explicit form. However, in practical applications, the specific form of this operator is usually unknown. One way to overcome this difficulty is to find the neural network operator, which approximates operator $\Phi\left(\mathbf{m}^{(1)}\right)$, and therefore $\mathbf{C}(\mathbf{m}^{(1)}, \mathbf{m}^{(2)})$ the best.

With this goal in mind, let us assume that we know pairs of vectors (the training set) representing these two model sets, $\mathbf{m}^{tr(1)}$ and $\mathbf{m}^{tr(2)}$, which correspond to the same target (e.g., density and conductivity of the same body).

According to the Universal Approximation Theorem, we can introduce a neural network operator, $\mathbf{N}_\Phi\left(\mathbf{m}^{(1)}, \mathbf{W}_J, \mathbf{B}_J\right)$, which approximates the continuous operator, Φ with the given accuracy ε:

$$\left\|\mathbf{N}_\Phi\left(\mathbf{m}^{(1)}, \mathbf{W}_J, \mathbf{B}_J\right) - \Phi\left(\mathbf{m}^{(1)}\right)\right\|_\infty = \left\|\mathbf{N}_\Phi\left(\mathbf{m}^{(1)}, \mathbf{W}_J, \mathbf{B}_J\right) - \mathbf{m}^{(2)}\right\|_\infty \leq \varepsilon. \tag{16.38}$$

We can apply the regularized neural network training algorithm described in Sect. 15.3.5 of Chap. 15, and determine the weights and biases $\mathbf{W}_J, \mathbf{B}_J$, of the $\mathbf{N}_\Phi$ operator using the training set $\mathbf{m}^{tr(1)}$ and $\mathbf{m}^{tr(2)}$.

The joint inverse problem for two model parameters, $\mathbf{m}^{(1)}$ and $\mathbf{m}^{(2)}$, can be now reformulated as a minimization of the following joint parametric functional:

$$P^\alpha(\mathbf{m}^{(1)}, \mathbf{m}^{(2)}) =$$

$$= \sum_{i=1}^{2} \left\|\mathbf{A}^{(i)}(\mathbf{m}^{(i)}) - \mathbf{d}^{(i)}\right\|_D^2 + \alpha \left\|\mathbf{N}_\Phi\left(\mathbf{m}^{(1)}, \mathbf{W}_J, \mathbf{B}_J\right) - \mathbf{m}^{(2)}\right\|_M^2 = \min. \tag{16.39}$$

The minimization problem (16.39) can be solved using any gradient-type method discussed in this book.

Finally, let us consider again a multimodal inverse problem described by the system of operator equations:

$$\mathbf{d}^{(i)} = A^{(i)}(\mathbf{m}^{(i)}), \quad i = 1, 2, 3, ..., n. \tag{16.40}$$

We assume now that there exists a relationship between all different model parameters, which in a general case, can be expressed in the following form:

$$\mathbf{m}^{(n)} = J\left(\mathbf{m}^{(1)}, \mathbf{m}^{(2)}, ..., \mathbf{m}^{(n-1)}\right), \tag{16.41}$$

where J is some unknown but continuous operator.

We also assume that we know the training sets of vectors representing all these models, $\mathbf{m}^{tr(1)}$, $\mathbf{m}^{tr(2)}$, ..., $\mathbf{m}^{tr(n)}$, corresponding to the same target. Based on the Universal Approximation Theorem, one can construct the neural network operator, $\mathbf{N}_j\left(\mathbf{m}^{(1)}, ..., \mathbf{m}^{(n-1)}, \mathbf{W}_J, \mathbf{B}_J\right)$, which approximates the operator relationship (16.41) with the given accuracy ε:

$$\left\|\mathbf{N}_J\left(\mathbf{m}^{(1)},...,\mathbf{m}^{(n-1)},\mathbf{W}_J,\mathbf{B}_J\right)-J\left(\mathbf{m}^{(1)},\mathbf{m}^{(2)},...,\mathbf{m}^{(n-1)}\right)\right\|_\infty$$

$$=\left\|\mathbf{N}_J\left(\mathbf{m}^{(1)},...,\mathbf{m}^{(n-1)},\mathbf{W}_J,\mathbf{B}_J\right)-\mathbf{m}^{(n)}\right\|_\infty\leq\varepsilon. \tag{16.42}$$

We call operator $\mathbf{N}_J$ a *joint neural network operator*, and the corresponding network *the joint neural network (JNN)*. The weights and biases $\mathbf{W}_J, \mathbf{B}_J$, of the JNN operator could be computed using the conventional machine learning method based on the known training sets $\mathbf{m}^{tr(1)}$, $\mathbf{m}^{tr(2)}$, ..., $\mathbf{m}^{tr(n)}$, and the gradient-type minimization with backpropagation algorithm presented in Chap. 15.

We can now introduce the following multimodal parametric functional:

$$P^\alpha(\mathbf{m}^{(1)},\mathbf{m}^{(2)},...,\mathbf{m}^{(n)})=$$

$$=\sum_{i=1}^{n}\left\|\mathbf{A}^{(i)}(\mathbf{m}^{(i)})-\mathbf{d}^{(i)}\right\|_D^2+\alpha\left\|\mathbf{N}_J\left(\mathbf{m}^{(1)},...,\mathbf{m}^{(n-1)},\mathbf{W}_J,\mathbf{B}_J\right)-\mathbf{m}^{(n)}\right\|_M^2. \tag{16.43}$$

The minimization of the parametric functional (16.43) delivers the solution to the joint multimodal inverse problem (16.40). Indeed, one can see that by minimizing parametric functional we determine the models, $\mathbf{m}^{(i)}$, $i=1,2,...,n$, which fit the observed data, while enforcing the complex relationships between different inverse models. The major advantage of this approach over the conventional joint inversion based on functional relationships between different model parameters is related to the use of the neural network operator as a stabilizer. The JNN-based stabilizer does not require exact knowledge about the specific functional relationships between different types of models. Instead, we use the relationships obtained by training the joint neural network over the training sets and extrapolate these relationships to the entire model space.

References and Recommended Reading to This Chapter

Aggarwal CC (2018) Neural networks and deep learning. Springer

Cover TM, Thomas JA (2006) Elements of information theory, 2nd edn. Willey, New York

Karniadakis GE, Kevrekidis IG, Lu L, Perdikaris P, Wang S, Yang L (2021) Physics-informed machine learning. Nat Rev Phys 3(6):422–440

Raissi M, Paris P, Karniadakis GE (2019) Physics-informed neural networks: a deep learning framework for solving forward and inverse problems involving nonlinear partial differential equations. J Comput Phys 378:686–707

Schuster G (2023) Machine learning methods in geoscience. Society of Exploration Geophysicists

Part V
Case Histories of Joint Inversion

Chapter 17
Modeling and Inversion of Potential Field Data

Abstract This chapter illustrates the inversion methods discussed in the book by joint inversion of potential field geophysical data. Methods of forward modeling the potential field data (gravity and magnetic, gravity and magnetic gradiometry) are discussed in detail. We also consider modeling and inversion of the total magnetic intensity (TMI) data. The general principles of standalone and joint inversion of potential field data are introduced. In the case of magnetic data, the inversion could be done for magnetic susceptibility and for the full magnetization vector, which is important for recovering the remanent magnetization.

Keywords Gravity field · Gravity potential · Magnetic field · Magnetic potential · Total magnetic intensity (TMI) · Magnetic susceptibility · Magnetization · Remanent magnetization

In this and the following chapters, I illustrate the inversion methods discussed in the book by case histories of joint inversion of potential field geophysical data. These data are widely used in the exploration of mineral resources and regional geological studies. I begin with reviewing the principles of modeling and inversion of gravity and magnetic data, followed by gravity and magnetic gradient tensors modeling and inversion.

17.1 Modeling and Inversion of the Gravity Fields

This section presents a short overview of modeling and inversion methods for gravity and gravity gradiometry data.

M. S. Zhdanov, *Advanced Methods of Joint Inversion and Fusion of Multiphysics Data*, Advances in Geological Science,
https://doi.org/10.1007/978-981-99-6722-3_17

17.1.1 Forward Modeling of the Vertical Component of the Gravity Field

The following equation describes the gravity forward modeling problem:

$$\mathbf{g}(\mathbf{r}') = \gamma \iiint_V \rho(\mathbf{r}) \frac{\mathbf{r} - \mathbf{r}'}{|\mathbf{r} - \mathbf{r}'|^3} dv, \tag{17.1}$$

where $\mathbf{r}$ is the source location; $\mathbf{r}'$ is the receiver location; $\rho(\mathbf{r})$ is the density distribution within some domain V; and γ is the universal gravitational constant. According to equation (17.1), forward modeling is reduced to calculating the integral over the domain occupied by masses with density $\rho(\mathbf{r})$.

It is important to note that the scalar components, $g_\alpha(\mathbf{r})$, $\alpha = x, y, z$, of the gravity field $\mathbf{g}(\mathbf{r})$, can be represented as the partial derivatives of the scalar gravity potential, $U(\mathbf{r})$, as follows:

$$g_\alpha(\mathbf{r}) = \frac{\partial}{\partial \alpha} U(\mathbf{r}), \ \alpha = x, y, z, \tag{17.2}$$

where

$$U(\mathbf{r}) = \gamma \iiint_V \frac{\rho(\mathbf{r}')}{|\mathbf{r}' - \mathbf{r}|} dv'. \tag{17.3}$$

We divide volume V, filled with the masses of density $\rho(\mathbf{r})$, into N_m small rectangular cells, V_k, $V = \cup_{k=1}^{N_m} V_k$, and assume that the density is constant within each cell, $\rho(\mathbf{r}) = \rho_k$, $\mathbf{r} \in V_k$. As a result, the vertical component of the gravity field, g_z, can be calculated using the following formula:

$$g_z(\mathbf{r}') = \gamma \sum_{k=1}^{N_m} \rho_k \iiint_{V_k} \frac{z - z'}{| \mathbf{r} - \mathbf{r}' |^3} dv. \tag{17.4}$$

Assuming a relatively small size of rectangular cells, V_k, we can use the point-mass approximation, which dramatically speeds up the processing time while yielding a very accurate result (Zhdanov 2009). We denote the coordinates of the cell centers as $\mathbf{r}_k = (x_k, y_k, z_k)$, $k = 1, ...N_m$, and the cell sides as Δx, Δy, Δz. Also, we have a discrete number of observation points $\mathbf{r}'_n = \left(x'_n, y'_n, z'_n\right)$, $n = 1, ...N_d$. Using discrete model parameters and discrete data, we can present the forward modeling operator for the gravity field (17.4) as follows:

$$g_z(\mathbf{r}'_n) \approx \sum_{k=1}^{N_m} A^{\rho}_{nk} \rho_k, n = 1, ...N_d, \tag{17.5}$$

where the gravity field kernels, A^{ρ}_{nk}, according to Eq. (17.4), are equal to

$$A^{\rho}_{nk} = \gamma \frac{\left(z_k - z'_n\right) \Delta x \Delta y \Delta z}{r^3_{nk}}, \tag{17.6}$$

and

$$r_{nk} = \sqrt{\left(x_k - x'_n\right)^2 + \left(y_k - y'_n\right)^2 + z^2_k}.$$

Thus, the discrete forward modeling operator for the gravity field can be expressed in matrix notations as follows:

$$\mathbf{d} = \mathbf{A}^g \mathbf{m}. \tag{17.7}$$

Here $\mathbf{m}$ is a vector of model parameters (densities, ρ_k) of the order N_m; $\mathbf{d}$ is a vector of observed data, g_z, of the order N_d; and $\mathbf{A}^g$ is a rectangular matrix of the size $N_d \times N_m$, formed by the gravity field kernels, Eq. (17.6).

17.1.2 Forward Modeling of the Full Tensor Gravity Gradiometry Data

The gravity gradient tensor is formed by the second spatial derivatives of the gravity potential $U(\mathbf{r})$,

$$g_{\alpha\beta}(\mathbf{r}) = \frac{\partial^2}{\partial\alpha\partial\beta} U(\mathbf{r}), \ \alpha, \beta = x, y, z. \tag{17.8}$$

The components, $g_{\alpha\beta}(\mathbf{r})$, can be organized in the form of *a gravity gradient tensor*,

$$\widehat{\mathbf{g}} = \begin{bmatrix} g_{xx} & g_{xy} & g_{xz} \\ g_{yx} & g_{yy} & g_{yz} \\ g_{zx} & g_{zy} & g_{zz} \end{bmatrix}. \tag{17.9}$$

They can be calculated based on formulas (17.8) and (17.3), as follows:

$$g_{\alpha\beta}(\mathbf{r}) = \gamma \iiint_V \rho(\mathbf{r}') \frac{1}{|\mathbf{r}' - \mathbf{r}|^3} K_{\alpha\beta} \left(\mathbf{r}' - \mathbf{r}\right) dv', \tag{17.10}$$

where the kernels, $K_{\alpha\beta}$, are equal to

$$K_{\alpha\beta}\left(\mathbf{r}' - \mathbf{r}\right) = \begin{cases} 3\frac{(\alpha-\alpha')(\beta-\beta')}{|\mathbf{r}'-\mathbf{r}|^2}, \ \alpha \neq \beta, \\ \\ 3\frac{(\alpha-\alpha')^2}{|\mathbf{r}'-\mathbf{r}|^2} - 1, \ \alpha = \beta, \end{cases} , \ \alpha, \beta = x, y, z. \tag{17.11}$$

In order to derive numerical expressions for the gravity tensor, we use the same discretization as above for the vertical component of the gravity field. Considering each cell as a point mass and using discrete model parameters and discrete data, we can present the forward modeling operator for the gravity tensor components, (17.10), as follows:

$$g_{\alpha\beta}(\mathbf{r}_n) \approx \sum_{k=1}^{N_m} A_{nk}^{\alpha\beta} \rho_k, \; n = 1, ...N_m; \; \alpha, \beta = x, y, z, \tag{17.12}$$

where the gravity tensor kernels, $A_{nk}^{\alpha\beta}$, according to (17.11) are expressed as follows

$$A_{nk}^{\alpha\beta} = \gamma \frac{\Delta x \Delta y \Delta z}{r_{nk}{}^3} K_{nk}^{\alpha\beta}, \tag{17.13}$$

where

$$K_{nk}^{\alpha\beta} = \begin{cases} 3\frac{(\alpha_k' - \alpha_n)(\beta_k' - \beta_n)}{r_{nk}{}^2}, \; \alpha \neq \beta, \\ \\ 3\frac{(\alpha_k' - \alpha_n)^2}{r_{nk}{}^2} - 1, \; \alpha = \beta, \end{cases} , \; \alpha, \beta = x, y, z; \tag{17.14}$$

and

$$r_{nk} = \sqrt{(x_k' - x_n)^2 + (y_k' - y_n)^2 + (z_k' - z_n)^2}.$$

Expression (17.12) represents a point-mass approximation of the gravity tensor components. It was demonstrated by Zhdanov (2009) that this approximation is very accurate while being much more computationally efficient than the conventional approach based on the exact analytical expression of the gravity gradient field produced by prismatic cells.

We can write Eq. (17.12) in matrix notations as follows:

$$\mathbf{d} = \mathbf{A}^{\widehat{\mathbf{g}}}\mathbf{m}, \tag{17.15}$$

where $\mathbf{m}$ is a vector of model parameters (densities, ρ_k); $\mathbf{d}$ is a vector of observed tensor data, $\widehat{\mathbf{g}}$; and $\mathbf{A}^{\widehat{\mathbf{g}}}$ is a rectangular matrix, formed by the gravity gradient field kernels, Eq. (17.13).

17.1.3 Inversion of the Gravity and Gravity Gradiometry Data

Gravity field and/or gravity tensor inversions are reduced to the solution of the linear matrix equations (17.7) or (17.15). In compact form, these equations can be written as follows:

$$\mathbf{d} = \mathbf{A}\mathbf{m}, \tag{17.16}$$

where $\mathbf{A}$ stands for the matrices of forward modeling operators, $\mathbf{A}^g$ or $\mathbf{A}^{\widehat{\mathbf{g}}}$; $\mathbf{m}$ is a vector of anomalous density distribution; and $\mathbf{d}$ is a vector formed by the observed gravity or gravity tensor data sets, g_z or $\widehat{\mathbf{g}}$.

To produce a stable solution of this problem, we can use the classical regularization approach based on the minimization of the Tikhonov parametric functional, described in detail in Chap. 4:

$$P^{\alpha}(\mathbf{m}) = \phi(\mathbf{m}) + \alpha\, s(\mathbf{m}) = \min, \tag{17.17}$$

where the misfit functional, $\phi(\mathbf{m})$, is specified by the least-square norm of the difference between the weighted predicted and observed data,

$$\phi(\mathbf{m}) = \|\mathbf{W}_d\,(\mathbf{Am} - \mathbf{d}))\|^{\,2}, \tag{17.18}$$

α is a regularization parameter, $\mathbf{W}_d$ is the data weighting matrix, and $s(\mathbf{m})$ is the corresponding stabilizer selected from the family of stabilizing functionals introduced in Chap. 4. We can also apply the model parameter weights to improve the depth resolution of the inversion, as discussed in Chap. 5.

The choice of stabilizing functional is usually based on the available knowledge about the targets. In Chap. 4, we described different smooth and focusing stabilizers in order to produce diffused or sharp images of the target. Examples of smooth stabilizers include minimum norm and spatial derivatives functionals.

A minimum norm (MN) stabilizer seeks to minimize the norm of the difference between the current model and an a priori model:

$$s_{MN}(\mathbf{m}) = \iiint_V (\mathbf{m} - \mathbf{m}_{apr})^2 dv, \tag{17.19}$$

and it usually produces a relatively smooth model.

The first derivative (FD) stabilizer implicitly introduces smoothness by minimizing the norm of spatial derivatives of the model parameters:

$$s_{FD}(\mathbf{m}) = \iiint_V (\nabla\mathbf{m} - \nabla\mathbf{m}_{apr})^2 dv. \tag{17.20}$$

The minimum support (MS) stabilizer,

$$s_{MS}(\mathbf{m}) = \iiint_V \frac{(\mathbf{m} - \mathbf{m}_{apr})^2}{(\mathbf{m} - \mathbf{m}_{apr})^2 + e^2} dv, \tag{17.21}$$

minimizes the volume with nonzero departures from the a priori model, effectively recovering compact bodies. Therefore, a smooth distribution of all model parameters with a small deviation from the a priori model is penalized.

The minimum gradient support (MGS) stabilizer,

$$s_{MGS}(\mathbf{m}) = \iiint_V \frac{\nabla m \cdot \nabla m}{\nabla m \cdot \nabla m + e^2} dv, \tag{17.22}$$

minimizes the areas where the big model parameter changes occur, thus emphasizing the sharp boundaries. There are several other choices of stabilizing functionals listed in this book.

The minimization problem (17.17) can be solved by the corresponding gradient-type methods of Chap. 7. Details of the numerical implementation of these methods can be found in Zhdanov (2014) and Čuma and Zhdanov (2014).

17.2 Modeling and Inversion of the Magnetic Fields for Magnetic Susceptibility

We now consider modeling and inversion methods for magnetic field data. Geophysicists study different components of the magnetic field. The most widely used are the total magnetic intensity (TMI) data based on measuring the magnitude of the magnetic field. One can also measure and analyze three scalar components of the magnetic field. Recently, the methods of measuring full tensor magnetic gradiometry (FTMG) data formed by the second derivatives of magnetic field potential have been introduced as well.

17.2.1 Forward Modeling of the Magnetic Field Data

The magnetic field of volume D, filled with magnetic masses with the intensity of magnetization $\mathbf{I}(\mathbf{r})$, can be expressed as a gradient of the magnetic potential, $U(\mathbf{r})$, as follows (Zhdanov 1988, 2002):

$$\mathbf{H}(\mathbf{r}') = \nabla' U(\mathbf{r}'), \tag{17.23}$$

where:

$$U(\mathbf{r}') = \iiint_V \mathbf{I}(\mathbf{r}) \cdot \nabla' \frac{1}{|\mathbf{r} - \mathbf{r}'|} dv. \tag{17.24}$$

The magnetic geophysical methods are often based on the assumption that there is no remanent magnetization, and the observed magnetic data are caused by induced magnetization only. Under such assumptions, the intensity of magnetization in the rock formation is linearly related to an inducing magnetic field, $\mathbf{H}^0$, through the magnetic susceptibility, $\chi(\mathbf{r})$:

$$\mathbf{I}(\mathbf{r}) = \chi(\mathbf{r})\,\mathbf{H}^0 = \chi(\mathbf{r})\,H^0\mathbf{l}, \tag{17.25}$$

where $\mathbf{r}$ is the radius-vector of a point within the volume V; H^0 is the magnitude of the inducing field; and $\mathbf{l} = (l_x, l_y, l_z)$ is a unit vector in the direction of this field. Assuming that the x-axis is directed eastward, the y-axis has a positive direction northward, and the z-axis is directed downward, one can calculate the direction of the inducing magnetic field as follows:

$$\begin{aligned} l_x &= cos(I)sin(D - A), \\ l_y &= cos(I)cos(D - A), \\ l_z &= sin(I), \end{aligned} \tag{17.26}$$

where I is the inclination, D is the declination, and A is the azimuth of the inducing field. The values of $I,\ D,\ A,\ H^0$ are variable with time and location on the Earth. The inclination, I, is given by an angle that can assume values between -90 ° (up) at the south magnetic pole to 90 ° (down) at the north magnetic pole. Declination, D, is positive for an eastward deviation of the field relative to true north. It could be –90° to 90 ° in most areas. The intensity of Earth's magnetic field H^0 ranges between approximately 25,000 and 65,000 nT. The values of $I,\ D,\ A,\ H^0$ can be found at International Geomagnetic Reference Field (IGRF) (Alken et al. 2021).

Thus, substituting formula (17.25) for the intensity of magnetization into equations (17.24) and (17.23), after some algebra, we arrive at the following integral representation of the magnetic field:

$$\mathbf{H}(\mathbf{r}') = -H_0 \iiint_V \frac{\chi(\mathbf{r})}{|\mathbf{r}-\mathbf{r}'|^3}\left[\mathbf{l} - \frac{3\left(\mathbf{l}\cdot(\mathbf{r}-\mathbf{r}')\right)(\mathbf{r}-\mathbf{r}')}{|\mathbf{r}-\mathbf{r}'|^2}\right]dv. \tag{17.27}$$

By discretizing the 3D earth model into a grid of N_m cells, each of constant magnetic susceptibility, we obtain the following discrete form of Eq. (17.27):

$$\mathbf{H}(\mathbf{r}') = -H_0 \sum_{k=1}^{N_m} \chi_k \iiint_{V_k} \frac{1}{|\mathbf{r}-\mathbf{r}'|^3}[\mathbf{l} - \frac{3(\mathbf{l}\cdot(\mathbf{r}-\mathbf{r}'))(\mathbf{r}-\mathbf{r}')}{|\mathbf{r}-\mathbf{r}'|^2}]dv, \tag{17.28}$$

where $\mathbf{r}' = (x', y', z')$ denotes the point of observation, $\mathbf{r} = (x, y, z)$ denotes the point of source location, $\mathbf{l} = (l_x, l_y, l_z)$, and H_0 are the direction and the absolute value of the inducing magnetic field, $\mathbf{H}^0$, respectively.

Closed-form solutions for the volume integral in Eq. (17.28) over right rectangular prisms of magnetic susceptibility have been previously presented (e.g., Bhattacharyya (1980)). We can also evaluate the volume integral numerically using single-point Gaussian integration with pulse basis functions in a similar way we considered above for the gravity field (Zhdanov 2009). In this case, $\mathbf{r} = (x, y, z)$ denotes the cell center. We assume constant discretization of Δx, Δy, and Δz in the x, y, and z directions, respectively.

Therefore, Eq. (17.28) can be simplified as follows:

$$\mathbf{H}(\mathbf{r}') = -H_0 \sum_{k=1}^{N_m} \chi_k \frac{1}{|\mathbf{r}-\mathbf{r}'|^3}[\mathbf{l} - \frac{3(\mathbf{l}\cdot(\mathbf{r}-\mathbf{r}'))(\mathbf{r}-\mathbf{r}')}{|\mathbf{r}-\mathbf{r}'|^2}]\Delta x \Delta y \Delta z. \quad (17.29)$$

From Eq. (17.29), we can derive discrete expressions for the scalar components of the magnetic field:

$$H_x(\mathbf{r}') = -H_0 \sum_{k=1}^{N_m} \chi_k \frac{1}{|\mathbf{r}-\mathbf{r}'|^3}[l_x - \frac{3t(x_k - x')}{|\mathbf{r}-\mathbf{r}'|^2}]\Delta x \Delta y \Delta z,$$

$$H_y(\mathbf{r}') = -H_0 \sum_{k=1}^{N_m} \chi_k \frac{1}{|\mathbf{r}-\mathbf{r}'|^3}[l_y - \frac{3t(y_k - y')}{|\mathbf{r}-\mathbf{r}'|^2}]\Delta x \Delta y \Delta z, \quad (17.30)$$

$$H_z(\mathbf{r}') = -H_0 \sum_{k=1}^{N_m} \chi_k \frac{1}{|\mathbf{r}-\mathbf{r}'|^3}[l_y - \frac{3t(y_k - y')}{|\mathbf{r}-\mathbf{r}'|^2}]\Delta x \Delta y \Delta z,$$

where

$$t = l_x(x_k - x') + l_y(y_k - y') + l_z(z_k - z'). \quad (17.31)$$

and $\mathbf{r} = \mathbf{r}_k = (x_k, y_k, z_k), k = 1, ..., N_m$; denotes the center of the cell, k. In compact operator form, we can write system of equations (17.30) as follows

$$\mathbf{d} = \mathbf{A}^{H\chi}(\mathbf{m}), \quad (17.32)$$

where $\mathbf{m}$ is a vector of model parameters (susceptibility, χ_l); $\mathbf{d}$ is a vector of observed magnetic field data $\mathbf{H}$; and $\mathbf{A}^{H\chi}$ is the discrete magnetic forward modeling operator described by equations (17.30).

17.2.2 Forward Modeling of the Total Magnetic Intensity Data

The standard geophysical surveys usually collect the total magnetic intensity (TMI) field data, which can be approximately represented as follows:

$$T\left(\mathbf{r}'\right) \approx \mathbf{l}\cdot\mathbf{H}\left(\mathbf{r}'\right) = -H^0 \iiint_V \frac{\chi(\mathbf{r})}{|\mathbf{r}-\mathbf{r}'|^3}\left[1 - \frac{3\left(\mathbf{l}\cdot(\mathbf{r}-\mathbf{r}')\right)^2}{\|\mathbf{r}-\mathbf{r}'\|^2}\right]dv, \quad (17.33)$$

where $\mathbf{r}$ is the source location; $\mathbf{r}'$ is the receiver location; $\chi(\mathbf{r})$ is the susceptibility distribution within domain V.

The discrete form of integral formula (17.33) is obtained by dividing domain V into N_m small rectangular cells, V_k, and assuming that the susceptibility is constant within each cell, χ_k. As a result, expression (17.33) for the TMI field takes the following form:

$$T\left(\mathbf{r}'\right) = -H^0 \sum_{k=1}^{N_m} \chi_k \iiint_{V_k} \frac{1}{|\mathbf{r}-\mathbf{r}'|^3}\left[1 - \frac{3\left(\mathbf{l}\cdot\left(\mathbf{r}-\mathbf{r}'\right)\right)^2}{|\mathbf{r}-\mathbf{r}'|^2}\right] dv. \quad (17.34)$$

Assuming a relatively small size of rectangular cells, V_k, we again can use the point-mass approximation, which dramatically speeds up the processing time while yielding very accurate results (Zhdanov 2009). The coordinates of the cell centers are $r_k = (x_k, y_k, z_k)$, $k = 1, ..., N_m$. and the cell sides are Δx, Δy, Δz. Also, we have a discrete number of observation points $\mathbf{r}'_n = (x'_n, y'_n, z'_n)$, $n = 1, ..., N_d$. Using discrete model parameters and discrete data, we can present the forward modeling equation for the magnetic field (17.34), as follows:

$$T(\mathbf{r}'_n) = \sum_{k=1}^{N_m} A^{\chi}_{nk}\chi_k, \quad (17.35)$$

where the magnetic field kernels, A^{χ}_{nk}, are calculated by the following formula:

$$A^{\chi}_{nk} = -H^0 \frac{1}{r^3_{nk}}\left[1 - \frac{3\left(l_x\left(x'_k - x_n\right) + l_y\left(y'_k - y_n\right) + l_z\left(z'_k - z_n\right)\right)^2}{r^2_{nk}}\right]\Delta x\Delta y\Delta z, \quad (17.36)$$

and

$$r_{nk} = \sqrt{\left(x'_k - x_n\right)^2 + \left(y'_k - y_n\right) + \left(z'_k - z_n\right)}.$$

Using the discrete model parameters introduced above, we can approximate the forward modeling operator for the TMI field produced by the volume distribution of magnetic rocks with susceptibility χ as follows:

$$\mathbf{d} = \mathbf{A}^{T\chi}(\mathbf{m}). \quad (17.37)$$

Here $\mathbf{m}$ is a vector of model parameters (susceptibility, χ_k) of the order N_m; $\mathbf{d}$ is a vector of observed TMI data T, of the order N_d; and $\mathbf{A}^{T\chi}$ is a rectangular matrix formed by the magnetic field kernels, Eq. (17.36). Note that the number of discretization cells, N_m, in the voxel-type inversion is obviously significantly greater than the number of observed data, N_d: $N_m \gg N_d$.

17.2.3 Forward Modeling of Magnetic Gradiometry Data

Another important representation of the magnetic field is given by the second derivatives of the magnetic potential:

$$H_{\alpha\beta}(\mathbf{r}) = \frac{\partial^2}{\partial\alpha\partial\beta}U(\mathbf{r}), \quad \alpha, \beta = x, y, z. \tag{17.38}$$

The second spatial derivatives of the magnetic potential form a symmetric tensor with zero trace:

$$\widehat{\mathbf{H}} = \begin{bmatrix} H_{xx} & H_{xy} & H_{xz} \\ H_{yx} & H_{yy} & H_{yz} \\ H_{zx} & H_{zy} & H_{zz} \end{bmatrix}, \quad H_{xx} + H_{yy} + H_{zz} = 0, \tag{17.39}$$

where:

$$H_{\alpha\beta} = \frac{\partial H_\alpha}{\partial\beta}, \quad \alpha, \beta = x, y, z. \tag{17.40}$$

This implies that of the nine tensor components, only five are independent.

After some algebra, we find from equations (17.30) and (17.40) the discrete forms of each component of the magnetic tensor:

$$H_{xx}(\mathbf{r}') = 3H_0 \sum_{k=1}^{N_m} \chi_k \{ \frac{[-l_x(x_k - x') - t]r^2 + 2(x' - x_k)^2 t}{r^7} + [l_x - \frac{3t(x_k - x')}{r^2}]\frac{(x' - x_k)}{r^5} \} \Delta x \Delta y \Delta z,$$

$$H_{xy}(\mathbf{r}') = 3H_0 \sum_{k=1}^{N_m} \chi_k \{ \frac{-l_y(x_k - x')r^2 - 2(y' - y_k)(x_k - x')t}{r^7} + [l_x - \frac{3t(x_k - x')}{r^2}]\frac{(y' - y_k)}{r^5} \} \Delta x \Delta y \Delta z,$$

$$H_{xz}(\mathbf{r}') = 3H_0 \sum_{k=1}^{N_m} \chi_k \{ \frac{-l_z(x_k - x')r^2 - 2(z' - z_k)(x_k - x')t}{r^7} + [l_x - \frac{3t(x_k - x')}{r^2}]\frac{(z' - z_k)}{r^5} \} \Delta x \Delta y \Delta z,$$

$$H_{yx}(\mathbf{r}') = 3H_0 \sum_{k=1}^{N_m} \chi_k \{ \frac{-l_x(y_k - y')r^2 - 2(x' - x_k)(y_k - y')t}{r^7} + [l_y - \frac{3t(y_k - y')}{r^2}] \frac{(x' - x_k)}{r^5} \} \Delta x \Delta y \Delta z,$$

$$H_{yy}(\mathbf{r}') = 3H_0 \sum_{k=1}^{N_m} \chi_k \{ \frac{[-l_y(y_k - y') - t]r^2 + 2(y' - y_k)^2 t}{r^7} + [l_y - \frac{3t(y_k - y')}{r^2}] \frac{(y' - y_k)}{r^5} \} \Delta x \Delta y \Delta z,$$

$$H_{yz}(\mathbf{r}') = 3H_0 \sum_{k=1}^{N_m} \chi_k \{ \frac{-l_z(y_k - y')r^2 - 2(z' - z_k)(y_k - y')t}{r^7} + [l_y - \frac{3t(y_k - y')}{r^2}] \frac{(z' - z_k)}{r^5} \} \Delta x \Delta y \Delta z,$$

$$H_{zx}(\mathbf{r}') = 3H_0 \sum_{k=1}^{N_m} \chi_k \{ \frac{-l_x(z_k - z')r^2 - 2(x' - x_k)(z_k - z')t}{r^7} + [l_z - \frac{3t(z_k - z')}{r^2}] \frac{(x' - x_k)}{r^5} \} \Delta x \Delta y \Delta z,$$

$$H_{zy}(\mathbf{r}') = 3H_0 \sum_{k=1}^{N_m} \chi_k \{ \frac{-l_y(z_k - z')r^2 - 2(y' - y_k)(z_k - z')t}{r^7} + [l_z - \frac{3t(z_k - z')}{r^2}] \frac{(y' - y_k)}{r^5} \} \Delta x \Delta y \Delta z,$$

$$H_{zz}(\mathbf{r}') = 3H_0 \sum_{k=1}^{N_m} \chi_k \{ \frac{[-l_z(z_k - z') - t]r^2 + 2(z' - z_k)^2 t}{r^7} + [l_z - \frac{3t(z_k - z')}{r^2}] \frac{(z' - z_k)}{r^5} \} \Delta x \Delta y \Delta z, \quad (17.41)$$

where t is defined by Eq. (17.31), and

$$r = [(x_k - x')^2 + (y_k - y')^2 + (z_k - z')^2]^{1/2}.$$

Equations (17.41) provide the basis for computing the full tensor magnetic gradiometry (FTMG) data for the models without the remanent magnetization. These equations can be written in compact operator form as follows:

$$\mathbf{d} = \mathbf{A}^{\widehat{\mathbf{H}}\chi}(\mathbf{m}), \tag{17.42}$$

where $\mathbf{m}$ is a vector of model parameters (susceptibility, χ_l); $\mathbf{d}$ is a vector of observed FTMG data $\widehat{\mathbf{H}}$; and $\mathbf{A}^{\widehat{\mathbf{H}}\chi}$ is the corresponding discrete FTMG forward modeling operator described by equations (17.41).

17.2.4 *Inversion of Magnetic Field Data into Susceptibility*

The regularized inversion of the magnetic field data is based on the same principles as gravity inversion discussed above. The fundamental inverse problem equation is written in a standard form, as follows:

$$\mathbf{d} = \mathbf{A}\mathbf{m}, \tag{17.43}$$

where $\mathbf{A}$ stands for the forward modeling operators $\mathbf{A}^{H\chi}$, $\mathbf{A}^{T\chi}$, or $\mathbf{A}^{\widehat{\mathbf{H}}\chi}$; $\mathbf{m}$ is a vector of anomalous susceptibility distribution, χ ; and $\mathbf{d}$ is a vector formed by the observed magnetic field, TMI or FTMG data sets, $\mathbf{H}$, T, or $\widehat{\mathbf{H}}$.

We introduce the Tikhonov parametric functional,

$$P^{\alpha}(\mathbf{m}) = \varphi(\mathbf{m}) + \alpha S_{MN,\ MS,\ MGS}(\mathbf{m}) \rightarrow \min, \tag{17.44}$$

where $\varphi(\mathbf{m})$ is a misfit functional specified by the least-square norm of the difference between the weighted predicted and observed data,

$$\phi(\mathbf{m}) = \|\mathbf{W}_d(\mathbf{A}\mathbf{m} - \mathbf{d})\|^{2}, \tag{17.45}$$

and $\mathbf{W}_d$ is the data weighting matrix.

The terms S_{MN}, S_{MS}, and S_{MGS} are the stabilizing functionals based on minimum norm, minimum support, and minimum gradient support constraints defined by equations (17.19), (17.21), and (17.22), respectively. The minimization problem of parametric functional (17.44) can be solved using a variety of optimization methods described in this book.

17.3 Magnetization-Based Modeling and Inversion of the Magnetic Fields

17.3.1 Magnetization-Based Modeling of the Magnetic Field Data

Remanent magnetization (or remanence) is a permanent magnetization of a rock that was obtained in the past when the Earth's magnetic field had a different magnitude and direction than what it has today. It follows that the total intensity of magnetization, $\mathbf{I}(\mathbf{r})$, is linearly related to both the induced, $\mathbf{M}_{ind}$, and remanent, $\mathbf{M}_{rem}$, magnetizations (Jorgensen et al. 2023):

$$\mathbf{I}(\mathbf{r}) = H^0 \left[\mathbf{M}_{ind}(\mathbf{r}) + \mathbf{M}_{rem}(\mathbf{r})\right], \tag{17.46}$$

where induced magnetization, $\mathbf{M}_{ind}$, is linear proportional to the inducing magnetic field, $\mathbf{H}^0(\mathbf{r})$, through the magnetic susceptibility, $\chi(\mathbf{r})$:

$$\mathbf{M}_{ind}(\mathbf{r}) = \chi(\mathbf{r})\mathbf{H}^0/H^0 = \chi(\mathbf{r})\mathbf{l}, \tag{17.47}$$

and $\mathbf{l}$ the unit vector in the direction of the inducing magnetic field, defined by equations (17.26).

We should note that we have defined the magnetization vectors (both the induced and remanent) as unitless for convenience of derivations.

The Koenigsberger ratio, Q, is the ratio of the absolute values of the remanent magnetization to the induced magnetization (Koenigsberger 1938):

$$Q = \frac{|\mathbf{M}_{rem}|}{|\mathbf{M}_{ind}|}. \tag{17.48}$$

For Koenigsberger ratios greater than 1, the remanent magnetization vector is the predominant contribution to the total intensity of magnetization.

We can rewrite Eq. (17.25) as follows:

$$\mathbf{I}(\mathbf{r}) = H_0\mathbf{M}, \tag{17.49}$$

where $\mathbf{M}$ is the magnetization vector:

$$\mathbf{M}(\mathbf{r}) = \mathbf{M}_{ind}(\mathbf{r}) + \mathbf{M}_{rem}(\mathbf{r}). \tag{17.50}$$

For modeling of the magnetic field data, we can use again the basic formulas (17.24) and (17.23). However, in the case of the arbitrary intensity of magnetization vector, $\mathbf{I}(\mathbf{r})$, we substitute expression (17.49) for $\mathbf{I}(\mathbf{r})$ into (17.24) and (17.23). As a result, we arrive at the following integral representation of the magnetic field:

$$\mathbf{H}(\mathbf{r}') = -H_0 \iiint_V \frac{1}{|\mathbf{r}-\mathbf{r}'|^3}\left[\mathbf{M}(\mathbf{r}) - \frac{3(\mathbf{M}(\mathbf{r})\cdot(\mathbf{r}-\mathbf{r}'))(\mathbf{r}-\mathbf{r}')}{|\mathbf{r}-\mathbf{r}'|^2}\right] dv, \tag{17.51}$$

where $\mathbf{M}(\mathbf{r})$ is the magnetization vector defined by Eq. (17.50).

We discretize the 3D earth model into a grid of N_m cells, each of a constant magnetization vector. Then integral representation (17.51) of the magnetic field can be expressed in discrete form as follows:

$$\mathbf{H}(\mathbf{r}') = -H_0 \sum_{k=1}^{N_m} \iiint_{V_k} \frac{1}{|\mathbf{r}-\mathbf{r}'|^3} [\mathbf{M}_k - \frac{3(\mathbf{M}_k \cdot (\mathbf{r}-\mathbf{r}'))(\mathbf{r}-\mathbf{r}')}{|\mathbf{r}-\mathbf{r}'|^2}] dv, \tag{17.52}$$

where $\mathbf{M}_k = (M_{xk}, M_{yk}, M_{zk})$ is the magnetization vector of the kth cell.

As discussed above, we evaluate the volume integral numerically using single-point Gaussian integration with pulse basis functions. This numerical solution is almost as accurate as the analytic solution provided the depth to the center of the cell exceeds twice the dimension of the cell (Zhdanov 2009). In this case, $\mathbf{r}_k = (x_k, y_k, z_k), k = 1; ::: N_m;$ denotes the center of the cell, k. We assume constant discretization of Δx, Δy, and Δz in the x, y, and z directions, respectively. Also, we have a discrete number of observation points $\mathbf{r}'_n = (x'_{xk}, y'_{yk}, z'_{zk}), n = 1, \ldots N_d$. Using discrete model parameters and discrete data, we can present (17.52) as follows:

$$\mathbf{H}(\mathbf{r}') = -H_0 \sum_{k=1}^{N_m} \frac{1}{|\mathbf{r}-\mathbf{r}'|^3} \left[\mathbf{M}_k - \frac{3(\mathbf{M}_k \cdot (\mathbf{r}-\mathbf{r}'))(\mathbf{r}-\mathbf{r}')}{|\mathbf{r}-\mathbf{r}'|^2} \right] \Delta x \Delta y \Delta z. \tag{17.53}$$

From Eq. (17.53), we obtain discrete expressions for the vector components of the magnetic field:

$$H_x(\mathbf{r}') = -H_0 \sum_{k=1}^{N_m} \frac{1}{|\mathbf{r}-\mathbf{r}'|^3} \left[M_{xk} - \frac{3t_k(x_k - x')}{|\mathbf{r}-\mathbf{r}'|^2} \right] \Delta x \Delta y \Delta z, \tag{17.54}$$

$$H_y(\mathbf{r}') = -H_0 \sum_{k=1}^{N_m} \frac{1}{|\mathbf{r}-\mathbf{r}'|^3} \left[M_{yk} - \frac{3t_k(y_k - y')}{|\mathbf{r}-\mathbf{r}'|^2} \right] \Delta x \Delta y \Delta z, \tag{17.55}$$

$$H_z(\mathbf{r}') = -H_0 \sum_{k=1}^{N_m} \frac{1}{|\mathbf{r}-\mathbf{r}'|^3} \left[M_{zk} - \frac{3t_k(y_k - y')}{|\mathbf{r}-\mathbf{r}'|^2} \right] \Delta x \Delta y \Delta z, \tag{17.56}$$

where indices n and k at the corresponding radius vectors of the observation and integration points are omitted to simplify the notations, and

$$\begin{aligned} |\mathbf{r}-\mathbf{r}'| = r_{nk} &= \sqrt{(x_k - x_n)^2 + (y_k - y_n)^2 + (z_k - z_n)^2}, \\ t_k &= M_{xk}(x_k - x'_n) + M_{yk}(y_k - y'_n) + M_{zk}(z_k - z'_n). \end{aligned} \tag{17.57}$$

We can write equations (17.54) to (17.56) in a compact form as follows:

$$H_\alpha(\mathbf{r}') = -H_0 \sum_{k=1}^{N_m} \frac{1}{|\mathbf{r}-\mathbf{r}'|^3}\left[M_{\alpha k} - \frac{3t_k(\alpha_k-\alpha')}{|\mathbf{r}-\mathbf{r}'|^2}\right]\Delta x \Delta y \Delta z,\ \alpha = x, y, z. \tag{17.58}$$

We can now write the magnetization-based modeling Eq. (17.58) in operator form as well:

$$\mathbf{d} = \mathbf{A}^{HM}(\mathbf{m}), \tag{17.59}$$

where $\mathbf{A}^{HM}$ is the discrete magnetic field forward modeling operator, $\mathbf{m}$ is a model parameter vector representing the discrete magnetization vector distribution $\mathbf{M}\left(M_x, M_y, M_z\right)$, and $\mathbf{d}$ is a vector of the corresponding magnetic field data, $\mathbf{H}$.

17.3.2 Magnetization-Based Modeling of TMI Data

For modeling the magnetic data, we project the magnetic field, Eq. (17.51), onto the direction, $\mathbf{l}$, of the inducing magnetic field, $\mathbf{H}_0$:

$$T\left(\mathbf{r}'\right) \approx \mathbf{l}\cdot\mathbf{H}\left(\mathbf{r}'\right) = -H_0\mathbf{l}(\mathbf{r}')\cdot \iiint_D \frac{1}{|\mathbf{r}-\mathbf{r}'|^3}\left[\mathbf{M} - \frac{3(\mathbf{M}\cdot(\mathbf{r}-\mathbf{r}'))(\mathbf{r}-\mathbf{r}')}{|\mathbf{r}-\mathbf{r}'|^2}\right]dv. \tag{17.60}$$

We again discretize the 3D earth model into a grid of N_m cells, each of constant magnetization vector. According to formula (17.60), the total magnetic intensity field can be computed approximately as follows:

$$T(\mathbf{r}') \approx \mathbf{l}(\mathbf{r}')\cdot\mathbf{H}(\mathbf{r}') =$$

$$= -H_0 \sum_{k=1}^{N_m} \mathbf{l}(\mathbf{r}')\cdot \iiint_{V_k} \frac{1}{|\mathbf{r}-\mathbf{r}'|^3}\left[\mathbf{M}_k - \frac{3(\mathbf{M}_k\cdot(\mathbf{r}-\mathbf{r}'))(\mathbf{r}-\mathbf{r}')}{|\mathbf{r}-\mathbf{r}'|^2}\right]dv, \tag{17.61}$$

where $\mathbf{M}_k = (M_{xk}, M_{yk}, M_{zk})$ is the magnetization vector of the kth cell.

As was done above for Eq. (17.33), we can also evaluate the volume integral numerically with sufficient accuracy using the point-mass approximation (Zhdanov 2009) (where indices n and k at the corresponding radius vectors of the observation and integration points are omitted to simplify the notations):

$$T(\mathbf{r}') \approx$$

$$\approx -H_0 \sum_{k=1}^{N_m} \mathbf{l}(\mathbf{r}')\cdot \frac{1}{|\mathbf{r}-\mathbf{r}'|^3}\left[\mathbf{M}_k - \frac{3(\mathbf{M}_k\cdot(\mathbf{r}-\mathbf{r}'))(\mathbf{r}-\mathbf{r}')}{|\mathbf{r}-\mathbf{r}'|^2}\right]\Delta x \Delta y \Delta z. \tag{17.62}$$

where $\mathbf{r} = \mathbf{r}_k = (x_k, y_k, z_k)$, $k = 1, ..., N_m$; denotes the center of the cell, k. As in the case of susceptibility-based modeling, Eq. (17.62) can be written in a compact form as well:

$$\mathbf{d} = \mathbf{A}^{TM}(\mathbf{m}), \tag{17.63}$$

where $\mathbf{A}^{TM}$ is the discrete TMI field forward modeling operator, $\mathbf{m}$ is a model parameter vector representing the discrete magnetization vector distribution $\mathbf{M}(M_x, M_y, M_z)$, and $\mathbf{d}^M$ is a vector of the corresponding TMI data, T.

17.3.3 Magnetization-Based Modeling of the Full Tensor Magnetic Gradiometry Data

The following formulas define the second spatial derivatives of the magnetic potential (Cai 2012; Jorgensen et al. 2023):

$$H_{\alpha\beta}(\mathbf{r}') = \frac{\partial^2}{\partial\alpha\partial\beta}U(\mathbf{r}) = \frac{\partial H_\alpha(\mathbf{r}')}{\partial\beta'}, \quad \alpha, \beta = x, y, z. \tag{17.64}$$

By introducing $r_k = r = [(x_k - x')^2 + (y_k - y')^2 + (z_k - z')^2]^{1/2}$ and differentiating Eq. (17.58), after some algebra, we find discrete forms of each component of the magnetic tensor:

$$H_{\alpha\beta}(\mathbf{r}') = -H_0\frac{\partial}{\partial\beta'}\sum_{k=1}^{N_m}\frac{1}{|\mathbf{r}-\mathbf{r}'|^3}[M_{\alpha k} - \frac{3t_k(\alpha_k - \alpha')}{|\mathbf{r}-\mathbf{r}'|^2}]\Delta x\Delta y\Delta z, \quad \alpha, \beta = x, y, z. \tag{17.65}$$

We now take the derivatives:

$$\frac{\partial}{\partial\beta'}\left(\frac{1}{|\mathbf{r}-\mathbf{r}'|^3}\left[M_{\alpha k} - \frac{3t_k(\alpha_k - \alpha')}{|\mathbf{r}-\mathbf{r}'|^2}\right]\right) =$$

$$\frac{3}{|\mathbf{r}-\mathbf{r}'|^5}\left[(\beta_k - \beta')M_{\alpha k} - \frac{5t_k(\alpha_k - \alpha')(\beta_k - \beta')}{|\mathbf{r}-\mathbf{r}'|^2} + M_{\beta k}(\alpha_k - \alpha') + t_k\delta_{\alpha\beta}\right], \tag{17.66}$$

where:

$$\delta_{\alpha\beta} = \begin{cases} 1, & \alpha = \beta \\ 0, & \alpha \neq \beta \end{cases}.$$

Substituting Eq. (17.66) into (17.65), we obtain:

$$H_{\alpha\beta}(\mathbf{r}') = -3H_0\sum_{k=1}^{N_m}\frac{1}{|\mathbf{r}-\mathbf{r}'|^5}\times$$

$$\left[\left(\beta_k-\beta'\right)M_{\alpha k}-\frac{5t_k(\alpha_k-\alpha')\left(\beta_k-\beta'\right)}{|\mathbf{r}-\mathbf{r}'|^2}+M_{\beta k}(\alpha_k-\alpha')+t_k\delta_{\alpha\beta}\right]\Delta x\Delta y\Delta z. \tag{17.67}$$

According to (17.57), we can write:

$$t_k=\sum_{\gamma=x,y,z}(\gamma_k-\gamma')M_{\gamma k}. \tag{17.68}$$

and

$$M_{\alpha k}=\sum_{\gamma=x,y,z}\delta_{\alpha\gamma}M_{\gamma k},\ \ M_{\beta k}=\sum_{\gamma=x,y,z}\delta_{\beta\gamma}M_{\gamma k}. \tag{17.69}$$

Substituting (17.68), and (17.69) into (17.67), we have:

$$H_{\alpha\beta}(\mathbf{r}')=-3H^0\sum_{k=1}^{N_m}\frac{\Delta x\Delta y\Delta z}{|\mathbf{r}-\mathbf{r}'|^5}\times$$

$$\sum_{\gamma=x,y,z}\left[\left(\beta_k-\beta'\right)\delta_{\alpha\gamma}+\left[\delta_{\alpha\beta}-\frac{5(\alpha_k-\alpha')\left(\beta_k-\beta'\right)}{|\mathbf{r}-\mathbf{r}'|^2}\right](\gamma_k-\gamma')+(\alpha_k-\alpha')\delta_{\beta\gamma}\right]M_{\gamma k}. \tag{17.70}$$

We introduce a sensitivity kernel for magnetic tensor as follows:

$$G^{\gamma}_{\alpha\beta k}=$$

$$\left(\beta_k-\beta'\right)\delta_{\alpha\gamma}+\left[\delta_{\alpha\beta}-\frac{5(\alpha_k-\alpha')\left(\beta_k-\beta'\right)}{|\mathbf{r}-\mathbf{r}'|^2}\right](\gamma_k-\gamma')+(\alpha_k-\alpha')\delta_{\beta\gamma}. \tag{17.71}$$

Using these notations, we can rewrite Eq. (17.70) in the following compact form:

$$H_{\alpha\beta}(\mathbf{r}')=-3H_0\sum_{k=1}^{N_m}\frac{\Delta x\Delta y\Delta z}{|\mathbf{r}-\mathbf{r}'|^5}\sum_{\gamma=x,y,z}G^{\gamma}_{\alpha\beta k}M_{\gamma k},\ \ \alpha,\beta=x,y,z, \tag{17.72}$$

where indices n and k at the corresponding radius vectors of the observation and integration points are omitted to simplify the notations and $M_{\gamma k}$ are the components of the magnetization vector:

$$\mathbf{M}_k=\left[M_{xk},M_{yk},M_{zk}\right]^T.$$

Equations (17.58) and (17.72) are the key equations that we need for solving both modeling and inversion problems for remanent magnetization.

Finally, we present equations (17.72) in compact operator form as follows:

$$\mathbf{d} = \mathbf{A}^{\widehat{\mathbf{H}}M}(\mathbf{m}), \tag{17.73}$$

where $\mathbf{A}^{\widehat{\mathbf{H}}M}$ is the discrete FTMG forward modeling operator described by equations (17.72); $\mathbf{m}$ is a model parameter vector representing the discrete magnetization vector distribution $\mathbf{M}\left(M_x, M_y, M_z\right)$, and $\mathbf{d}$ is a vector of the corresponding observed FTMG data, $\widehat{\mathbf{H}}$.

17.3.4 Inversion of Magnetic Field Data into Magnetization Vector

The inverse problem for magnetization vector can be written in compact form as follows:

$$\mathbf{d} = \mathbf{A}(\mathbf{m}), \tag{17.74}$$

where $\mathbf{A}$ represents the magnetic field, TMI or FTMG forward modeling operators, $\mathbf{A}^{HM}$, $\mathbf{A}^{TM}$, or $\mathbf{A}^{\widehat{\mathbf{H}}M}$, described by equations (17.59), (17.63), or (17.73); $\mathbf{m}$ is a model parameter vector representing the discrete magnetization vector distribution $\mathbf{M}\left(M_x, M_y, M_z\right)$, and $\mathbf{d}$ is a vector of the corresponding observed magnetic, TMI, or FTMG data sets, $\mathbf{H}$, T or $\widehat{\mathbf{H}}$.

Inverting for the magnetization vector is a more challenging problem than inverting for scalar magnetic susceptibility because we have three unknown scalar components of the magnetization vector for every cell. We should notice that there is an inherent correlation between the different components of the magnetization vector. The different scalar components have similar spatial variations and represent the same zones of anomalous magnetization. Therefore, it is possible to expect that the different components of the magnetization vector should be mutually correlated (Zhu et al. 2015). It was demonstrated in Chap. 12 that one could enforce the correlation between the different model parameters by using the Gramian constraints. Following this idea, Zhu et al. (2015) and Jorgensen and Zhdanov et al. (2021) included the Gramian constraint in Eq. (17.44) as follows:

$$P^{\alpha}(\mathbf{m}) = \varphi(\mathbf{m}) + \alpha c_1 S_{MN,\ MS,\ MGS}(\mathbf{m}) + \alpha c_2 \sum_{\beta=x,y,z} S_G(\mathbf{m}_\beta, \chi_{eff}), \tag{17.75}$$

where α is the regularization parameter; c_1 and c_2 are the weights defining the relative contributions of the focusing and Gramian stabilizers; $\mathbf{m}$ is the $3N_m$ vector of magnetization vector components; $\mathbf{m}_\beta$ is the N_m length vector of the β component of magnetization vector, $\beta = x, y, z$; $\boldsymbol{\chi}_{eff}$ is the N_m length vector of the effective magnetic susceptibility, defined as the magnitude of the magnetization vector,

$$\chi_{eff} = \sqrt{M_x^2 + M_y^2 + M_z^2}; \tag{17.76}$$

and S_G is the Gramian stabilizer,

$$S_G(m_\beta, \chi_{eff}) = \begin{vmatrix} (m_\beta, m_\beta) & (m_\beta, \chi_{eff}) \\ (\chi_{eff}, m_\beta) & (\chi_{eff}, \chi_{eff}) \end{vmatrix}. \tag{17.77}$$

Using the Gramian constraint (17.77), we enhance a direct correlation between the scalar components of the magnetization vector with χ_{eff}, which is computed at the previous iteration of an inversion and is updated on every iteration. The advantage of using the Gramian constraint, Eq. (17.77), is that it does not require any a priori information about the magnetization vector (e.g., direction, the relationship between different components, etc.).

The minimization problem (17.75) can be solved using one of the regularized methods described in this book.

References and Recommended Reading to this Chapter

Alken P, Thébault E, Beggan CD (2021) International geomagnetic reference field: the thirteenth generation. Earth Planets Space, 73:49.

Bhattacharyya BK (1980) A generalized multibody model for inversion of magnetic anomalies. Geophysics 29:517–531.

Cai H (2012) Migration and inversion of magnetic and magnetic gradiometry data: Master's Thesis, The University of Utah, Salt Lake City, UT, USA.

Čuma M, Zhdanov MS (2014) Massively parallel regularized 3D inversion of potential fields on CPUs and GPUs. Comput Geosci 62:80–87.

Jorgensen M, Zhdanov MS (2021) Recovering magnetization of rock formations by jointly inverting airborne gravity gradiometry and total magnetic intensity data. Minerals 11:366.

Jorgensen M, Zhdanov MS, Parsons B (2023) 3D focusing inversion of full tensor magnetic gradiometry data with Gramian regularization. Minerals 13, 851.

Koenigsberger JG (1938) Natural residual magnetism of eruptive rocks. Terr Magn Atmos Electr 43(3):299–320.

Meyer B, Saltus R, Chulliat A (2016) EMAG2: Earth magnetic anomaly grid (2-arc-minute resolution) version 3. National Centers for Environmental Information, NOAA.

Zhu Y, Zhdanov MS, Čuma M (2015) Inversion of TMI data for the magnetization vector using Gramian constraints. In: Expanded abstracts, proceedings of the 85th SEG international exposition and annual meeting, New Orleans, LA, USA, 18–23 October 2015. Society of Exploration Geophysicists. Tulsa, OK, USA, 2015; pp 1602–1606.

Zhdanov MS, Ellis R, Mukherjee S (2004) Three-dimensional regularized focusing inversion of gravity gradient tensor component data. Geophysics 69(4):925–937.

Zhdanov MS (1988) Integral transforms in geophysics. Springer, New York, Berlin, London, Tokyo, pp 367.

Zhdanov MS (2002) Geophysical inverse theory and regularization problems. Elsevier, pp 628.

Zhdanov MS (2009) New advances in regularized inversion of gravity and electromagnetic data. Geophys Prospect 57:463–478.

Chapter 18
Case Histories of Joint Inversion of Gravity and Magnetic Data

Abstract This chapter presents several case histories of joint inversion of potential field geophysical data as an illustration of the joint inversion methods introduced in this book. We describe, as an example, the joint inversion of the airborne gravity gradient (AGG) and total magnetic intensity (TMI) data collected by Fugro in the Ring of Fire area of northwestern Ontario, Canada. A comparison of the standalone inverted density and magnetization vector models versus the jointly inverted models demonstrates that the latter recovers more compact bodies with more structural correlation and more geologically reasonable models than the standalone inverse solutions. Another example illustrates the joint Gramian inversion of the gravity full tensor gradiometry (FTG) and total magnetic intensity (TMI) data collected within the Nordkapp basin in the Barents Sea, offshore Norway. It shows the improved correlation between the density and magnetization in the vertical sections of the corresponding inverse models produced by the joint probabilistic Gramian inversion. Finally, this chapter presents a case history of a joint inversion of gravity and magnetic data covering the US state of Alaska and the Canadian province of Yukon. A computationally effective algorithm for joint inversion of the gravity and magnetic data on a continental scale based on a Gramian stabilizer has been developed. It is demonstrated that the utility of a joint inversion approach on a continental scale lies in a single coupled density and susceptibility model of the whole continent, where a geoscientist can focus on areas of geological interest without the need to perform separate inversions in these areas.

Keywords Total magnetic intensity (TMI) · Full tensor gradiometry (FTG) · Gramian stabilizer · Continental scale inversion

This final chapter presents several case histories of joint inversion of potential field geophysical data, as an illustration of the joint inversion methods introduced in this book. The interested reader can find many other examples of joint inversion published in geophysical literature over the last decade.

M. S. Zhdanov, *Advanced Methods of Joint Inversion and Fusion of Multiphysics Data*, Advances in Geological Science,
https://doi.org/10.1007/978-981-99-6722-3_18

18.1 Joint Inversions of Airborne Gravity Gradient (AGG) and Magnetic Data in the Ring of Fire, Ontario, Canada

This section[1] describes, as an example, the joint inversion of the airborne gravity gradient (AGG) and total magnetic intensity (TMI) data, collected by Fugro in the Ring of Fire area of northwestern Ontario, Canada (Fig. 18.1). The Ring of Fire comprises mafic metavolcanic flows, felsic metavolcanic flows, pyroclastic rocks, and a suit of layered mafic to ultramafic intrusions that trend subparallel with and obliquely cut the westernmost part of the belt, close to a large granitoid batholith lying west of the belt. The significant layered intrusion at its base hosts Ni-Cu-PGE deposits of exceptional grade and overlying stratiform chromite deposits further east and higher in the layered intrusion stratigraphy (Ontario Geological Survey and Geological Survey of Canada 2011).

Jorgensen and Zhdanov (2021) studied the Thunderbird deposit consisting of semi-massive vanadium and titanium-enriched magnetite, corresponding to strong gravity and magnetic anomalies. Panel A of Fig. 18.2 shows the vertical gradient of the gravity field, Gzz, observed by AGG data acquisition system. Panel B shows the observed TMI data map. The location of profile AA' is shown in black. We filtered

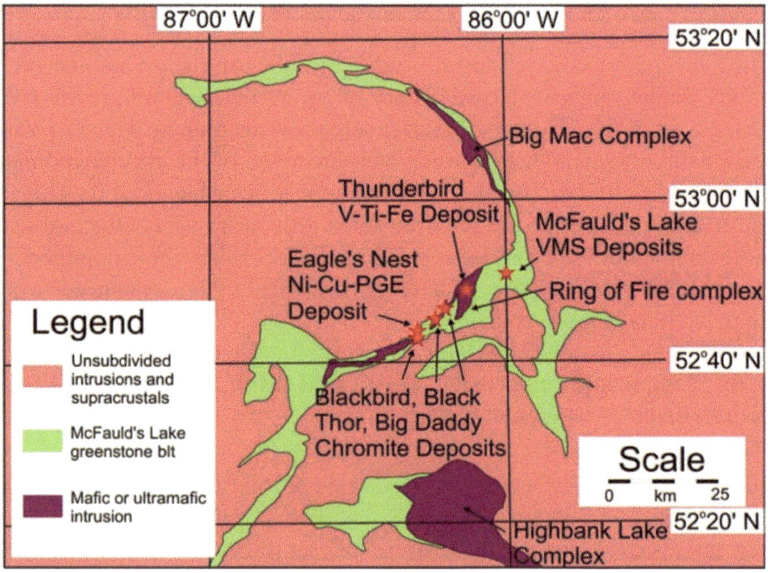

Fig. 18.1 Geological map of the Ring of Fire area with marked known deposits (from Mungall et al. 2010)

[1] This section was written in collaboration with M. Jorgensen.

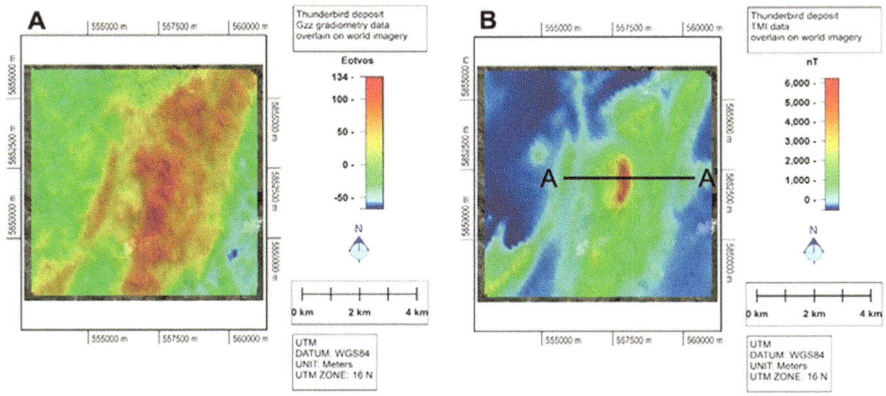

Fig. 18.2 Panel **A** presents the G_{zz} component of the observed AGG data shown in UTM coordinates. Panel **B** presents the observed TMI data map. The location of profile AA' is shown by black line

the observed data by a spatial filter with a wavelength longer than 10 km to remove the regional anomaly.

We have inverted the observed data separately and jointly on the same 50 m by 50 m horizontal grid with a logarithmic depth discretization ranging from 25 at the top to 150 m at the bottom. The total grid size was about 250,000 cells. The joint inversion of the gravity and magnetic data was run using two methods. One was based on the joint focusing stabilizing functional, and the other used the Gramian stabilizer.

Figure 18.3 shows, as an example, a comparison between the observed and predicted Gzz component of the gravity gradient field and TMI data produced by standalone and joint inversions. One can see an excellent fit of the observed data with the data computed for the inverse models using all three approaches—standalone, joint Gramian, and focusing inversions.

Figure 18.4 shows the vertical sections of the inverse density and magnetization vector models produced by standalone, joint focusing and joint Gramian inversions. One can see that the jointly inverted images have sharper boundaries and more structural correlation, while maintaining the same level of data misfit as the standalone inversions.

Figure 18.5 presents model parameter cross plots for the different inversion scenarios. Additionally, the correlation coefficient was calculated for the density model versus the vertical component of the magnetization vector. In the case of the standalone inversions, the correlation coefficient was about 0.8; for joint focusing inversion, it increased to 8.5. The models produced by the joint Gramian inversion have the highest value of the correlation coefficient of 8.9.

Comparison of the standalone inverted density and magnetization vector models versus the jointly inverted models demonstrates that the latter recovers the more compact bodies, with more structural correlation and more geologically reasonable

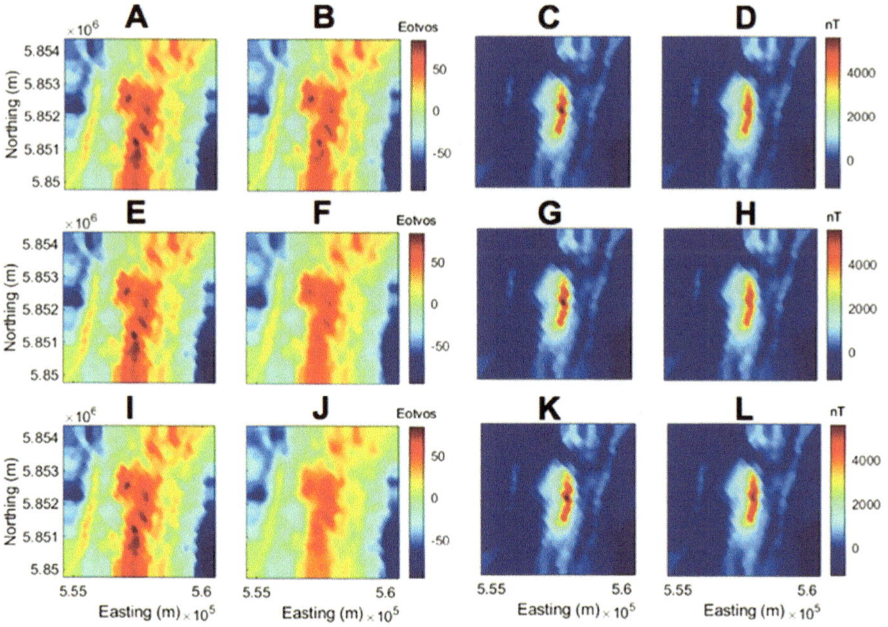

Fig. 18.3 Panels **A** and **B** show observed and predicted G_{zz} component of gravity gradient field data from standalone inversion, respectively. Panels **C** and **D** show observed and predicted TMI data from standalone inversion, respectively. Panels **E** and **F** show observed and predicted G_{zz} component of gravity gradient field data from Gramian inversion, respectively. Panels **G** and **H** show observed and predicted TMI data from Gramian inversion, respectively. Panels **I** and **J** show observed and predicted G_{zz} component of gravity gradient field data from joint focused inversion, respectively. Panels **K** and **L** show observed and predicted TMI data from joint focused inversion, respectively

models than the standalone inverse solutions. The Gramian inversion does provide the highest level of structural correlation.

18.2 Joint Inversion of Marine Gravity Gradient and Magnetic Data

This section[2] illustrates the joint Gramian inversion of the gravity full tensor gradiometry (FTG) and total magnetic intensity (TMI) data collected within the Nordkapp basin in the Barents Sea, offshore Norway (Fig. 18.6). The Nordkapp basin can be divided into two parts—the south-western part (SWP) and the northeastern part (NEP). The SWP sub-basin (Obelix survey location) is a narrow, northeast-trending 150 km long and 25–50 km wide geological structure. It contains some 17 salt

[2] This section was written in collaboration with Jorgensen and Tao (Zhdanov et al. 2023).

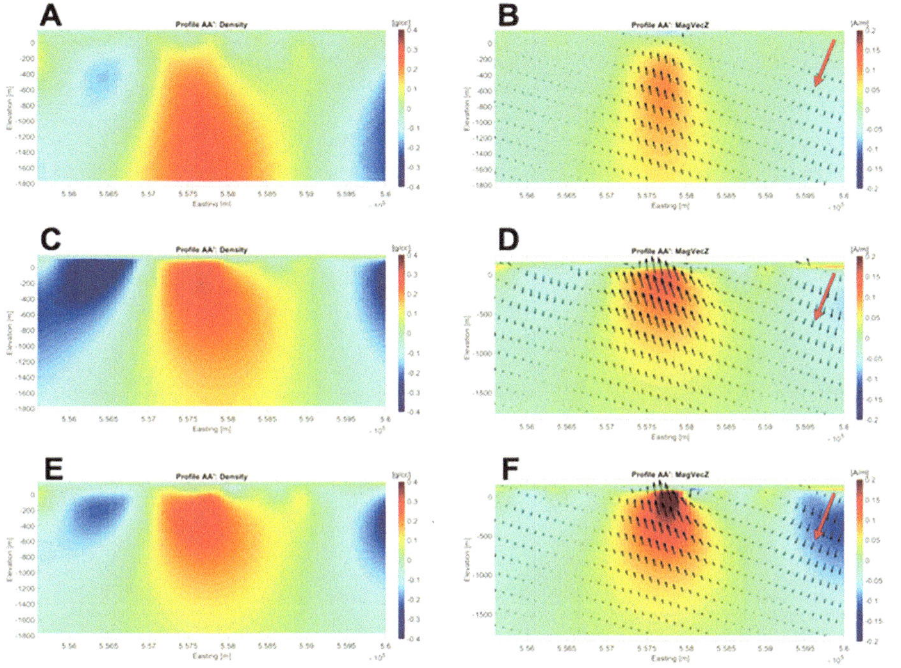

Fig. 18.4 Panels **A** and **B** show vertical sections of the standalone inverted density and magnetic vector models, respectively. Panels **C** and **D** present vertical sections of the Gramian jointly inverted density and magnetic vector models, respectively. Panels **E** and **F** show vertical sections of the jointly focused inverted density and magnetic vector models, respectively. The color map in panels (**B**), (**D**), and (**F**) is the vertical component of the magnetic vector, the black arrows are the full magnetic vector, and the red arrows in the upper right corner indicate the direction of the inducing field

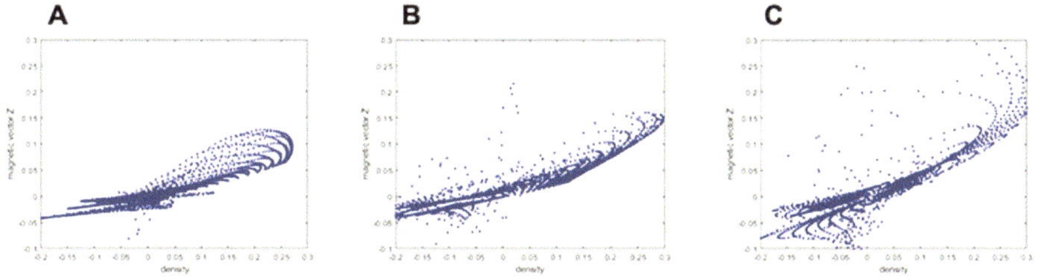

Fig. 18.5 Model parameter cross plots shown in Panels **A**, **B**, and **C** correspond to the standalone inversions, joint Gramian inversion, and joint focusing inversion, respectively

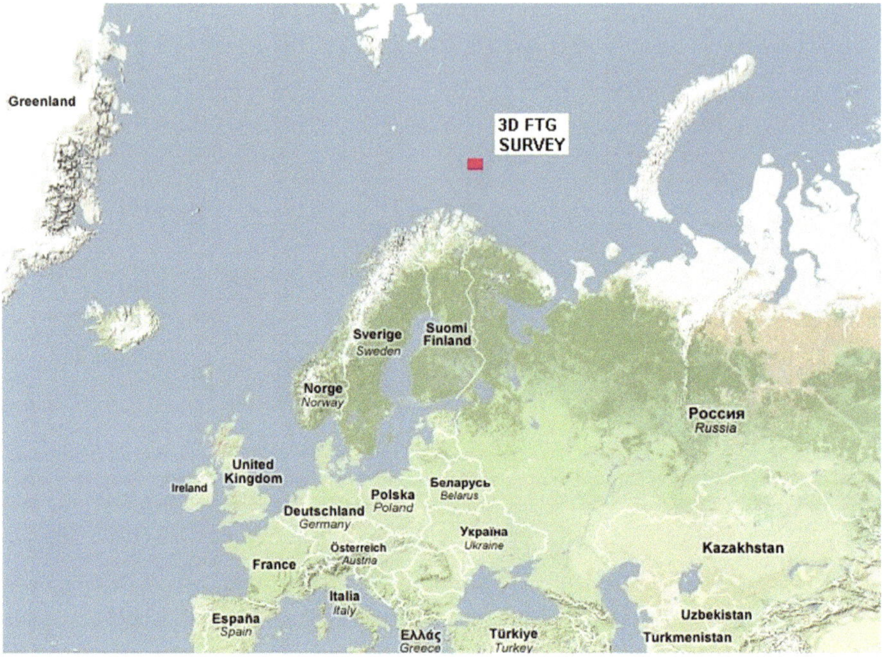

Fig. 18.6 Location of the multiphysics survey area in the Nordkapp basin in the Barents Sea, offshore Norway

diapirs located along the basin's axis. The NEP sub-basin is about 200 km long and 50–70 km wide, and it contains more than 16 salt domes. The hydrocarbon (HC) exploration in the Nordkapp basin started in the 1980s. Several wells have been drilled to date, all on the flanks of the basin. The recent results of geological and geophysical exploration and the discovery of hydrocarbons in wells outside the basin indicate that there is a potential for HC reservoir discovery within the Nordkapp basin.

The complex salt diapirs, however, represent the major geological structures known in this area. Much of the present uncertainty and exploration risk associated with these salt features results from severe seismic imaging/distortion problems and subsequent interpretation ambiguity of the salt isopach (specifically the ability to define/map base of salt seismically). The multiphysics survey aimed to provide additional information for evaluating these complex salt overhang geometries. FTG, by its very nature, is very well suited to solve this problem. It can be used to define the geological boundaries with the strong density contrasts typical for salt dome structures. In addition, the salt diapirs are characterized by diamagnetic properties, which makes it possible to use magnetic data to delineate the salt. There were a number of publications dedicated to standalone inversions of the FTG and TMI data (Wan and Zhdanov 2008; Gernigon et al. 2011; Stadtler et al. 2014; Paoletti et al. 2020; Tu and Zhdanov 2021; Tao et al. 2021).

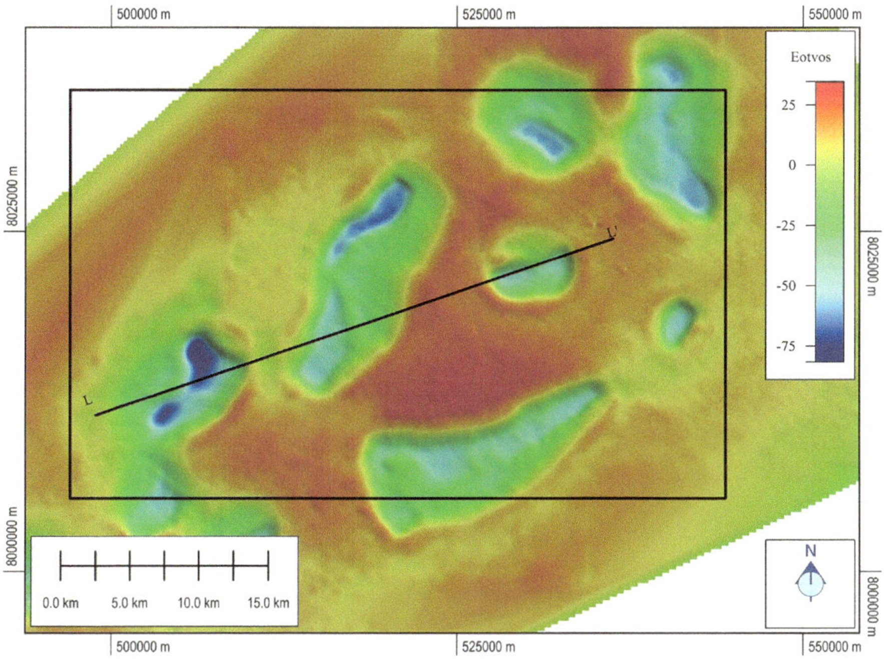

Fig. 18.7 Map of the vertical gradient of the gravity field, G_{zz}, over the area of inversion. The salt diapirs are manifested by the relatively low values of G_{zz}. The location of profile LL' is shown by black line

However, to produce well-resolved images of the salt structure, conducting a joined inversion of the FTG and TMI data is advantageous. In this section, I present the results of the joint inversion of the G_{zz} component of the gravity tensor and TMI field using the Gramian stabilizers.

Figure 18.7 shows the map of the G_{zz} component observed in the Nordkapp survey area. The black rectangular outlines the inversion area. One can clearly see the lateral position of the salt diapirs in this map shown by cool colors, which corresponds to the relatively low values of the vertical gradient of the gravity field, G_{zz}. The black line shows profile II' traversing the salt diapirs.

Figure 18.8 presents the map of the TMI component over the survey area. The diapirs are not as clearly shown in the magnetic map as in the gravity gradient data (Fig. 18.7). Nevertheless, some local lows in the magnetic data can be observed along the same profile II' on the magnetic field map as on the gravity gradient map data (Fig. 18.7).

We have applied three types of inversion to the observed gravity gradient and magnetic data: (1) separate (standalone) inversions of each data set; (2) joint inversion using the deterministic Gramian stabilizer; (3) joint inversion using the probabilistic Gramian stabilizer.

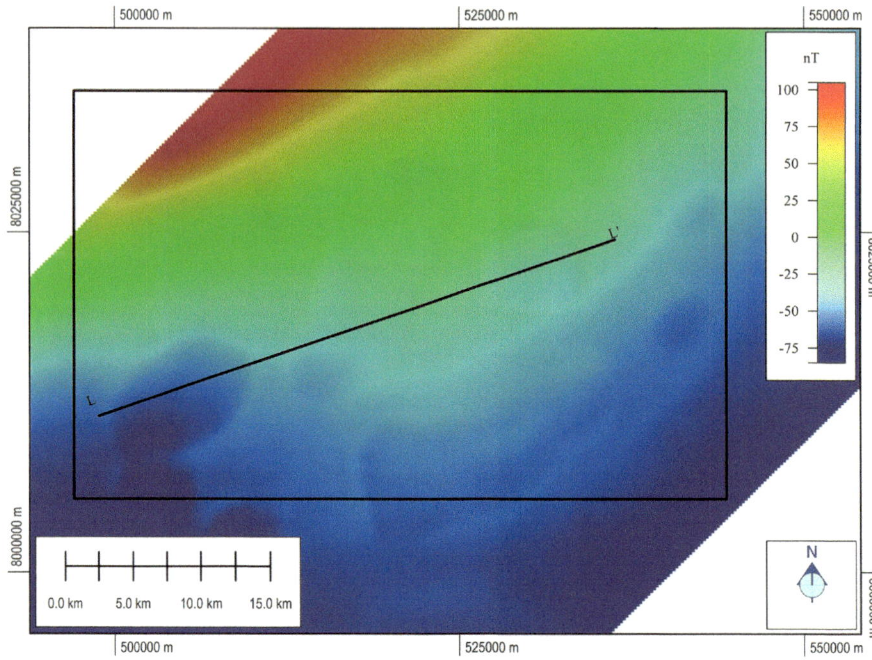

Fig. 18.8 Map of the total magnetic intensity field over the area of inversion. The salt diapirs are manifested by the relatively low magnetization. The location of profile LL' is shown by black line

The iterative conjugate-gradient method was used in all inversions. Figure 18.9 shows the convergence plots of the iterative process for the standalone inversions (gravity and magnetic) and joint inversions with deterministic and probabilistic Gramian, respectively. We use the same level of 5% for the normalized misfit between the observed and predicted data to terminate all iterative inversions. One can see that the standalone inversions converge very fast in less than ten iterations. The deterministic Gramian requires a relatively large number of iterations (about 110). The inversion based on probabilistic Gramian converges much faster (about 60 iterations).

Figures 18.10, 18.11, and 18.12 show the vertical sections along profile II' of the inverse density and magnetization images produced by standalone, deterministic Gramian, and probabilistic Gramian, respectively. One can clearly see the salt diapirs in these images manifested by low density and magnetization. A typical density of the base tertiary rocks in the area of investigation is within 2.30–2.38 g/cm^3. The salt diapirs are usually characterized by negative density anomalies, which is clearly seen in top panels of Figs. 18.10, 18.11, and 18.12. The negative magnetization within the salt diapir indicates that it is opposite to the direction of the inducing magnetic field, which corresponds to a diamagnetic property of the salt structures. At the same time, the magnetization is positive outside the diapir, which is typical for paramagnetic minerals present in Cretaceous sea-bottom layers of the host formations (Paoletti et al. 2020; Tao et al. 2021). Thus, the volume distribution of the density and magnetization

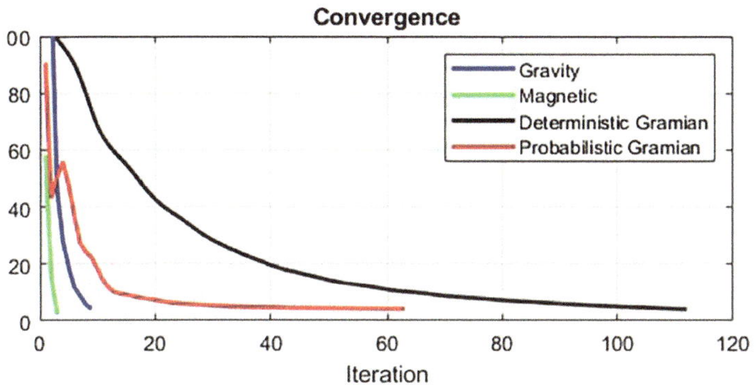

Fig. 18.9 Convergence plots of the iterative process for the standalone inversions are shown by blue line for gravity inversion and by green line for magnetic inversion. The corresponding convergence plots for joint inversions are shown by black and red lines for deterministic and probabilistic Gramian, respectively

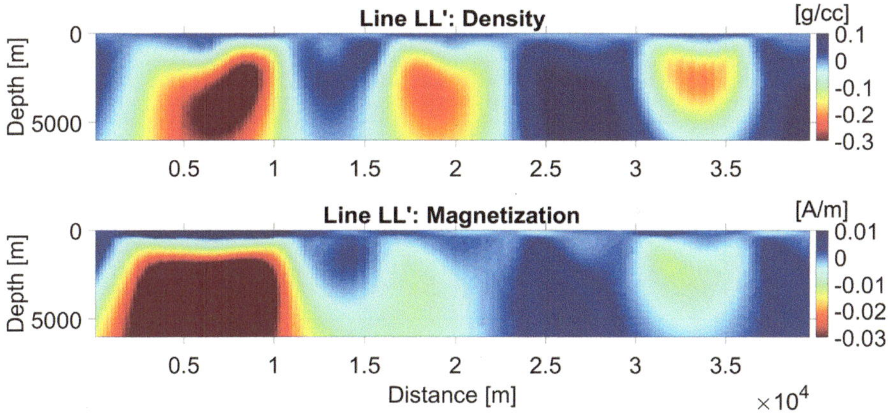

Fig. 18.10 Vertical sections of the inverse density (top panel) and magnetization (bottom panel) models produced by standalone inversions

produced by inversion indicates the salt diapir structure in the Nordkapp basin. Note that all three inversions generate the density and magnetization anomalies in the respective models with approximately the same locations. However, the structure of diapers mapped by the separate gravity and magnetic inversions is quite different. At the same time, we observe a robust structural correlation between density and magnetization models generated by the Gramian inversions.

Figure 18.13 presents the cross-correlation plots of density versus magnetization for the models produced by each of the three inversions discussed above. The cross-plot produced by the results of the separate inversions (upper panel in Fig. 18.13) represents a cloud with a weak correlation (the calculated correlation coefficient, η is equal 0.7). The joint inversion results with deterministic Gramian have a slightly

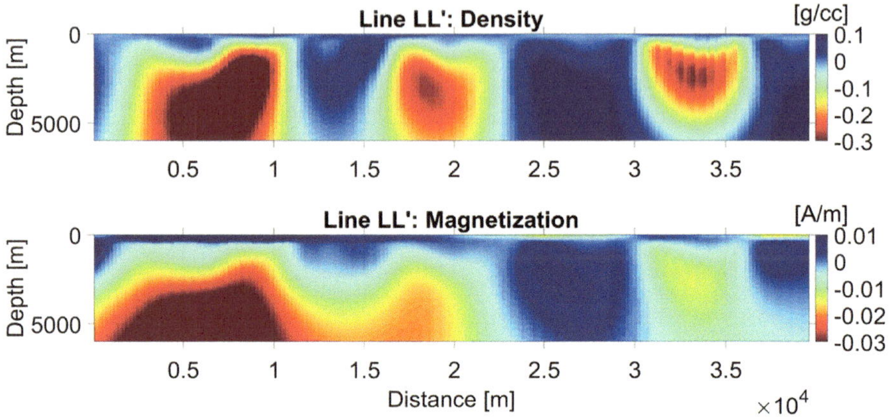

Fig. 18.11 Vertical sections of the inverse density (top panel) and magnetization (bottom panel) models produced by the joint deterministic Gramian inversion

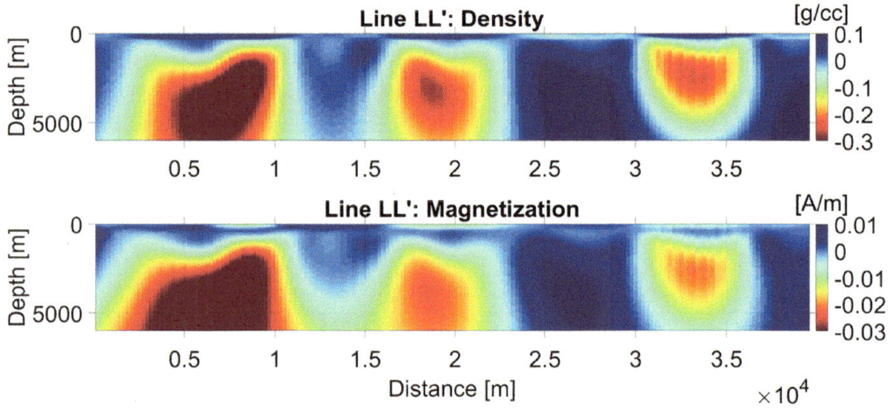

Fig. 18.12 Vertical sections of the inverse density (top panel) and magnetization (bottom panel) models produced by the joint probabilistic Gramian inversion

improved correlation with $\eta = 0.73$ (middle panel in Fig. 18.13). The joint inversion with the probabilistic Gramian shows the best structural correlation with $\eta = 0.93$ (bottom panel in Fig. 18.13). One can also observe the improved correlation between the density and magnetization in the vertical sections of the corresponding inverse models produced by the joint probabilistic Gramian inversion (Fig. 18.12).

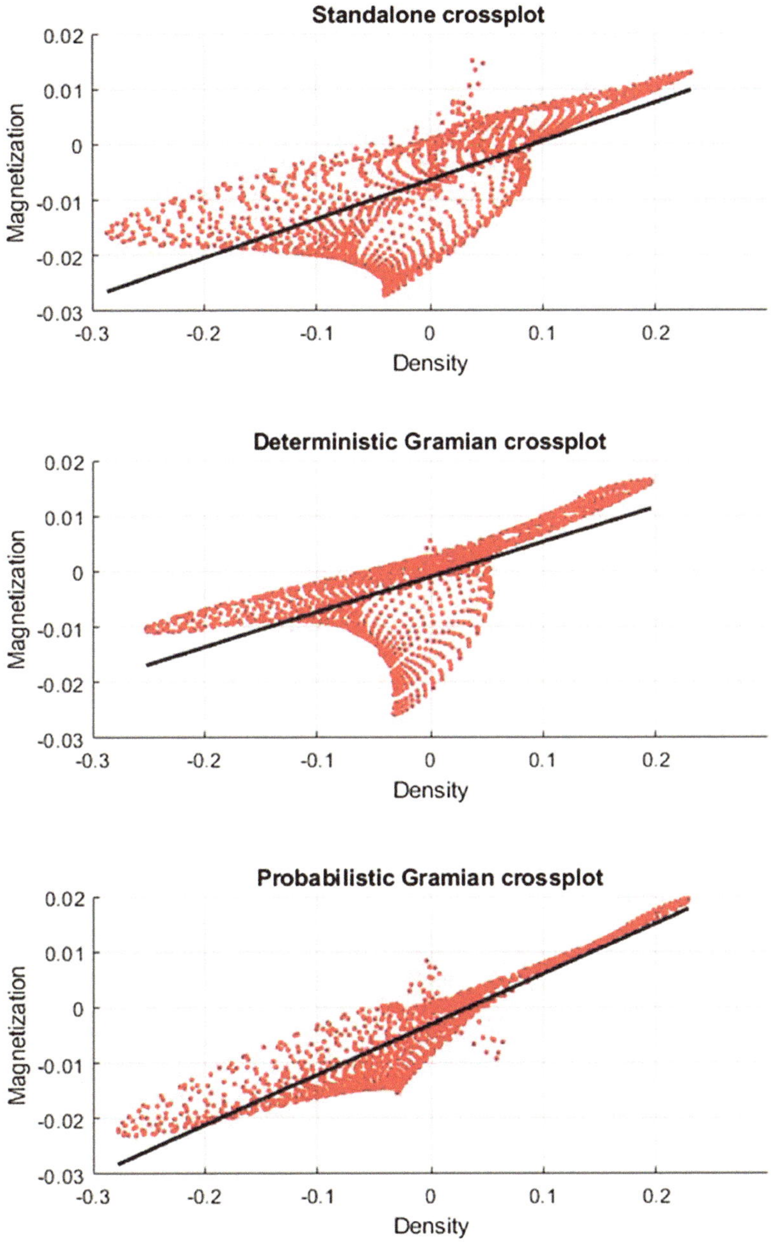

Fig. 18.13 Cross-correlation plots of density versus magnetization for the models produced by standalone inversions (top panel), joint deterministic Gramian inversion (middle panel) and joint probabilistic Gramian inversion (bottom panel)

18.3 Joint Gramian Inversion of Alaska and Yukon Gravity and Magnetic Data

We now discuss the case history[3] representing a joint inversion of gravity and magnetic data covering the US state of Alaska and the Canadian province of Yukon. Čuma and Zhdanov (2014) developed a computationally effective algorithm for joint inversion of the gravity and magnetic data on a continental scale based on a Gramian stabilizer. The developed algorithm was applied to the gravity and magnetic data collected over the US state of Alaska and the Canadian province of Yukon, covering an area of approximately 2,300,000 km^2 (Čuma and Zhdanov 2017).

Figure 18.14 presents the geological map of Alaska and Yukon, with the area of investigation outlined by the red line. Geologically, Alaska and Yukon are highly complex. Fundamentally, the geology of this vast region can be viewed as a collage

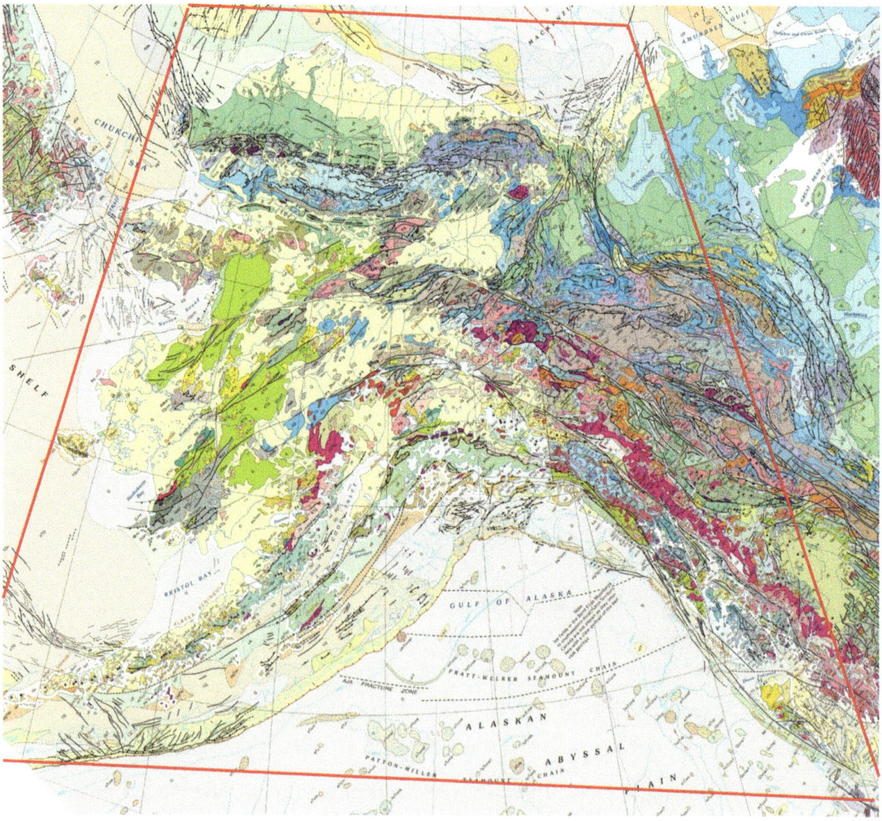

Fig. 18.14 The geological map of Alaska and Yukon. The red line outlines the area of investigation

[3] This section was written in collaboration with M. Čuma.

of terranes pieced together about 65 to 100 million years ago. Still, the process of their formation and assembly continues today. That is why using geophysical methods is extremely important to understand the geological structure of the vast region rich with mineral resources.

The gravity and magnetic data over Alaska and Yukon are available as global products, with the base data typically obtained by satellite observations and augmented with the ground or airborne data, where available.

In this case study, Čuma and Zhdanov (2017) used the Alaska ground-based gravity data from the United States Geological Survey (USGS), with a total of 91,547 stations covering an area of approximately 1,800,000 km^2. Natural Resources Canada assembled the Yukon gravity data based on 12,211 stations distributed over an area of roughly 500,000 km^2. We have interpolated the corresponding Bouguer gravity anomaly data into a 1 km by 1 km grid resulting in 2,504,665 observation points. In addition, these gravity data were then processed with a high-pass filter with a cutoff of 10,000 m to remove regional effects. Figure 18.15 presents the map of the Bouguer anomaly gravity data used in the inversion. The black dots denote the coastline.

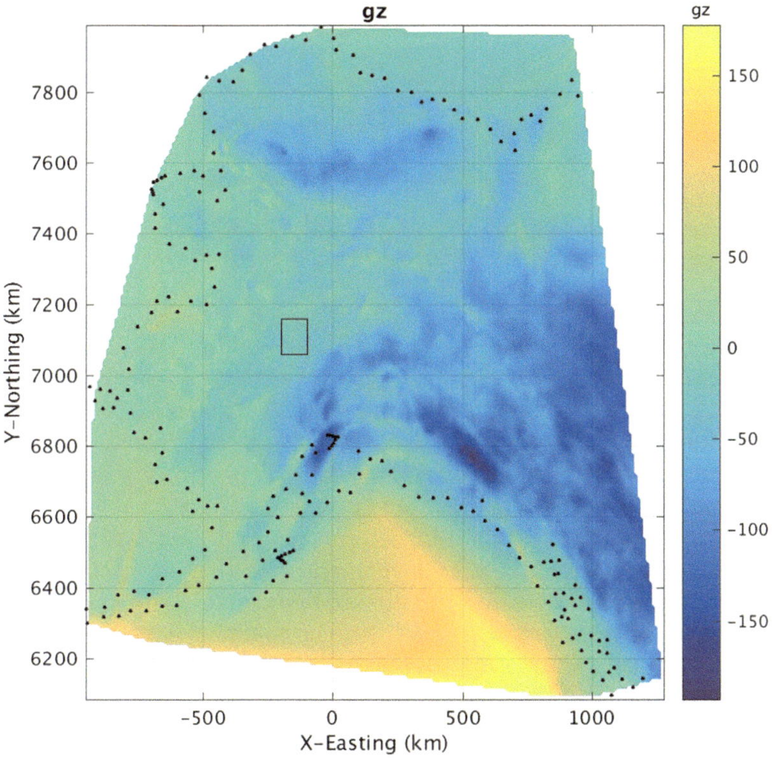

Fig. 18.15 Map of the Bouguer Anomaly gravity data used in the inversion. The black dots denote the coastline, and the black box outlines the Minchumina Basin area analyzed in more detail

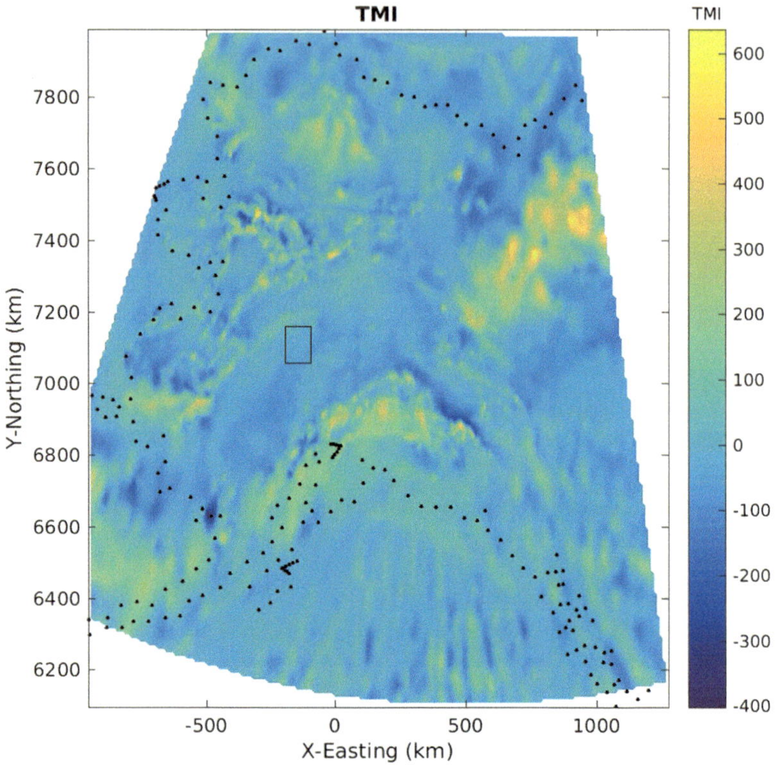

Fig. 18.16 Map of the total magnetic intensity data used in the inversion. The black dots denote the coastline and the black box outlines the Minchumina Basin area analyzed in more detail

Magnetic data for this study were based on EMAG2_V3 model (Meyer et al. 2016), which was compiled from satellite, ship, and airborne magnetic measurements and delivered on a 2-arc-minute grid (approximately 2 km) in WGS94 coordinate system. We extracted the data covering the area of Alaska and Yukon and analytically continued the data upward to an elevation of 4 km. Figure 18.16 shows the map of magnetic data on a 2 km × 2 km grid used in the inversion.

For both independent and joint inversions on the continental scale, we have discretized the inversion domain by a rectangular grid with a horizontal cell size of 2 km by 2 km and with the vertical discretization started with a 200 m vertical cell size and increased by 3% at every horizontal layer up to a total depth of the inversion domain of 27 km. Thus, the total number of cells in the discretization grid was about 60 million. The inversion was done with a parallel OpenACC GPU-enabled program (Čuma and Zhdanov 2014).

I present the results of independent and joint inversions of the gravity and magnetic data here.

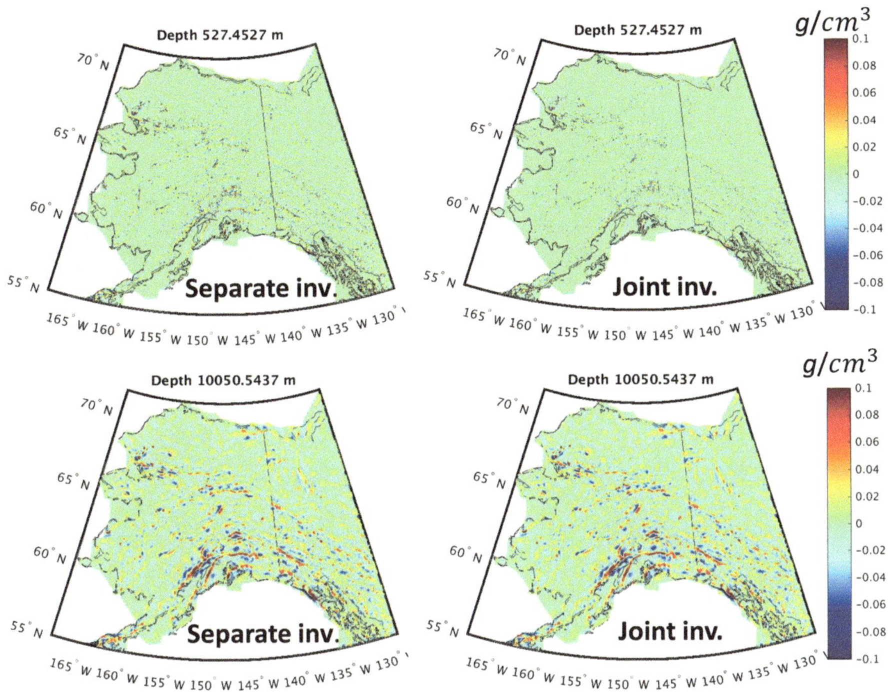

Fig. 18.17 Horizontal cross sections of recovered density by separate (left panels) and joint (right panels) inversions at 500 m depth (upper two panels) and 10 km depth (lower two panels)

Figures 18.17 and 18.18 show horizontal cross sections of density and susceptibility obtained with separate and joint continental-scale inversions. The differences between the results of the individual and joint inversions are hard to notice due to the large scale of the inversion domain, but overall, they are relatively small due to the aforementioned weak coupling enforcement.

To assess the utility of continental-scale inversion for regional anomaly exploration and to evaluate the effect of density and susceptibility coupling at a reasonably observable scale, we also focused on a regional section of about 100 km by 100 km in the Minchumina Basin in central Alaska. There is a number of underexplored basins in Alaska that have hydrocarbon potential, and this is one of them. We ran this regional scale inversion on 500 m by 500 m horizontal cell size and vertical size starting with 150 m and increasing by 3% up to the depth of 27 km. In this basin, the correlation between density and susceptibility is relatively strong, and therefore we were able to increase the strength of the Gramian stabilizer. As in the whole domain case, we ran the inversion to a normalized L_2 norm misfit of 1%.

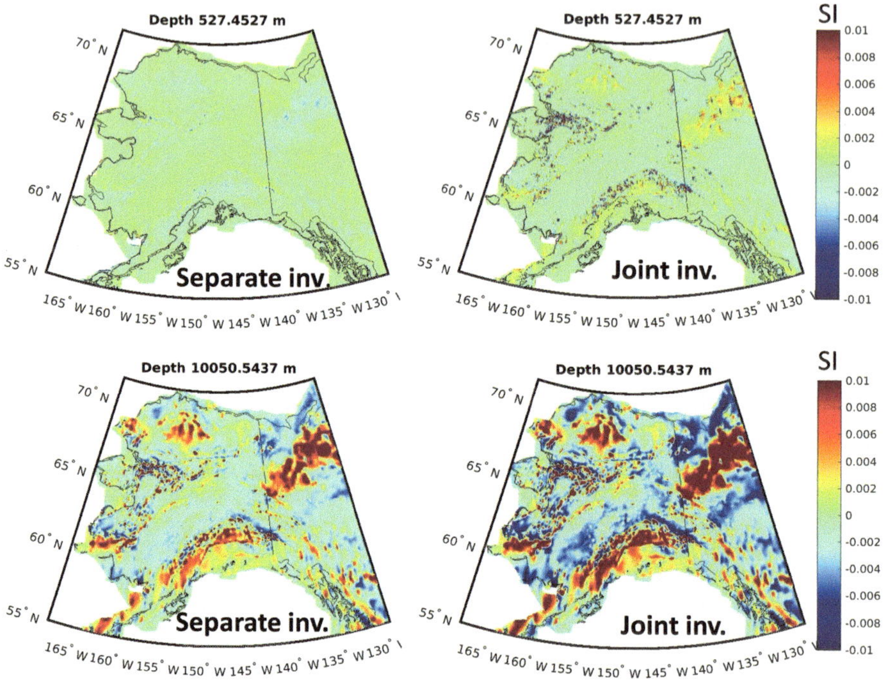

Fig. 18.18 Horizontal cross sections of recovered susceptibility by separate (left panels) and joint (right panels) inversions at 500 m depth (upper two panels) and 10 km depth (lower two panels)

In Figs. 18.19 and 18.20, we compare the results of the separate and joint inversions of the Minchumina Basin subset for a few horizontal cross sections obtained from the continental-scale inversions. The density maps shown in the figures are fairly similar between the separate and joint inversions, but we observe noticeable focusing of the susceptibility, particularly near the surface. For these particular upward continued TMI data, the main advantage of the joint inversion is the considerably improved susceptibility resolution at the near-surface.

Thus, the utility of a joint inversion approach on a continental scale lies in a single coupled density and susceptibility model of the whole continent, where a geoscientist can focus on areas of geological interest without the need to perform separate inversions on these areas.

We should note that the cited paper Čuma and Zhdanov (2017) did not account for the variability of relationships between materials' properties. In large-scale interpretation like this, one can expect different correlations between rock properties, varying with the area. In this case, one can apply a concept of localized Gramian discussed in Chap. 12 (Sect. 12.5). This concept allows the researcher to recover multiple litho-

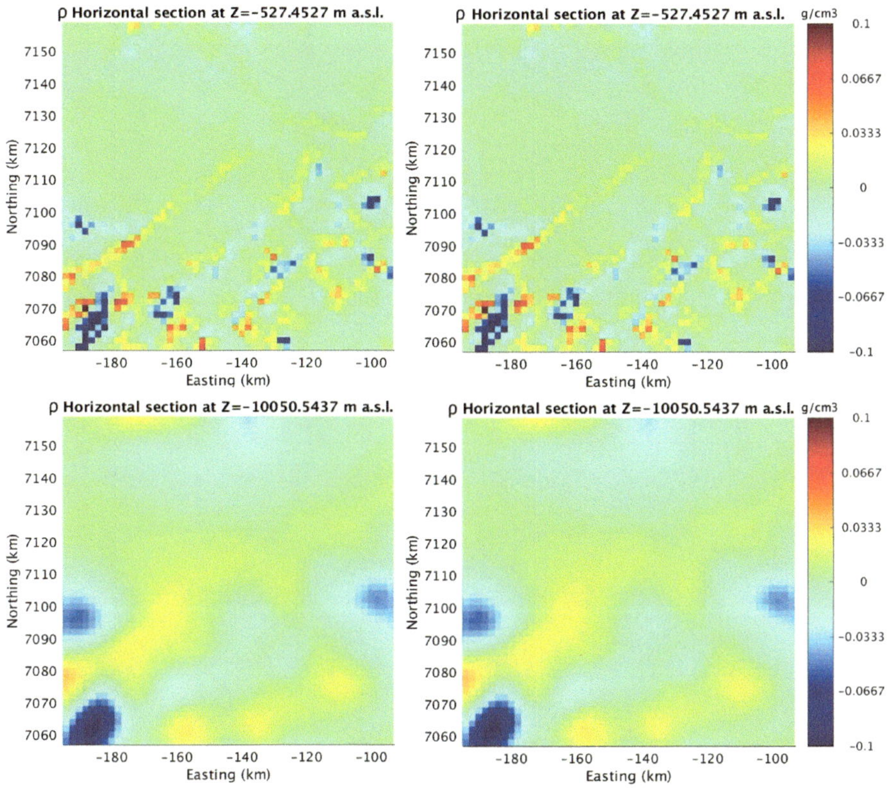

Fig. 18.19 Minchumina Basin area: horizontal cross sections of density recovered by the independent continental-scale inversion (left panels) and joint continental-scale inversion (right panels) at 500 m depth (upper panels) and 10 km depth (lower panels)

logic relationships between the different physical properties, which can vary over the survey area. Future research should be directed toward implementing this approach in the framework of continental-scale inversion. Also, it was shown above that inversion for the magnetization vector instead of susceptibility is more flexible as it allows accounting for irregularly magnetized rocks such as those with remanence or strong magnetization.

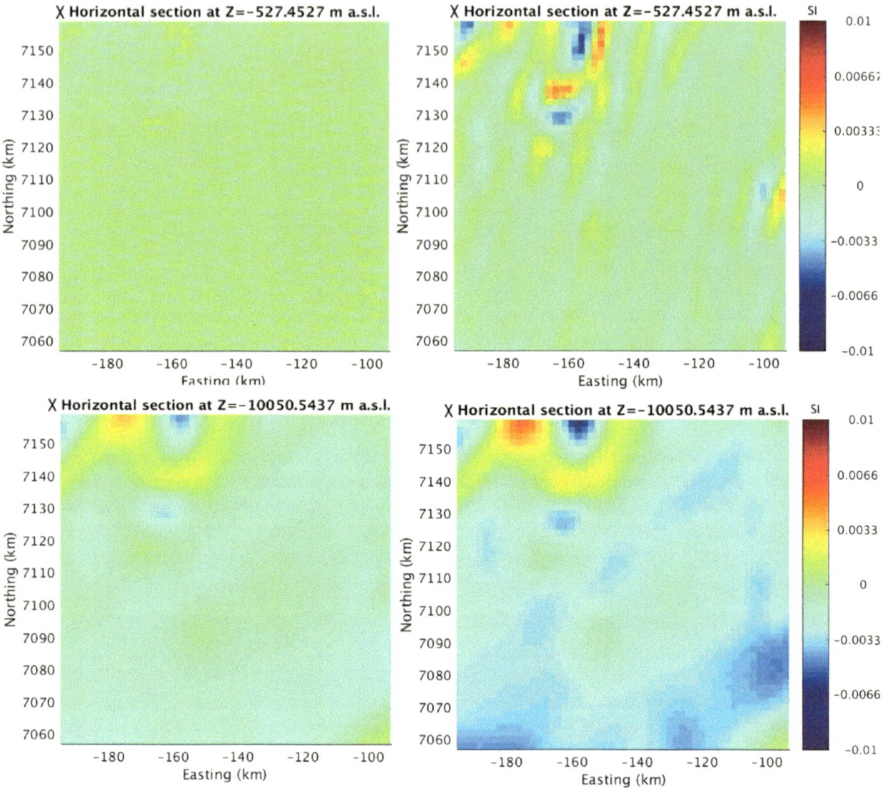

Fig. 18.20 Minchumina Basin area: horizontal cross sections of susceptibility recovered by the independent continental-scale inversion (left panels) and joint continental-scale inversion (right panels) at 500 m depth (upper panels) and 10 km depth (lower panels)

References and Recommended Reading

Čuma, M., and M.S. Zhdanov, 2014, Massively parallel regularized 3D inversion of potential fields on CPUs and GPUs: Computers and Geosciences, **62**, 80-87.

Čuma, M., and M. S. Zhdanov, 2017, Continental-scale joint inversion of Alaska and Yukon gravity and magnetic data: First Break, **35** (9).

Gernigon, L., M. Bronner, C. Fichler, L. Lovas, L. Marello, and O. Olesen, 2011, Magnetic expression of salt diapir marine related structures in the Nordkapp Basin, western Barents Sea: Geology, 39 (2), 135–138.

Jorgensen, M. and M. S. Zhdanov, 2021, Recovering magnetization of rock formations by jointly inverting airborne gravity gradiometry and total magnetic intensity data: Minerals, **11**, 366.

Meyer, B., R. Saltus, A. Chulliat, 2016, EMAG2: Earth Magnetic Anomaly Grid (2-arc-minute resolution) Version 3: National Centers for Environmental Information, NOAA.

Mungall, J. E., J. D. Harvey, S. J. Balch, B. Azar, J. Atkinson, and M. A. Hamilton, 2010), Eagle's nest: a magmatic Ni-sulfide deposit in the James Bay Lowlands, Ontario, Canada: SEG Special Publication, **15**, 539–557.

Ontario Geological Survey and Geological Survey of Canada, 2021, Ontario airborne geophysical surveys, gravity gradiometer and magnetic data, grid and profile data (ASCII and Geosoft formats) and vector data, McFaulds Lake area. In Ontario Geological Survey, Geophysical Data Set 1068.

Paoletti, V., M. Milano, J. Baniamerian and M. Fedi, 2020, Magnetic field imaging of salt structures at Nordkapp Basin, Barents Sea: Geophysical Research Letters, **47** (18)

Stadtler, C., C. Fichler, K. Hokstad, E. A. Myrlund, S. Wienecke, and B. Fotland, 2014, Improved salt imaging in a basin context by high resolution potential field data: Nordkapp Basin, Barents Sea: Geophysical Prospecting, **62** (3), 615–630.

Tao, M., M. Jorgensen, and M. S. Zhdanov, 2021, Mapping the salt structures from magnetic and gravity gradiometry data in Nordkapp Basin, Barents Sea: SEG/AAPG International Meeting for Applied Geoscience & Energy, Houston, Texas, USA, 874–878.

Tu, X., and M. S. Zhdanov, 2020, Enhancement and sharpening the migration images of the gravity field and its gradients: Pure and Applied Geophysics, **177** (6), 2853–2870.

Wan, L., and M. S. Zhdanov, 2008, Focusing inversion of marine full tensor gradiometry data in offshore geophysical exploration: In 78th SEG Technical Program Expanded Abstracts,.Tulsa, OK, 751–754.

Zhu, Y. M. S. Zhdanov, and M. Čuma, 2015: Inversion of TMI data for the magnetization vector using Gramian constraints. In Expanded Abstracts, Proceedings of the 85th SEG International Exposition and Annual Meeting, New Orleans, LA, USA, 18–23 October 2015; Society of Exploration Geophysicists: Tulsa, OK, USA, 2015; pp. 1602–1606.

Zhdanov, M. S., M. Jorgensen, and M. Tao, 2023, Probabilistic approach to Gramian inversion of multiphysics data: Frontiers in Earth Science, 11:1127597.

Bibliography

Ables JG (1974) Maximum entropy spectral analysis. Astron Astrophys Suppl Ser 15:383
Acar R, Vogel CR (1994) Analysis of total variation penalty methods. Inverse Probl 10:1217–1229
Aczel J, Daroczy Z (1975) On measures of information and their characterizations. Academic, New York
Aggarwal CC (2018) Neural networks and deep learning. Springer
Alken P, Thébault E, Beggan CD (2021) International geomagnetic reference field: the thirteenth generation. Earth Planets Space 73:49
Amato U, Hughes W (1991) Maximum entropy regularization of Fredholm integral equations of the first kind. Inverse Probl 7:793–808
Aster RC, Borchers B, Thurber CH (2013) Parameter estimation and inverse problems. Academic
Backus GE (1970a) Inference from inadequate and inaccurate data, I. Proc Natl Acad Sci 65:1–7
Backus GE (1970b) Inference from inadequate and inaccurate data, II. Proc Natl Acad Sci 65:281–287
Backus GE (1970c) Inference from inadequate and inaccurate data, III. Proc Natl Acad Sci 67:282–289
Backus GE, Gilbert TI (1967) Numerical applications of a formalism for geophysical inverse problems. Geophys J R Astr Soc 13:247–276
Bhattacharyya BK (1980) A generalized multibody model for inversion of magnetic anomalies. Geophysics 29:517–531
Barth N (1999) The Gramian and k-volume in n-space: some classical results in linear algebra. J Young Investig 2
Bishop CM (2006) Pattern recognition and machine learning. Springer
Burg JP (1975) Maximum entropy spectral analysis. Stanford University
Cai H (2012) Migration and inversion of magnetic and magnetic gradiometry data. Master's Thesis, The University of Utah, Salt Lake City, UT, USA
Cary PW, Chapman CH (1988) Automatic 1D waveform inversion of marine seismic refraction data. Geophys J 93:527–46
Colombo D, De Stefano M (2007) Geophysical modeling via simultaneous joint inversion of seismic, gravity, and electromagnetic data. Appl prestack Depth Imaging: Lead Edge 26:326–331

M. S. Zhdanov, *Advanced Methods of Joint Inversion and Fusion of Multiphysics Data*, Advances in Geological Science,
https://doi.org/10.1007/978-981-99-6722-3

Constable SC, Parker RC, Constable GG (1987) Occam's inversion: a practical algorithm for generating smooth models from EM sounding data. Geophysics 52:289–300
Corana A, Marchesi M, Martini C, Ridella S (1987) Minimizing multimodal functions of continuous variables with the "Simulated Annealing algorithm". ACM Trans Math Softw 13:262–280
Cover TM, Thomas JA (2006) Elements of information theory, 2nd edn. Willey, New York
Čuma M, Zhdanov MS (2014) Massively parallel regularized 3D inversion of potential fields on CPUs and GPUs. Comput Geosci 62:80–87
Čuma M, Zhdanov MS (2017) Continental-scale joint inversion of Alaska and Yukon gravity and magnetic data. First Break 35(9)
Cybenko (1989) Approximation by superpositions of sigmoidal function. Math Control Signals Syst 2:304–314
Dell'Aversana P (2013) Cognition in Geosciences - The feeding loop between geo-disciplines, cognitive sciences and epistemology. EAGE Publications, 204 pp
De Stefano M, Andreasi FG, Re S, Virgilio M, Snyder FF (2011) Multiple-domain, simultaneous joint inversion of geophysical data with application to subsalt imaging. Geophysics 76:R69–R80
Dmitriev VI, Editor in chief (1990) Computational mathematics and techniques in exploration geophysics (in Russian). Nedra, Moscow, 498 pp
Droske M, Rumpf M (2003) A variational approach to nonrigid morphological image registration. SIAM J Appl Math 64(2):668–687
Ekstrom MP (2012) Image recovery: theory and applications. Academic
Everitt WN (1958) Some properties of Gram matrices and determinants. Q J Math 9(1):87–98
Fletcher R (1995) Practical methods of optimization. Willey, Chichester, New-York, p 436
Foster M (1961) An application of the Wiener-Kolmogorov smoothing theory to matrix inversion. J Soci Ind Appl Math 9:387–392
Franklin JN (1970) Well-posed stochastic extensions of ill-posed linear problems. J Math Anal Appl 31:682–716
Fregoso E, Gallardo LA (2009) Cross-gradients joint 3D inversion with applications to gravity and magnetic data. Geophysics 74:L31–L42
Gao Y (1998) An upper bound on the convergence rates of canonical genetic algorithms. Complex Int 5
Gallardo LA (2007) Multiple cross-gradient joint inversion for geospectral imaging. Geophys Res Lett 34:L19301
Gallardo LA, Meju MA (2003) Characterization of heterogeneous near-surface materials by joint 2D inversion of DC resistivity and seismic data. Geophys Res Let 30:1658–1661
Gallardo LA, Meju MA (2004) Joint two-dimensional DC resistivity and seismic travel time inversion with cross-gradients constraints. J Geophys Res: Solid Earth 109(B3)
Gallardo LA, Meju MA (2007) Joint two-dimensional cross-gradient imaging of magnetotelluric and seismic traveltime data for structural and lithological classification. Geophys J Int 169(3):1261–1272
Gallardo LA, Meju MA (2011) Structure-coupled multi-physics imaging in geophysical sciences. Rev Geophys 49:RG1003
Garcia AG (2000) Orthogonal sampling formulas: a unified approach. SIAM Rev 42(3):499–512
Gernigon L, Bronner M, Fichler C, Lovas L, Marello L, Olesen O (2011) Magnetic expression of salt diapir marine related structures in the Nordkapp Basin, western Barents Sea. Geology 39(2):135–138
Giusti E (1984) Minimal surfaces and functions of bounded variations. Birkhauser-Verlag, 240 pp
Goldberg DE (1989) Genetic Algorithms in search, optimization, and machine learning. Addison-Wesley, New York
Golub GH, Van Loan CF (2013) Matrix computations, 4th edn. The Johns Hopkins University Press, Baltimore and London, p 753
Gonzalez RC (2009) Digital image processing. Pearson Education, India
Greenhalgh D, Marshall S (2000) Convergence criteria for genetic algorithms. SIAM J Comput 30:269–82

Haber E, Gazit MH (2013) Model fusion and joint inversion. Surv Geophys 34:675–695
Haber E, Modersitzki J (2007) Intensity gradient based registration and fusion of multimodal images. Methods Inf Med 46:292–299
Haber E, Oldenburg D (1997) Joint inversion: a structural approach. Inverse Probl 13:63–67
Hadamard J (1902) Sur les problèmes aux derivées partielles et leur signification physique. Princeton Univ Bull 13:49–52. Reprinted in his Oeuvres, vol III, Centre Nat Recherche Sci Paris 1968:1099–1105
Hansen C (1998) Rank-deficient and discrete ill-posed problems. Department of mathematical modeling, Technical University of Denmark, Lyngby, Numerical aspects of linear inversion, p 247
Hansen PC (2010) Discrete inverse problems: insight and algorithms. SIAM Press
Hansen C, Pereyra V, Scherer G (2013) Least squares data fitting with applications. Johns Hopkins University Press, 328 pp
Hjelt S-E (1992) Pragmatic inversion of geophysical data. Springer, Heidelberg, New York, Berlin, p 262
Holland JH (1975) Adaptation in natural and artificial systems. University of Michigan Press, Ann Arbor
Hornik K, Stinchcombe M, Halbert H (1989) Multilayer feedforward networks are universal approximators. Neural Netw 2:359–366. Pergamon Press
Hoversten GM, Gritto R, Washbournez J, Daley T (2003) Pressure and fluid saturation prediction in a multicomponent reservoir using combined seismic and electromagnetic imaging. Geophysics 68:1580–1591
Hoversten GM, Cassassuce F, Gasperikova E, Newman GA, Chen J, Rubin Y, Hou Z, Vasco D (2006) Direct reservoir parameter estimation using joint inversion of marine seismic AVA and CSEM data. Geophysics 71:C1–C13
Hu WY, Abubakar A, Habashy TM (2009) Joint electromagnetic and seismic inversion using structural constraints. Geophysics 74:R99–R109
Jackson DD (1972) Interpretation of inaccurate, insufficient and inconsistent data. Geophys J Roy Astron Soc 28:97–110
Jegen MD, Hobbs RW, Tarits P, Chave A (2009) Joint inversion of marine magnetotelluric and gravity data incorporating seismic constraints: preliminary results of sub-basalt imaging off the Faroe Shelf. Earth Planet Sci Lett 282:47–55
Jorgensen M, Zhdanov MS (2021) Recovering magnetization of rock formations by jointly inverting airborne gravity gradiometry and total magnetic intensity data. Minerals 11:366
Jorgensen M, Zhdanov MS, Parsons B (2023) 3D focusing inversion of full tensor magnetic gradiometry data with Gramian regularization. Minerals 13:851
Jupp DLB, Vozoff K (1975) Joint inversion of geophysical data. Geophys J Roy Astron Soc 42:977–991
Isakov V (1993) Uniqueness and stability in multi-dimensional inverse problem. Inverse Probl 6:389–414
Kapur JN (1989) Maximum-entropy models in science and engineering. Wiley Eastern Limited, New Delhi
Kapur JN, Kesavan HK (1992) Entropy optimization principles with applications. Academic, New York
Keilis-Borok VI, Yanovskaya TB (1967) Inverse problems of seismology. Geophys J 13:223–34
Karniadakis GE, Kevrekidis IG, Lu L, Perdikaris P, Wang S, Yang L (2021) Physics-informed machine learning. Nat Rev Phys 3(6):422–440
Khan A, Mosegaard K, Rasmussen KL (2000) A new seismic velocity model for the Moon from a Monte Carlo inversion of the Apollo Lunar seismic data. Geophys Res Lett 27:1591–1594
Khan A, Mosegaard K (2001) New information on the deep lunar interior from an inversion of lunar free oscillation periods. Geophys Res Lett 28:1791
Khinchin AYa (1957) Mathematical foundations of information theory. Dover, New York

Kirkpatrick SC, Gelatt D, Vecchi MP (1983) Optimization by simulated annealing. Science 220:671–680
Koenigsberger JG (1938) Natural residual magnetism of eruptive rocks. Terr Magn Atmos Electr 43(3):299–320
Kopec S (1991) Properties of maximum entropy approximate solutions to Fredholm integral equations. J Math Phys 32:1269–1272
Kreyszig E (1989) Introductory functional analysis with applications. Wiley, 688 pp
Lanczos C (1961) Linear differential operators. D. van Nostrand Co
Last BJ, Kubik K (1983) Compact gravity inversion. Geophysics 48:713–721
Lavrent'ev MM, Romanov VG, Shishatskii SP (1986) Ill-posed problems of mathematical physics and analysis. Translations of Mathematical Monographs, vol 64. American Mathematical Society, Providence, Rhode Island, p 290
Levenberg K (1944) A method for the solution of certain nonlinear problems in least squares: Quart Appl Math 2:164–168
Marquardt DW (1963) An algorithm for least-squares estimation of nonlinear parameters. J Soc Ind Appl Math 11:431–441
Marquardt DW (1970) Generalized inverses, ridge regression, biased linear estimation, and nonlinear estimation. Technometrics 12:591–612
McCulloch WS, Pitts W (1943) A logical calculus of the ideas immanent in nervous activity. Bull Math Biophys 5(4):115–133
Meju MA (2011) Joint multi-geophysical inversion: effective model integration, challenges and directions for future research. Presented at international workshop on gravity, Electrical and Magnetic Methods and their Applications, Bejing, China
Meju MA, Gallardo LA (2016) Structural coupling approaches in integrated geophysical imaging. In: Integrated imaging of the earth, Chap 4. American Geophysical Union (AGU), pp 49–67
Mehlig B (2021) Machine learning with neural networks: an introduction for scientists and engineers. Cambridge University Press
Menke W (2018) Geophysical data analysis: discrete inverse theory, 4th edn. Elsevier
Metropolis N, Ulam SM (1949) The Monte Carlo method. J Amer Stat Assoc 44:335–341
Metropolis N, Rosenbluth MN, Rosenbluth AW, Teller AH, Teller E (1953) Equation of state calculations by fast computing machines. J Chem Phys 21:1087–1092
Meyer B, Saltus R, Chulliat A (2016) EMAG2: earth magnetic anomaly grid (2-arc-minute resolution) Version 3. NOAA, National Centers for Environmental Information
Michalewicz Z, Schoenauer M (1996) Evolutionary algorithms for constrained parameter optimization problems. Evolut Comput 4(1):1–32
Minsky ML, Papert SA (1969) Perceptrons, MIT Press (Expanded Edition, 1990)
Mitrinović DS, Pečarić JE, Fink AM (1993) Gram's inequality. In: Classical and new inequalities in analysis. Mathematics and its applications (East European Series), vol 61. Springer
Mohri M, Rostamizadeh A, Talwalkar A (2012) Foundations of machine learning. MIT Press
Molodtsov DM, Troyan V (2017) Multiphysics joint inversion through joint sparsity regularization. In: Expanded Abstracts: Proceedings of 87th annual international meeting. Society of Exploration Geophysicists: Tulsa, OK, USA, pp 1262–1267
Moorkamp M, Heincke B, Jegen M, Robert AW, Hobbs RW (2011) A framework for 3-D joint inversion of MT, gravity and seismic refraction data. Geophys J Int 184:477–493
Morozov VA (1993) Regularization methods for ill-posed problems. CRC Press
Mosegaard K, Sambridge M (2002) Monte Carlo analysis of inverse problems. Inverse Probl 18:R29–R54
Muller JL, Siltanen S (2012) Linear and nonlinear inverse problems with practical applications. SIAM Press
Mungall JE, Harvey JD, Balch SJ, Azar B, Atkinson J, Hamilton MA (2010) Eagle's nest: a magmatic Ni-sulfide deposit in the James Bay Lowlands. Ontario, Canada, vol 15. SEG Special Publication, pp 539–557
Neto FDM, Neto AJ (2013) An introduction to inverse problems with applications. Springer

Oliveira JP, Bioucas-Dias JM, Figueiredo MA (2009) Adaptive total variation image deblurring: a majorization-minimization approach. Signal Proc 89(9):1683–1693

Ontario Geological Survey and Geological Survey of Canada (2021) Ontario airborne geophysical surveys, gravity gradiometer and magnetic data, grid and profile data (ASCII and Geosoft formats) and vector data. Geophysical Data Set, McFaulds Lake area, In Ontario Geological Survey, p 1068

Paoletti V, Milano M, Baniamerian J, Fedi M (2020) Magnetic field imaging of salt structures at Nordkapp Basin, Barents Sea. Geophys Res Lett 47(18)

Parker RL (1994) Geophysical inverse theory. Princeton University Press, Princeton, NJ, p 386

Portniaguine O, Zhdanov MS (1999) Focusing geophysical inversion images. Geophysics 64(3):874–887

Portniaguine ON, Zhdanov MS (2005) Method of digital image enhancement and sharpening. US Patent No. 6,879,735

Pratt WK (2007) Digital image processing: PIKS Scientific inside, vol 4. Wiley-Interscience, Hoboken, New Jersey

Press F (1968) Earth models obtained by Monte Carlo inversion. J Geophys Res 73:5223–34

Press F (1970) Earth models consistent with geophysical data. Phys Earth Planet Inter 3:3–22

Press F (1970) Regionalized earth models. J Geophys Res 75:6575–81

Press WH, Flannery BP, Teukolsky SA, Vettering WT (1987) Numerical recipes, the art of scientific computing, vol I and II. Cambridge University Press, Cambridge, 1447 pp

Raissi M, Paris P, Karniadakis GE (2019) Physics-informed neural networks: a deep learning framework for solving forward and inverse problems involving nonlinear partial differential equations. J Comput Phys 378:686–707

Ramos FM, Campos Velho HF, Carvalho JC, Ferreira NJ (1999) Novel approaches to entropic regularization. Inverse Probl 15:1139–1148

Rawlinson N, Hauser J, Sambridge M (2008) Seismic ray tracing and wavefront tracking in laterally heterogeneous media. Adv Geophys 49:203–273

Reddy BD (1998) Introductory functional analysis. Springer, 472 pp

Rosenblatt F (1957) The Perceptron—a perceiving and recognizing automaton. Report 85-460-1. Cornell Aeronautical Laboratory

Rosenblatt F (1962) Principles of neurodynamics. Spartan Books, DC Washington

Ross S (2010) A first course in probability, 8th edn. Printice Hall, Upper Saddle River, New Jersey, p 07458

Rudin LI, Osher S, Fatemi E (1992) Nonlinear total variation based noise removal algorithms. Physica D 60:259–268

Sabatier PC (1977) On geophysical inverse problems and constraints. J Geophys Res 43:115–137

Schuster G (2023) Machine learning methods in geoscience. Society of Exploration Geophysicists

Sambridge M, Mosegaard K (2002) Monte Carlo methods in geophysical inverse problems. Rev Geophys 40(3):1–29

Shannon CE (1948) A mathematical theory of communication. Bell Sys Tech J 27(379–423):623–656

Smith JT, Booker JR (1991) Rapid inversion of two- and three-dimensional magnetotelluric data. J Geophys Res 96:3905–3922

Smith RT, Zoltani CK, Klem GJ, Coleman MW (1991) Reconstruction of the tomographic images from sparse data sets by a new finite element maximum entropy approach. Appl Opt 30:573–582

Stadtler C, Fichler C, Hokstad K, Myrlund EA, Wienecke S, Fotland B (2014) Improved salt imaging in a basin context by high resolution potential field data: Nordkapp Basin. Barents Sea. Geophys Prospect 62(3):615–630

Stark H (2013) Image recovery: theory and applications. Academic

Strakhov VN (1968) Numerical solution of incorrect problems representable by integral equations of convolution type (in Russian). DAN SSSR 178(2):299

Strakhov VN (1969) Theory of approximate solution of the linear ill-posed problems in a Hilbert space and its application in applied geophysics, Part I (in Russian). Izvestia AN SSSR, Fizika Zemli, No 8:30–53

Strakhov VN (1969) Theory of approximate solution of the linear ill-posed problems in a Hilbert space and its application in applied geophysics, Part II (in Russian). Izvestia AN SSSR, Fizika Zemli, No 9:64–96

Tao M, Jorgensen M, Zhdanov MS (2021) Mapping the salt structures from magnetic and gravity gradiometry data in Nordkapp Basin. SEG/AAPG International meeting for applied geoscience & energy, Houston, Texas, USA, Barents Sea, pp 874–878

Tarantola A (1987) Inverse problem theory. Elsevier, Oxford, New York, Tokyo, Amsterdam, p 613

Tarantola A (2005) Inverse problem theory and methods for model parameter estimation. SIAM, 344 pp

Tarantola A, Valette B (1982) Generalized nonlinear inverse problem solved using the least squares criterion. Rev Geophys Space Phys 20:219–232

Tikhonov AN (1943) On the stability of inverse problems (in Russian). Doklady AN SSSR 39(5):195–198

Tikhonov AN (1999) Mathematical geophysics (in Russian). Moscow State University, 476 pp

Tikhonov AN, Arsenin VY (1977) Solution of ill-posed problems. V. H. Winston and Sons

Tu X, Zhdanov MS (2020) Enhancement and sharpening the migration images of the gravity field and its gradients. Pure Appl Geophys 177(6):2853–2870

Tu X, Zhdanov MS (2021) Joint Gramian inversion of geophysical data with different resolution capabilities: case study in Yellowstone. Geophys J Int 226(2):1058–1085

Tu X, Zhdanov MS (2022) Joint focusing inversion of marine controlled-source electromagnetic and full tensor gravity gradiometry data. Geophysics 87(5):K35–K47

Vogel CR, Oman ME (1998) Fast total variation based reconstruction of noisy, blurred images. IEEE Trans Image Proc 7:813–824

Vogel CR (2002) Computational methods for inverse problems. Society for Industrial and Applied Mathematics

Wan L, Zhdanov MS (2008) Focusing inversion of marine full tensor gradiometry data in offshore geophysical exploration. 78th SEG technical program expanded abstracts. Tulsa, OK, pp 751–754

Wang R, Tao D (2014) Recent progress in image deblurring. arXiv:1409.6838

Wang Z, Bovik A, Sheikh H, Simoncelli E (2004) Image quality assessment: from error visibility to structural similarity. Proc IEEE Trans Image Proc 13(4):600–612

Wernecke SJ, D'Addario LR (1977) Maximum entropy image reconstruction. IEEE Trans Comput 26:351–364

Whitley DL (1994) A genetic algorithm tutorial. Stat Comput 4:65–85

Wolberg J (2006) Data analysis using the method of least squares. Springer, 250 pp

Zhdanov MS, Fang S (1996) 3-D quasi-linear electromagnetic inversion. Radio Sci 3(4):741–754

Zhdanov MS (1988) Integral transforms in geophysics. Springer, Berlin, London, Tokyo, New York, p 367

Zhdanov MS (1993) Tutorial: regularization in inversion theory. CWP-136, Colorado School of Mines, 47 pp

Zhdanov MS (2002) Geophysical inverse theory and regularization problems. Elsevier, 628 pp

Zhdanov MS (2009) New advances in 3D regularized inversion of gravity and electromagnetic data. Geophys Prospect 57(4):463–478

Zhdanov MS (2015) Inverse theory and applications in geophysics. Elsevier, 704 pp

Zhdanov MS (2018) Foundations of geophysical electromagnetic theory and methods. Elsevier, 770 pp

Zhdanov MS (2019) Method of simultaneous imaging of different physical properties using joint inversion of multiple datasets. U.S. Patent 10,242,126

Zhdanov MS (2022) Joint minimum entropy method for simultaneous processing and fusion of multi-physics data and images. U.S. Patent Application 17/343,218

Zhdanov MS, Čuma M (2018) Joint inversion of multimodal data using focusing stabilizers and Gramian constraints: In: Expanded Abstracts, proceedings of the 88th SEG international exposition and annual meeting. Society of Exploration Geophysicists, Tulsa, OK, USA, pp 1430–1434

Zhdanov MS, Ellis R, Mukherjee S (2004) Three-dimensional regularized focusing inversion of gravity gradient tensor component data. Geophysics 69(4):925–937

Zhdanov MS, Gribenko AV, Wilson G (2012) Generalized joint inversion of multimodal geophysical data using Gramian constraints. Geophys Res Lett 39(L09301):1–7

Zhdanov MS, Gribenko AV, Wilson G, Funk C (2012) 3D joint inversion of geophysical data with Gramian constraints: a case study from the Carrapateena IOCG deposit. The Leading Edge, November, South Australia, pp 1382–1388

Zhdanov MS, Jorgensen M, Cox L (2021) Advanced methods of joint inversion of multiphysics data for mineral exploration. Geosciences 11:262

Zhdanov MS, Jorgensen M, Tao M (2023) Probabilistic approach to Gramian inversion of multiphysics data. Front Earth Sci 11:1127597

Zhdanov MS, Tu X, Čuma M (2022) Cooperative inversion of multiphysics data using joint minimum entropy constraints. Near Surf Geophys 1:1–14

Zhu YM, Zhdanov S, Čuma M (2015) Inversion of TMI data for the magnetization vector using Gramian constraints. In: Expanded Abstracts, proceedings of the 85th SEG international exposition and annual meeting, New Orleans, LA, USA, 18–23 October 2015. Society of Exploration Geophysicists, Tulsa, OK, USA, pp 1602–1606

Index

M. S. Zhdanov, *Advanced Methods of Joint Inversion and Fusion of Multiphysics Data*, Advances in Geological Science,
https://doi.org/10.1007/978-981-99-6722-3